Lecture Notes in Physics

The Lecture Notes in Physics

The series Lecture Notes in Physics (LNP), founded in 1969, reports new developments in physics research and teaching – quickly and informally, but with a high quality and the explicit aim to summarize and communicate current knowledge in an accessible way. Books published in this series are conceived as bridging material between advanced graduate textbooks and the forefront of research and to serve three purposes:

- to be a compact and modern up-to-date source of reference on a well-defined topic

- to serve as an accessible introduction to the field to postgraduate students and nonspecialist researchers from related areas

- to be a source of advanced teaching material for specialized seminars, courses and schools

Both monographs and multi-author volumes will be considered for publication. Edited volumes should, however, consist of a very limited number of contributions only. Proceedings will not be considered for LNP.

Volumes published in LNP are disseminated both in print and in electronic formats, the electronic archive being available at springerlink.com. The series content is indexed, abstracted and referenced by many abstracting and information services, bibliographic networks, subscription agencies, library networks, and consortia.

Proposals should be sent to a member of the Editorial Board, or directly to the managing editor at Springer:

Christian Caron
Springer Heidelberg
Physics Editorial Department I
Tiergartenstrasse 17
69121 Heidelberg / Germany
christian.caron@springer.com

L. Vlahos
P. Cargill (Eds.)

Turbulence in Space Plasmas

 Springer

Loukas Vlahos
Aristotle University of Thessaloniki
Dept. Physics
540 06 Thessaloniki
Greece
vlahos@astro.auth.gr

Peter Cargill
Imperial College
The Blackett Laboratory
Space and Atmospheric Physics
London SW7 2BW
United Kingdom
p.cargill@imperial.ac.uk

Vlahos, L., Cargill, P. (Eds.), *Turbulence in Space Plasmas*, Lect. Notes Phys. 778 (Springer, Berlin Heidelberg 2009), DOI 10.1007/978-3-642-00210-6

ISBN 978-3-642-00209-0   e-ISBN 978-3-642-00210-6

DOI 10.1007/978-3-642-00210-6

Springer Dordrecht Heidelberg London New York

Lecture Notes in Physics ISSN 0075-8450   e-ISSN 1616-6361

Library of Congress Control Number: 2009920101

Cover design: Integra Software Services Pvt. Ltd., Pondicherry

Printed on acid-free paper

Springer is part of Springer Science+Business Media (www.springer.com)

To Dennis Papadopoulos

Preface

The basis of this book lies in a collaboration between researchers at six European institutions (Imperial College London, Universita di Firenze, Aristotle University of Thessaloniki, Institute Astrophysique Spatial—Orsay, University of Oslo, and Universita di Calabria). Beginning in the late 1990s, they began to develop a research program targeted at Framework 5 of the European Commission. A proposal for a Research Training Network (RTN) was submitted in May 2001, was funded later that year, and ran from 2002 to 2006. The network title "Theory, Observations and Simulations of Turbulence in Space Plasmas" encapsulated the network aim: to address key questions of plasma turbulence at the Sun, in the solar wind, and in planetary magnetospheres through a multi-pronged approach.

The two most important aspects of the network were firstly, the training of young researchers in the methods needed to study turbulence in space plasmas. This involved considerable collaboration between institutes and has, we believe, helped in the development of several young researchers who will become familiar names in the future. Secondly, the network sponsored three summer schools aimed at young researchers. The first, in Halkidiki, Greece, in September 2003 focused on basic principles of turbulent plasmas. The second, in Calabria in October 2004 addressed analysis techniques. The final one, at Montegufoni, Italy, in October 2005 addressed applications to solar, space, and astrophysical environments. At each school we were privileged to have lectures from senior scientists both from and outwith the network.

It was apparent at the first school that it was worthwhile considering the publication of a book based on the schools. Not every lecture would be written up, but by careful selection, we were able to obtain a good coverage of the topics discussed. The aim is to provide readers with a general flavor of the topic ranging from introductory reviews to descriptions of current research. The intended audience is starting postgraduate students and postdoctoral researchers changing their field of research, though more senior researchers will doubtless find much of interest as well. Final year undergraduates will find some material useful for any major project they are undertaking.

The book begins with a chapter by Dennis Papadopoulos that develops the theory of high-frequency plasma waves from linear waves through to the introduction of non-linear effects, and subsequent turbulence, outlining in a simple way the new physics that must be introduced at each step. Next, Chap. 2 by Claudio Chiuderi and Marco Velli presents an introduction to the solar atmosphere from the viewpoint of magnetohydrodynamics (MHD) and discuss briefly the origin of the hot solar corona. Chapter 3 by Vincenzo Carbone and Annick Pouquet is an introduction to the theory of first hydrodynamic and subsequently MHD turbulence. Particular emphasis is placed on the key phenomenon of "intermittency," and on the importance of anisotropy in MHD turbulence. In Chap. 4, Viggo Hansteen and Mats Carlsson address turbulence in the outer solar atmosphere. Observations of simulations of the hierarchy of scales present in the photosphere, chromosphere, and corona, with particular emphasis on 3D modeling. The solar atmosphere is discussed further in Chap. 5 by Loukas Vlahos, Sam Krucker, and Peter Cargill. They demonstrate the formation of multiple dissipation regions in a turbulent corona, and show that such sites are potentially very effective particle accelerators. The following two chapters turn to the solar wind. In Chap. 6, Karine Issautier discussed measurement techniques for turbulent solar wind plasmas, especially in the more distant parts of the heliosphere traversed by the Ulysses mission. Chapter 7 by Marco Velli outlines the "classic" Parker solar wind theory, modifications thereof, and the inverse accretion problem. He also discusses MHD turbulence in the specific context of the wind. Finally, in Chap. 8, Manuel Guedel discusses the above in the context of stellar coronae. Using in particular radio and X-ray observations, the presence of hot, dynamic plasmas can be inferred, suggesting the ubiquity of turbulent dissipation and acceleration processes.

Thessaloniki, Greece   *Loukas Vlahos*
London, UK   *Peter Cargill*

Contents

1 **Waves and Instabilities in Space Plasmas** 1
K. Papadopoulos
  1.1   Introduction 1
  1.2   Plasma Description – Response Function – Generalized Ohm's Law 2
  1.3   Examples of Plasma Waves in Space – Isotropic Plasmas 4
  1.4   Properties of Plasma Waves in Isotropic Plasmas 8
  1.5   Generation of Plasma Waves 14
  1.6   Plasma Instabilities 22
  1.7   Electromagnetic Emission from Isotropic Plasmas 29
  1.8   Type III Radio-Bursts – Puzzles and Resolution – The Triumph
        of Strong Turbulence Theory 34
  1.9   Epilogue 41
  References 42

2 **Solar MHD: An Introduction** 45
C. Chiuderi and M. Velli
  2.1   Introduction 45
  2.2   Plasma Physics and Magnetohydrodynamics 46
  2.3   A Quick Tour of the Sun 51
  2.4   Plasma Instabilities 54
  2.5   Coronal Heating, Waves, and Turbulence 60
  References 68

3 **An Introduction to Fluid and MHD Turbulence for Astrophysical
Flows: Theory, Observational and Numerical Data, and Modeling** 71
V. Carbone and A. Pouquet
  3.1   Introduction 71
  3.2   Fundamental Equations 74
  3.3   The Problem of Intermittency in Turbulence 85
  3.4   Modeling Turbulence 92
  3.5   Theoretical Approaches 111

3.6   Conclusions and Perspectives................................. 117
Appendix A: Data Analysis of Real Turbulent Field.................... 121
Appendix B: Transition to Chaos in Low-Order Galerkin Approximations . 123
Appendix C: About the Word *Intermittency* Encountered in Literature 125
References 126

4   The Solar Atmosphere 129
V.H. Hansteen and M. Carlsson
4.1   Introduction 129
4.2   The Photosphere 131
4.3   The Chromosphere 136
4.4   Coronal Heating.. 143
References 154

5   The Solar Flare: A Strongly Turbulent Particle Accelerator 157
L. Vlahos, S. Krucker, and P. Cargill
5.1   Introduction 157
5.2   Observational Constraints................................... 159
5.3   Models for Impulsive Energy Release......................... 176
5.4   Particle Acceleration in Turbulent Electromagnetic Fields 189
5.5   Discussion of the Global Consideration of Particle Acceleration ... 214
5.6   Summary 216
References 217

6   Diagnostics of the Solar Wind Plasma 223
K. Issautier
6.1   Introduction 223
6.2   Measuring the Key Parameters in the Inner Heliosphere.......... 224
6.3   Measuring the Key Parameters of the Solar Wind 227
6.4   Comparison of Solar Wind Plasma Parameters 236
6.5   Large-Scale Variations of the Heliosphere 240
6.6   Summary and Perspectives 244
References 244

7   Physical Processes in the Solar Wind 247
M. Velli
7.1   The Solar Wind 247
7.2   Hydrodynamics of a Featureless Solar Wind Expansion: Why the
         Solar Wind Is Supersonic 248
7.3   A Wind-Accretion Hysteresis Cycle 254
7.4   Solar Wind Energetics and Empirical Solar Wind Models 257
7.5   Alfvénic Fluctuations and the Solar Wind 260
References 267

8   Physics of Stellar Coronae 269
     M. Güdel
     8.1   Introduction 269
     8.2   Stellar Coronae – Defining the Theme 271
     8.3   The Coronal Hertzsprung–Russell Diagram 271
     8.4   Non-flaring Radio Emission from Stellar Coronae 272
     8.5   Thermal X-ray Emission from Stellar Coronae 277
     8.6   The Structure of Stellar Coronae 286
     8.7   Stellar Radio Flares 300
     8.8   Stellar X-ray Flares 303
     8.9   The Statistics of Flares 315
     8.10   A Flare-Heating Approach 319
     References 322

Chapter 1
Waves and Instabilities in Space Plasmas

K. Papadopoulos

1.1 Introduction

Electromagnetic plasma waves are a ubiquitous feature of space plasmas. They are important agents in propagating energy across different space regions, in providing plasma transport in the absence of collisions in the form of anomalous resistivity, viscosity and isotropization, in accelerating particles to high energies, and in transmitting diagnostic information for the local plasma properties from regions not accessible to in situ measurements. Waves are generated by thermal and non-thermal particle distributions of the plasma populations, by spontaneous or stimulated emission or by instabilities driven by free energy sources in the plasma.

This tutorial lecture does not attempt to provide a comprehensive coverage of the topic that has been the subject of numerous books and articles [2, 22, 6, 21, 10, 7, 9, 3]. It is an eclectic review of wave processes of importance to space plasmas; it reflects my personal style as a practitioner of plasma physics in various space and laboratory settings. The style emphasizes simplicity and physics intuition over strict mathematical rigor (that I liked to leave to my graduate students!). An important ingredient of the tutorial is a parallel exposition of the basic plasma characteristics of the modes, including polarization, phase and group velocities, refractive index surfaces, interaction with particles, mode conversion and transport properties, weak and strong turbulence theories, coupled to observations from space and the laboratory and computer simulations. Emphasis is placed on concepts rather than detailed analysis. Thus most of the review deals with unmagnetized or equivalently weakly magnetized plasmas. Extension to magnetized plasmas is straightforward but laborious.

K. Papadopoulos (✉)
University of Maryland, Department of Physics, College Park, MD 20742, USA
kp@astro.umd.edu

Papadopoulos, K.: *Waves and Instabilities in Space Plasmas*. Lect. Notes Phys. **778**, 1–43 (2009)
DOI 10.1007/978-3-642-00210-6_1   © Springer-Verlag Berlin Heidelberg 2009

1.2 Plasma Description – Response Function – Generalized Ohm's Law

The basic set of equations that provides a complete description of waves in a plasma is

$$\nabla \times \mathbf{B}(\mathbf{r}, t) = \mu_0[\mathbf{J}_0(\mathbf{r}, t) + \mathbf{J}_p(\mathbf{r}, t)] + \mu_0 \varepsilon_0 \frac{\partial \mathbf{E}(\mathbf{r}, t)}{\partial t}, \tag{1.1}$$

$$\nabla \times \mathbf{E}(\mathbf{r}, t) = -\frac{\partial \mathbf{B}(\mathbf{r}, t)}{\partial t}, \tag{1.2}$$

$$\nabla \cdot \mathbf{B}(\mathbf{r}, t) = 0, \tag{1.3}$$

$$\nabla \cdot \mathbf{E}(\mathbf{r}, t) = \frac{1}{\varepsilon_0}[\rho_0(\mathbf{r}, t) + \rho_p(\mathbf{r}, t)], \tag{1.4}$$

where $\mathbf{J}_0(\mathbf{r}, t)$ and $\rho_0(\mathbf{r}, t)$ are current and charge densities due to external sources (e.g., coils, and internal currents of the earth or planets) and $\mathbf{J}_p(\mathbf{r}, t)$ and $\rho_p(\mathbf{r}, t)$ are current and charge densities induced in the plasma.

It is important to note the physical meaning of the electric and magnetic field $\mathbf{E}(\mathbf{r}, t)$ and $\mathbf{B}(\mathbf{r}, t)$. In both vacuum and inside plasma the fields are defined by the force they exert on a test charge e moving with velocity $\mathbf{V}$, i.e.,

$$\mathbf{F}(\mathbf{r}, t) = e[\mathbf{E}(\mathbf{r}, t) + \mathbf{V} \times \mathbf{B}(\mathbf{r}, t)]. \tag{1.5}$$

As in all traditional electrodynamics only the first two of Maxwell's equations are required, while the last two are essentially initial conditions that provide self-consistency at $t = 0$. Furthermore, only the current densities are required since the charge densities can be found from the continuity of charge equation.

In order to solve plasma wave problems we need a model that expresses $\mathbf{J}_p(\mathbf{r}, t)$ as a function of the fields $\mathbf{E}(\mathbf{r}, t)$ and $\mathbf{B}(\mathbf{r}, t)$. The relationship $\mathbf{J}_p(\mathbf{r}, t) = f(\mathbf{E}, \mathbf{B})$ is called the internal response function and does not have to be a linear function of $\mathbf{E}$ and $\mathbf{B}$, although most often is taken as linear. A major part of plasma physics is devoted to the development and justification of plasma models that allow us to compute the plasma currents that are created by the external excitation of an electric or magnetic field. For any given plasma model a self-consistent description can be developed that, at least in principle, solves the problem of waves in plasmas.

A common approach to the computation of the plasma response function is to introduce a conductivity tensor $\bar{\bar{\sigma}}$ that connects the plasma current density to the electric field that drives it. It is a generalized Ohm's law. The most general form of the conductivity tensor should be **non-linear** in E and non-local in space–time (i.e., it should express the dependence of the plasma current on the space–time history of the electric field). However, here as in most plasma wave analysis, we restrict ourselves to situations that the conductivity tensor is independent of the electric field amplitude. This is called the **small signal or linear theory**. In this case, the generalized Ohm's law is written in the following form:

$$\mathbf{J}(\mathbf{x}, t) = \int_V d\mathbf{x}' \int_{-\infty}^t dt' \bar{\bar{\sigma}}(\mathbf{x}, \mathbf{x}'; t, t') \cdot \mathbf{E}(\mathbf{x}', t'). \tag{1.6}$$

The tensor $\bar{\bar{\sigma}}(\mathbf{x}, \mathbf{x}'; t, t')$ describes the propagation characteristics of a disturbance applied at $(\mathbf{x}', t')$ as an electric field and observed at $(\mathbf{x}, t)$ as current. The above relationship represents a linear response function in that the conductivity tensor is determined solely by the properties of the system without the perturbation. In addition to linearity the above equation contains the concept of causality. It is reflected in the choice of the time integration limits $[-\infty, t]$. It states clearly that the effect $\mathbf{J}(\mathbf{x}, t)$ cannot proceed the cause $\mathbf{E}(\mathbf{x}', t')$.

A final simplification in the plasma wave analysis is introduced by assuming that the plasma is stationary in time and homogeneous in space. In this case, $\bar{\bar{\sigma}}$ becomes a function of $\mathbf{x} - \mathbf{x}'$ and $t - t'$ only and the generalized Ohm's law becomes

$$\mathbf{J}(\mathbf{x}, t) = \int_V d\mathbf{x}' \int_{-\infty}^t dt' \bar{\bar{\sigma}}(\mathbf{x}', t') \cdot \mathbf{E}(\mathbf{x} - \mathbf{x}', t - t'). \tag{1.7}$$

This is an expression of translational invariance in space–time. This assumption should be tested carefully. It is valid only if the spatial plasma inhomogeneity has scale lengths much larger than the perturbation scale lengths and the temporal variation of the medium much slower than the perturbation timescale. These will become clearer in the examples presented in the next section.

The major analytical advantage of space–time translational invariance derives from the fact that the natural representation of the system's physical variables are in terms of plane wave states, as can be seen by taking the Fourier transform of (1.7) and using the convolution theorem. Within the framework of spatially homogeneous, temporally stationary, small signal theory, the strategy of investigating plasma response to perturbations is illustrated in Fig. 1.1.

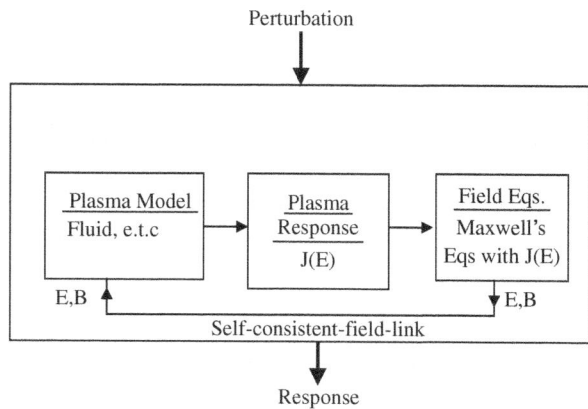

Fig. 1.1 Investigation strategy for plasma wave response to perturbations

1.3 Examples of Plasma Waves in Space – Isotropic Plasmas

We proceed next to apply the above considerations to particular plasma waves relevant to space plasmas. The choice of the examples is not comprehensive but reflects the lecturer's taste and availability of clear data. In this section we focus on high-frequency plasma waves that do not involve ion motion and isotropic plasmas in the sense that they are weakly magnetized.

1.3.1 Electron Plasma Oscillations

Electron plasma oscillations represent the simplest as well as the most fundamental plasma mode. The plasma model used for their description ignores the motion of the ions and describes the electron motion in the presence of the perturbing electric field $E(x, t)$ by the so-called collisionless cold fluid equation of motion so that all electrons move according to the equation

$$m\frac{dv}{dt} = -eE. \tag{1.8}$$

The problem can be described using one spatial dimension (the x-coordinate say), so we can write the perturbed quantities as scalars. The next step is to derive the generalized Ohm's law. We find this by considering the definition of the current density in terms of the ambient plasma density n_0,

$$J(x, t) = -en_0v(x, t). \tag{1.9}$$

From (1.8) and (1.9) we find that

$$\frac{dJ}{dt} = \varepsilon_0\omega_e^2 E, \tag{1.10}$$

where ω_e is the plasma frequency given by

$$\omega_e = \sqrt{\frac{n_0e^2}{\varepsilon_0 m}}. \tag{1.11}$$

Equation (1.10) is the generalized Ohm's law for this particular example. It connects the current and the applied electric field. Notice that it is indicative of reactive response with the phase of the current lagging the phase of the electric field by $\pi/2$. Furthermore by analogy with the equations for AC circuits, $(1/\varepsilon_0\omega_e)$ is equivalent to inductance per unit length.

The plasma behavior can be found by combining (1.10) with Maxwell's equations. Before this step we simplify Maxwell's equations by using the **electrostatic approximation**, i.e., $\nabla \times \mathbf{E} = 0$. It is easy to see that in this case the wave magnetic field is negligible and the set of Maxwell's equations is replaced by

$$J = -\varepsilon_0 \frac{\partial E}{\partial t}. \tag{1.12}$$

From (1.10) and (1.12), we find the equation describing the temporal plasma response to an applied small electric field as

$$\frac{d^2 E}{dt^2} + \omega_e^2 E = 0. \tag{1.13}$$

The plasma electric fields, as well as the associated current and density perturbation oscillate at the plasma frequency. The behavior is similar to that of an harmonic oscillator with mass m and spring constant $n_0 e^2/\varepsilon_0$ or of an LC circuit with capacitance $(n_0 e^2/\varepsilon_0)^{-1}$ and inductance m.

Introduction of dissipation in terms of a collision frequency v modifies (1.13) to a damped harmonic oscillator of the form

$$\frac{d^2 E}{dt^2} + v\frac{dE}{dt} + \omega_e^2 E = 0. \tag{1.14}$$

It should be emphasized that plasma oscillations are just oscillations, not propagating waves and in this sense they do not transport energy or information.

In deriving the equations describing the plasma oscillations we have made several approximations (cold plasma, electrostatic waves, neglect of the magnetic field, small collisionality, homogeneity, and time stationarity). Is there any evidence that they exist in space and where?

An example of such plasma oscillations can be found in the wave data acquired by the wave instrument on the Voyager 1 spacecraft when it crossed Jupiter's bow shock. An excellent description of the measurements can be found in W. Kurth's website (www-pw.physics.uiowa.edu/plasma-wave/tutorial/voyager1/jupiter/bow shock) from where the figures below are taken.

Bow shocks are present in the front of all magnetized planets. Their presence slows down the solar wind flow by deflecting it around the obstacle. In the process, the energy in the bulk motion of the solar wind is converted into thermal energy at the shock. That is, the solar wind observed downstream of the shock is found to be much hotter than the unperturbed supersonic solar wind. In many respects, these same processes occur in the atmosphere around a supersonic aircraft. To be sure, there are several complications associated with the fact that the solar wind is plasma and not simply neutral gas, but the analogy is still good in many qualitative ways.

Bow shocks are excellent sites for the study of plasma waves since any perturbation to the ambient plasma results in the generation of plasma waves. In some cases it may be that the plasma waves act as one mechanism for heating the plasma, turning the energy of bulk flow of the supersonic plasma into the more chaotic downstream flow of heated plasma. In other cases, the waves generated at the shock may simply

Fig. 1.2 Plasma oscillations in front of Jupiter's bow shock. The plot shows a frequency–time spectrogram with an associated color table. The shock is located at approximately 80, and the unshocked(shocked) plasma is to the *left* (*right*) of the shock

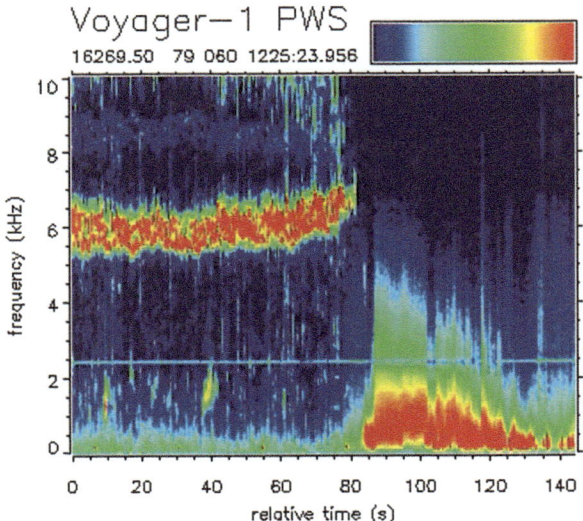

be a result of the heating process and be of more interest for their diagnostic utility or particle acceleration.

Voyager 1 made a remarkable set of observations of the Jovian bow shock as it crossed the boundary on February 28, 1979, at approximately 72 Jovian radii from the planet. The wave measurements are most often studied in the form of frequency–time spectrograms such as the one depicted in Fig. 1.2 (courtesy of W. Kurth).

This spectrogram shows the variation of wave intensity and frequency with time as Voyager 1 passed through the shock. The ordinate or vertical axis depicts the frequency of waves detected by the Voyager plasma wave instrument. The horizontal axis (abscissa) represents time and the entire time for this spectrogram is 144 s (2.4 min). The color scheme is employed in order to convey information about the intensity of the waves. Red implies the most intense waves and blue the least intense waves.

The spectrogram shows narrowband emission near 6 kHz during the early part of the time interval. This emission is seen when the spacecraft is still in the supersonic solar wind upstream of the shock. These emissions correspond to electron plasma oscillations and represent the oscillatory motion of electrons about their equilibrium positions. From the frequency of this band of emissions we can say that the plasma density of the solar wind just upstream of the bow shock is about 0.44 particles/cm^3. The cause of these oscillations as well as the nature of the waves shown downstream of the shock will be discussed later.

1.3.2 Propagating Electron Plasma Waves

The physics of transforming electron plasma oscillations to electron plasma waves is similar to the propagating waves generated by transitioning from a single pendulum to a set of pendulums coupled by springs or from an LC circuit to a network of coupled capacitances and inductances. The spring or the inductive or capacitive coupling transmits any oscillatory motion of the first pendulum or LC circuit to the subsequent pendulums or circuits, generating propagating waves. For plasmas the coupling agent is the random thermal velocity of the electrons. To determine the properties of plasma waves we have to improve the plasma model over the one given by (1.8) by including the effect of the thermal motion of the electrons that can be described as a fluid pressure P. The new plasma model is given by

$$m\frac{\partial v}{\partial t} = -eE - \frac{\nabla P}{n_0}. \tag{1.15}$$

Notice that it is a partial differential equation since it now includes a spatial derivative. For plasmas the electron pressure as a function of the electron thermal speed V_e is given by

$$P = \alpha(n_0 + n)mV_e^2, \tag{1.16}$$

where n is the density perturbation. The value of α depends on the degrees of freedom of the electron motion. For one-dimensional plasmas (i.e., strongly magnetized) it is $1/2$, while for unmagnetized or weekly magnetized plasmas it is $3/2$.

Using (1.15) and (1.16) in conjunction with (1.4) we find that

$$\nabla P = \alpha mV_e^2\nabla n = -\left(\frac{\alpha\varepsilon_0}{e}\right)mV_e^2\nabla^2 E. \tag{1.17}$$

Equations (1.15, 1.16, 1.17) and (1.9) give

$$\frac{\partial J}{\partial t} = \varepsilon_0(\omega_e^2 E - \alpha V_e^2\nabla^2 E). \tag{1.18}$$

Equation (1.18) is the plasma response to an electric field E in the form of a generalized Ohm's law similar to (1.10) but including electron thermal motion. Using the electrostatic approximation of Maxwell's equations (1.12) with (1.18) we find

$$\frac{\partial^2 E}{\partial t^2} - \alpha V_e^2\nabla^2 E + \omega_e^2 E = 0. \tag{1.19}$$

Equation (1.19) has the form of a wave equation that describes propagating electron plasma waves.

1.3.3 Electromagnetic Waves in Isotropic Plasmas

Sections 3.1 and 3.2 deal with electrostatic oscillations and waves that are mainly longitudinal waves similar to sound waves in a gas. In the absence of external currents ($J_0 = 0$) and eliminating B from (1.1) and (1.2) we find

$$\frac{\partial^2 E}{\partial t^2} - c^2 \nabla^2 E = \varepsilon_0^{-1} \frac{\partial J}{\partial t}. \tag{1.20}$$

In the right-hand side of (1.20) we have dropped the subscript p from J_p. With the exception of [1] the symbol J represents plasma currents. For $J = 0$ (1.20) is the usual wave equation that describes propagation of electromagnetic waves in vacuum at the speed of light. To find the propagation of electromagnetic waves in isotropic plasmas, we need to replace the right-hand side by either the cold or thermal plasma Ohm's law given by (1.10) or (1.18). It turns out that for non-relativistic plasmas (1.10) gives a very accurate – to order $(V_e/c)^2$ – representation. From (1.10) and (1.20) we find

$$\frac{\partial^2 E}{\partial t^2} - c^2 \nabla^2 E + \omega_e^2 E = 0. \tag{1.21}$$

1.4 Properties of Plasma Waves in Isotropic Plasmas

1.4.1 Electrostatic Waves – Dispersion Relations – Phase and Group Velocity – Dielectric Constant

It was noted earlier that a major advantage in the analysis of translationally invariant systems derives from the fact that the natural representation of the system's physical variables is in terms of plane waves. We assume next that the electric field of the wave is of the form

$$E(r, t) \sim \exp[i(kx - \omega t)] \equiv \exp[i\varphi(x, t)]. \tag{1.22}$$

Substituting (1.22) into (1.13) and (1.19) we find for the plasma oscillations and the electrostatic plasma waves correspondingly the relationships

$$\omega^2 = \omega_e^2, \tag{1.23}$$

$$\omega^2 = \omega_e^2 + \frac{3}{2} k^2 V_e^2. \tag{1.24}$$

Relationships of the form of (1.23) and (1.24) are known as **dispersion relations**. They give the relation between the frequency and the wavelength of the normal modes of the plasma. Notice that dispersion relations are asymptotic. They give the wavelength of a perturbation as a function of frequency after the initial transients have died-down. Notice that for the plasma oscillations all wavelengths have the same frequency. This, however, is not true for the plasma waves. Equations (1.23) and (1.24) plotted in diagrams of ω vs. k are known as dispersion diagrams and characterize the behavior of electrostatic waves in isotropic plasmas (see Fig. 1.3).

Fig. 1.3 Dispersion diagrams for electrostatic (*lower curve*) and electromagnetic plasma waves (*upper curve*) in isotropic plasmas. The *solid lines* are the actual dispersion relations, the *dashed* ones correspond to the labeled dispersion relation

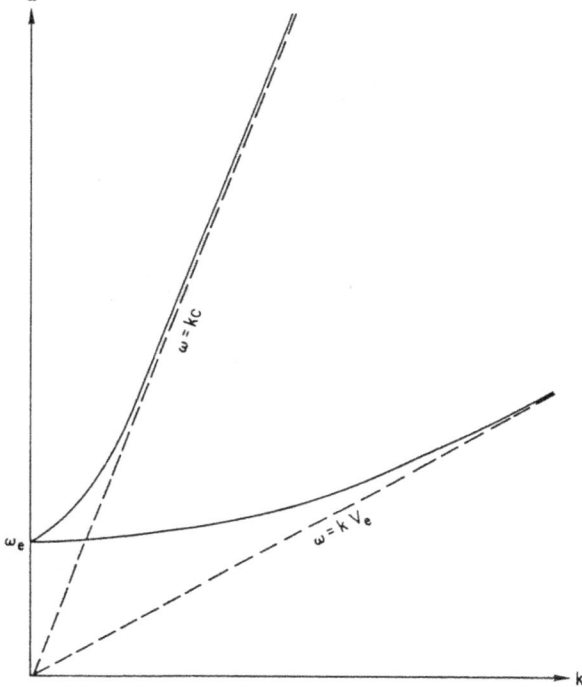

An equivalent and most important description of plasmas is through their dielectric properties. Using as an example the electrostatic plasma oscillations described in Sect. 3.1, we note that the introduction of a perturbed electric field E led to a plasma current given by (1.10). This in turn leads to the development of a polarization charge ρ that relates to the plasma current through the current continuity equation.

Equations (1.23) and (1.24) allow us to compute the two characteristic velocities of plasma waves, the phase velocity V_p and the group velocity V_g. The phase velocity of a harmonic wave is defined as the speed at which a specific phase $\phi(x, t)$ of the particular wave propagates in the plasma. The locus of points $d\phi(x, t) = 0$ with constant phase can be found by taking the total differential. From the definition of phase for plane waves as given in (1.22) we find

$$d\varphi(x, t) = k dx - \omega dt. \tag{1.25}$$

This is zero provided that dt and dx are related by

$$V_p = \left(\frac{dx}{dt} \right)_{[d\varphi=0]} = \frac{\omega}{k}. \tag{1.26}$$

Equation (1.26) is the definition of the phase velocity. It is the speed at which the crests and troughs of the wave progress. From Eq. (1.23) we find that the phase velocity of plasma oscillations can be anything – in fact it is meaningless. For $\alpha = 3/2$ the phase velocity of electrostatic plasma waves is given by

$$V_p^2 = \frac{\omega_e^2}{k^2} + \frac{3}{2} V_e^2 > V_e^2, \tag{1.27}$$

Notice that it is always larger than the thermal velocity of the plasma. By comparing (1.23) and (1.24) we find that the cold plasma approximation that describes electron plasma oscillations is valid when

$$\sqrt{3/2} \frac{k V_e}{\omega_e} = \sqrt{3} k \lambda_d << 1. \tag{1.28}$$

The right-hand side of the inequality (1.28) includes the definition of the Debye length λ_d. The physical meaning of the cold plasma approximation is that the thermal velocity of the plasma is low; the particles move less than a wavelength during one plasma time $1/\omega_e$ and therefore cannot propagate any disturbance.

The second characteristic velocity is the group velocity. It is the velocity that a modulation (wave packet) imposed on a plasma wave propagates. It is the velocity at which energy and information propagate and is defined as

$$V_g = \frac{d\omega}{dk}. \tag{1.29}$$

The group velocity is a real velocity and is always bounded by the speed of light. As we will discuss in the next section this is not the case for the phase velocity.

For electrostatic plasma waves we find that

$$V_g = \left(\frac{3}{2}\right)\left(\frac{k V_e}{\omega}\right) V_e = \left(\frac{3}{2}\right)\left(\frac{V_e}{V_p}\right) V_e < V_e. \tag{1.30}$$

From (1.30) we see that the group velocity of the electrostatic plasma waves is smaller than the thermal velocity of the plasma and for $V_e \to 0$ it becomes zero as expected from cold plasma oscillations. Another characteristic of (1.30) is the occurrence of dispersion. Namely disturbances with different wave numbers k propagate at different speeds and the original wave packet disperses as it propagates.

The plane wave description given by (1.22) facilitates the introduction of another important plasma function, the dielectric constant $\varepsilon(k, \omega)$ of a plasma when it is viewed as a dielectric medium. In this description equation (1.1) with $J_0 = 0$ is replaced by

$$\nabla \times (B) = \mu_0 \varepsilon_0 \varepsilon \frac{\partial(E)}{\partial t}. \tag{1.31}$$

The above equation is essentially the definition of the plasma dielectric constant ε. Comparing this equation with Eq. (1.1) we find that

$$(\varepsilon - 1)\frac{\partial^2 E}{\partial t^2} = (1/\varepsilon_0)\frac{\partial J}{\partial t}. \tag{1.32}$$

From (1.22), (1.32) and (1.10) we find the dielectric constant for cold plasmas is

$$\varepsilon(\mathbf{k}, \omega) = 1 - \frac{\omega_e^2}{\omega^2}. \tag{1.33}$$

If instead of (1.10) we use (1.19) for thermal plasmas we find

$$\varepsilon(\mathbf{k}, \omega) = 1 - \frac{\omega_e^2 + (3/2)k^2 V_e^2}{\omega^2}. \tag{1.34}$$

Notice that for electrostatic waves the dispersion relations given by (1.23 and 1.24) correspond to $\varepsilon(\mathbf{k}, \omega) = 0$.

Before closing this section we should note that calling the disturbances shown in Fig. 1.2 plasma oscillations is erroneous. Plasma oscillations have zero bandwidth, while the observed bandwidth is approximately 1 kHz, corresponding to $(1/6)\omega_e$, too large to attribute it to Doppler shift due to motion of the satellite. They are in fact plasma waves with wave number spread Δk of the order $(1/6)\lambda_d$. This is an important observation that will guide us when we attempt to understand their origin.

1.4.2 Electromagnetic Plasma Waves – Index of Refraction – Cutoff

Applying (1.22) to (1.21) and following the procedures and definitions of Sect. 4.1 we find electromagnetic waves described by

$$\omega^2 = \omega_e^2 + k^2 c^2, \tag{1.35}$$

$$V_p = c^2 + \frac{\omega_e^2}{k^2} > c^2, \tag{1.36}$$

$$V_g = c^2 / V_p < c. \tag{1.37}$$

Notice that the phase velocity of electromagnetic waves is larger than the speed of light. The cause of the super-luminous speed is the presence of the term ω_e^2. If this term is zero, that is, in the absence of plasma, the waves have the same group and phase velocity c and the wave packets do not disperse as expected for wave propagation in free space. The dispersion diagram for (1.35) is shown in Fig. 1.3.

Equation (1.36) can be written as

$$\eta^2 = \frac{k^2 c^2}{\omega^2} = 1 - \frac{\omega_e^2}{\omega^2}. \tag{1.38}$$

In (1.38) we have incorporated the definition of the index of refraction η. In general we can write

$$\eta^2 \equiv \frac{k^2 c^2}{\omega^2} = \varepsilon(k, \omega). \tag{1.39}$$

In this respect (1.38) corresponds to the cold plasma model. For thermal plasmas one should use (1.34). However, as long as $V_e \ll c$ the use of the cold plasma model is fully justified.

Equation (1.38) exhibits a phenomenon known as **cutoff**. If we inject into the plasma an electromagnetic wave with frequency ω, the wave number k will acquire a value given by (1.35). As the plasma density, and hence ω_e, increases the wave number k decreases, while the wavelength becomes longer. When the wave number becomes zero, which occurs for a frequency equal to the plasma frequency, the wave cannot propagate and is reflected. This is known as a **cutoff**. Frequencies below the plasma frequency cannot propagate since the value of the wave number becomes imaginary and the wave becomes **evanescent**.

Notice that both the super-luminous phase velocity and the reflection at the critical frequency can be understood in an elementary fashion by considering an electromagnetic wave incident upon a thin one-dimensional plasma slab with thickness Δz. The transmitted radiation is the sum of the incident wave and the radiation produced by a thin infinite slab of electrons whose radiation is anti-phased to the original wave. For $\omega \gg \omega_e$ there is only partial cancelation of the incident field and the main effect is a phase advance of the wave in steady state resulting in super-luminous phase speed, when the electron number density reaches the critical value (i.e., $\omega = \omega_e$) total field cancelation occurs and the incident wave is reflected.

1.4.3 Examples of Plasma Waves in Space

Besides the electrostatic plasma waves that we described previously there are many sites where electromagnetic plasma waves are measured in space. Of particular importance are the type II and type III solar radio-bursts shown in Figs. 1.4 and 1.5. These waves are generated locally at the plasma frequency and its harmonics and detected either on the ground or on orbiting spacecrafts. They appear to be driven by an agent that propagates at a speed of either about 1/3 of the speed of light (type III) or at about 10000 km/s (type II). The agent of the type III bursts is an energetic electron beam, while the type II bursts originate at a shock waves moving through the solar corona and the interplanetary medium. The physics underlying

Fig. 1.4 An example of a type III radio-burst observed in space. The *lower panel* is a time–frequency spectrogram. The *upper panel* interprets this in terms of a sequence of type III bursts, each with rapidly decreasing frequency, superimposed on which are bursts of Auroral Kilometric Radiation (AKR)

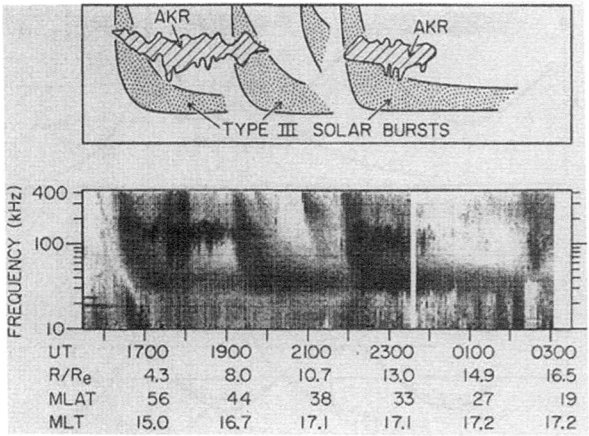

these radio-bursts was among the most challenging of the 1960s and 1970s and their eventual resolution was a triumph of strong plasma turbulence theories ([15] and references therein). We will comment on these later on in this chapter.

Figure 1.4 shows a type III radio-burst detected by orbiting satellites. The emission frequency decreases with time as the electron beam propagates from regions of high plasma density near the corona to lower plasma densities.

Figure 1.5 shows the emission at the local plasma frequency and its second and third harmonic as measured from the ground at the Meudon observatory. As will be discussed later the presence of the third harmonic can only be a consequence of strong turbulence.

The next example of plasma waves in space comes from ionospheric physics and refers to an instrument known as ionosonde that measures the plasma density as a function of altitude for 90–400 km heights. The measurement principle is illustrated in Fig. 1.6. The ionosonde is an instrument that sends pulses of electromagnetic waves vertically upward while sweeping the frequency from 0.5 to 10 MHz. By measuring the time delay between the pulses and echoes received on the ground

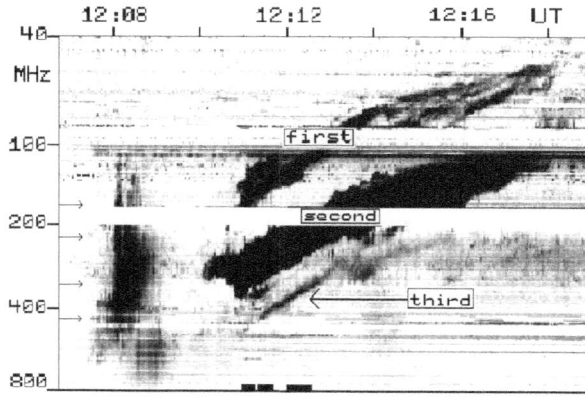

Fig. 1.5 Radio-bursts measured at the Meudon observatory shown in a time – frequency spectrogram. Notice the presence of the third harmonic

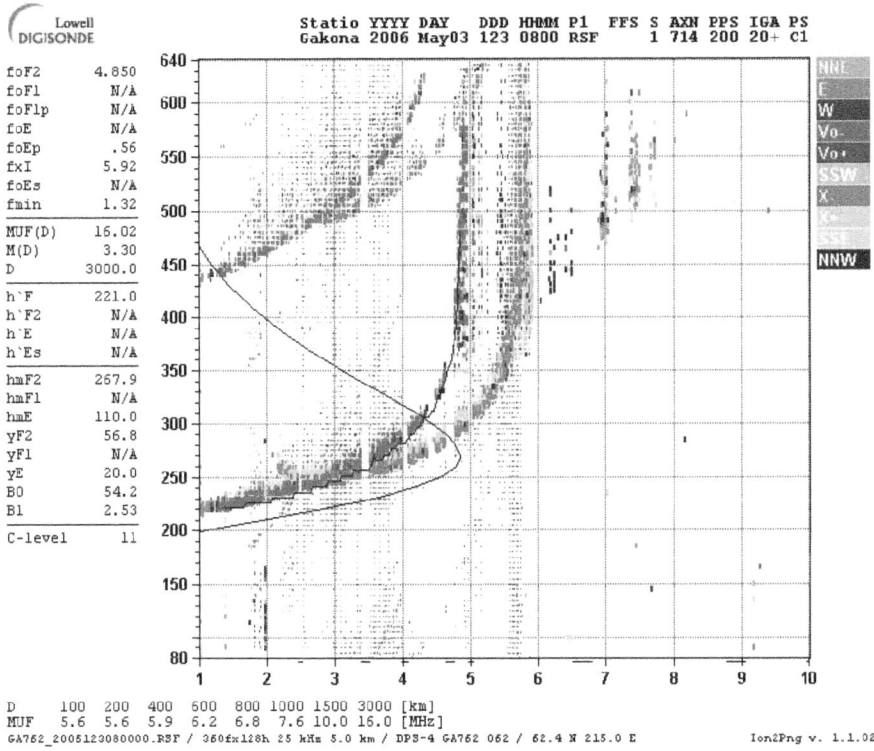

Fig. 1.6 Ionosonde measurement of the ionospheric density profile. The *right-hand panel* shows
how a density profile can be built up by measuring delay times. [Multiple echoes are due to multiple
reflections.] The *left-hand panel* shows a typical ionospheric density profile

we can find the reflection height as a function of frequency. Since the reflection
occurs at the critical frequency (i.e., the local plasma frequency), the density at the
reflection height is given by

$$n \approx 1.25 \times 10^4 (f/\text{MHz})^2 \text{particles/cm}^3.$$

We end this section by commenting that AM radio-wave propagation is entirely due
to wave reflection from the ionospheric plasma. Furthermore, satellite communication and navigation systems use frequencies in the GHz range, which are much
higher than the maximum plasma frequency in the ionosphere that is usually below
10–12 MHz and are thus minimally affected by the ionospheric plasma.

1.5 Generation of Plasma Waves

There are many ways to generate plasma waves. In fact any perturbation of the
plasma from its equilibrium state results in the generation of normal plasma wave
modes during return to equilibrium. We describe below some of the most important
as well as common ways to excite waves in plasmas.

1.5.1 Externally Applied Sources – Antenna Radiation in Plasmas

The problem of plasma wave generation by externally applied oscillating current sources, such as antennas, can be addressed by retaining in the wave equations the source terms $J_0(r, t)$ of (1.1). The plasma wave equations (1.19) and (1.21) become

$$\frac{\partial^2 E}{\partial t^2} - \alpha V_e^2 \nabla^2 E + \omega_e^2 E = \varepsilon_0^{-1} \frac{\partial J_0(r, t)}{\partial t}, \tag{1.40}$$

$$\frac{\partial^2 E}{\partial t^2} - c^2 \nabla^2 E + \omega_e^2 E = \varepsilon_0^{-1} \frac{\partial J_0(r, t)}{\partial t}. \tag{1.41}$$

The terms on the right-hand side of the equations represent the antenna currents. For a harmonic external current of the form $\exp[-i\omega t]$ (1.40) becomes

$$\nabla^2 E + k_0^2 E = \frac{i\omega J_0(r)}{\varepsilon_0 \alpha V_e^2}, \tag{1.42}$$

where

$$k_0^2 = \frac{\omega^2 - \omega_e^2}{\alpha V_e^2}. \tag{1.43}$$

Equation (1.42) is an inhomogeneous Helmholz equation. The right-hand side is the externally forced term due to the antenna current. The expression for the wave number k_0 indicates the dispersion relation for electron plasma waves. The solution is given by

$$\mathbf{E}(r, t) = \hat{\mathbf{k}} \frac{i\omega \exp(-i\omega t)}{\varepsilon_0 \alpha V_e^2} \int_{V'} dV' \frac{J_0(r') \exp(-ik|r - r'|)}{|r - r'|}, \tag{1.44}$$

The prime coordinate refers to the external source, V is a volume, and $\hat{\mathbf{k}}$ is the unit vector in the k direction. For a point or highly localized source (1.44) becomes

$$\mathbf{E}(r, t) = \hat{\mathbf{k}} \frac{i\omega J_0 \exp(-i\omega t + ik_0 R)}{\varepsilon_0 \alpha R V_e^2}. \tag{1.45}$$

In (1.45) R is the distance from the source to the observation point. The equation indicates omni-directional spherical plasma waves radiated from a point source with wave number k_0 and spatial attenuation $1/R$. The wavefronts are shown in Fig. 1.7. The value of α is 3/2 for this case.

Radiation from dipole sources for both electrostatic and electromagnetic waves can be found by using similar methodology as above. For dipole radiation sources in isotropic plasmas the electromagnetic radiation is radiated mostly in the direction perpendicular to dipole direction, while the electrostatic plasma wave in the direction of the dipole (see Fig. 1.7). This is a typical difference between longitudinal and transverse plasma waves. In fact by determining the directivity of plasma waves one can distinguish their character.

Fig. 1.7 Wavefronts for an isotropic (*upper*) and dipole (*lower*) radiation source

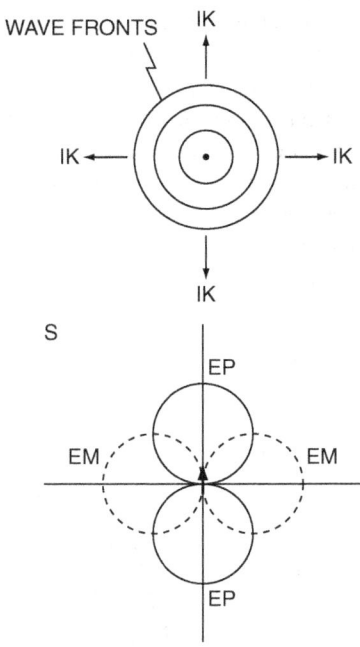

1.5.2 Particle Beam Antennas

Particle beam antennas consist of charged particle (electron or ion) beams. Many space experiments have been performed using frequency-modulated particle beams to excite plasma waves. Such radiators have the advantage of giving large equivalent antennas with small beam sources and avoiding plasma contact with metallic structures. On the other hand, there are issues related to charge and current neutralization when the beams are injected into the plasma. Excellent discussions on beam antennas can be found in [11] from which this discussion has been abstracted. An advantage of beam antennas is the possibility to control the wavefronts of the radiating waves by adjusting the frequency and the ratio of beam to the phase velocity of the excited mode. An example of electron plasma waves radiated from electron beam antennas is shown in Figs. 1.8 and 1.9.

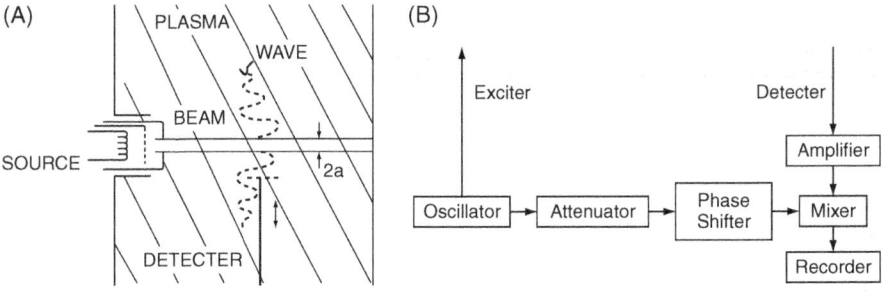

Fig. 1.8 An overview of the electron beam antenna (**A**) and the measurement system (**B**) for generating and detecting plasma waves

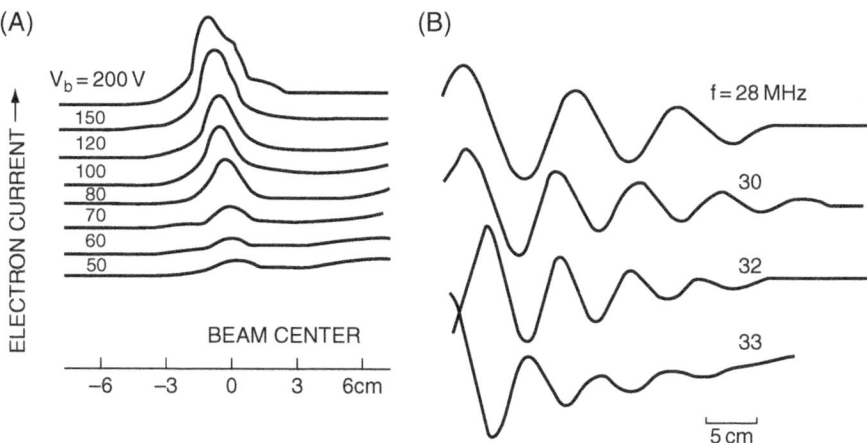

Fig. 1.9 The *upper panel* shows the current profile of the electron beam described in Fig. 1.8. The *lower panel* shows the profiles of the radiated plasma waves

Figure 1.8 shows the experimental setup for launching the waves along with the measurement system. Figure 1.9A shows the current distributions across the electron sheet beam. The zero level corresponds to a beam voltage of 50-V, with larger voltages shifted successively. Figure 1.9B shows the wave patterns of plasma waves radiated perpendicular to the sheet beam.

1.5.3 Cerenkov Emission

Another way to generate plasma waves is Cerenkov emission. Cerenkov discovered this process in 1934 as a bluish glow from water in the presence of β-ray emitters. The cause of the emission is the presence of electrons in the water moving faster than the phase speed of the normal modes, in this case electromagnetic waves in the water, that since $\eta > 1$ they have $V_p = c/\eta < c$. A process analogous to Cerenkov emission is the sonic boom created by supersonic planes. Referring to (1.38) for a homogeneous plasma $\eta < 1$ and as a result there **is no Cerenkov emission of electromagnetic waves**. Notice, however, that the phase velocity of electrostatic plasma waves is of the order of a few times V_e and as a result super-thermal electrons satisfying the condition $V > \omega/k$ can emit plasma waves even if they move at constant speed. For example, electrons in the tail of a Maxwellian distribution can emit plasma waves. The condition for Cerenkov emission is

$$\omega - \mathbf{k} \cdot \mathbf{V} = 0. \tag{1.46}$$

The angle $\theta_c = \arccos(\omega/kv)$ is known as Cerenkov angle. As shown in Fig. 1.10, a particle at O at $t = 0$ emits waves with $V_p = \omega/k$ [8]. At t the particle is at P at a distance $OP = vt$ from O and the wavefront reaches A at a distance

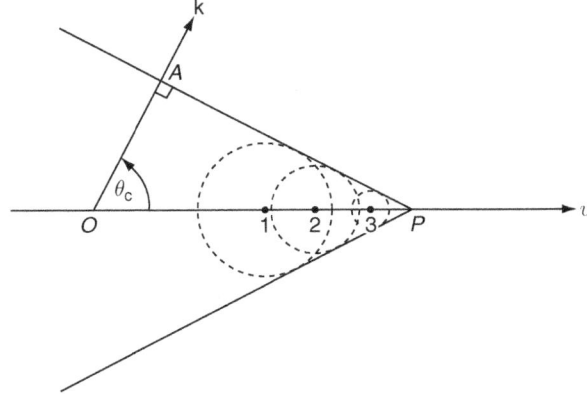

Fig. 1.10 The mechanism responsible for Cerenkov emission from a super-thermal particle. Details can be found in the text

$OA = V_p t$ from O. The wavefronts emitted at the various times between zero and t form a bow wave PA.

The rate of Cerenkov emission by a particle distribution $f(\mathbf{v})$ is given by

$$\frac{\partial E(k)}{\partial t} = \frac{\pi e^2 \omega_e^4}{2k^2} \int d\mathbf{v} f(\mathbf{v}) \delta(\omega_e - \mathbf{k} \cdot \mathbf{v}). \tag{1.47}$$

1.5.4 Landau Damping and Stimulated Emission

While Cerenkov emission describes emission of a wave by a particle moving in phase synchronism, Landau damping describes the opposite; absorption of plasma waves by a particle in phase synchronism. Landau damping is probably the most subtle and important property of collisionless plasmas.

A simple but physical description of Landau damping can be given by referring to the one-dimensional interaction of a sinusoidal wave with a particle moving at a constant speed v. The particle motion is then given by

$$\frac{dv}{\partial t} = \frac{eE}{m} \exp[i(\omega t - kx(t))], \tag{1.48}$$
$$x(t) = vt.$$

The solution of the above equations is

$$\Delta v = \frac{eE}{m} \int dt \exp[i(\omega - \mathbf{k} \cdot \mathbf{v})t]. \tag{1.49}$$

Notice that for $\omega \gg kv$ and $\omega \ll kv$ the interaction results in an oscillatory quivering of the electron in the presence of the wave, similar to the sloshing of boats in the presence of small waves with no energy exchange. This, however, is not the case at the resonance $\omega - \mathbf{k} \cdot \mathbf{v} = 0$. The solution of (1.49) gives a secular behavior in time

$$\Delta v = (eE/m)t. \tag{1.50}$$

Under these circumstances the proper way to handle the interaction between waves and resonant particles is to introduce in the kinetic description of the interaction a resonant denominator of the form $1/(\omega - \mathbf{k} \cdot \mathbf{v} + i\varepsilon)$. This is known as the Landau prescription. Using the Landau prescription we find that the damping of a plasma wave by resonant particles is given by

$$\gamma = -\frac{\pi \omega_e^3}{2k^2} \left[\frac{\partial F(u)}{\partial u} \right]_{u=\frac{\omega}{k}}, \tag{1.51}$$

$$F(\mathbf{u}) = \int f(\mathbf{v})d\mathbf{v}_T,$$

where $\mathbf{v}_T$ is the component of velocity transverse to $\mathbf{k}$ and γ is the damping (or growth) rate of the wave. Notice that (1.51) gives wave damping for $\frac{\partial F(u)}{\partial u} > 0$. In the opposite case it describes stimulated emission of plasma waves. To quote Chen [4]

> the theoretical discovery of wave damping without energy dissipation by collisions is perhaps the most astounding result of plasma physics research. It is a real effect having been demonstrated in the laboratory many times. Although a physical explanation of this damping is now available, it is a triumph of applied mathematics that this unexpected effect was first discovered purely mathematically in the course of a careful analysis of a contour integral. Landau damping is a characteristic of collisionless plasmas, but it may have applications to other fields. For instance, in the kinetic treatment of galaxy formation, stars can be considered as atoms of a plasma interacting via gravitational rather than electromagnetic forces. Instabilities of the gas of stars can cause spiral arms to form but this process is limited by Landau damping.

The simplest way to understand Landau damping is by referring to Fig. 1.11 that shows a surfer trying to catch an ocean wave. If the surfboard is not moving it simply sloshes up and down in a way equivalent to Δv of (1.49) for $\omega >> kv$ and $\omega << kv$. However, if the surfboard has similar velocity with the phase velocity

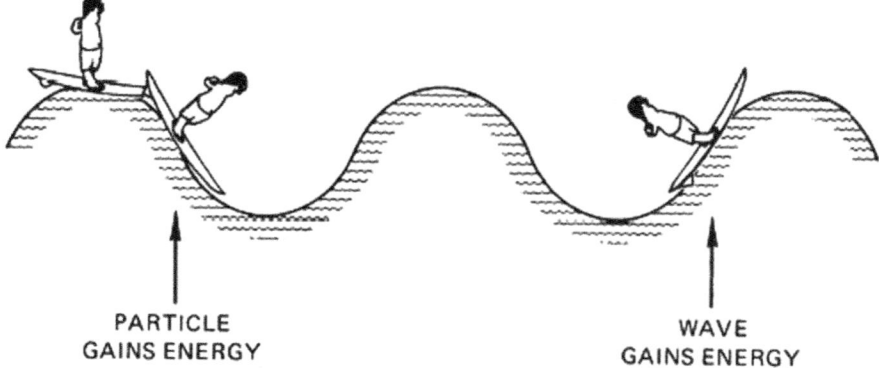

PARTICLE   WAVE
GAINS ENERGY   GAINS ENERGY

Fig. 1.11 A customary physical picture of single particle Landau damping and stimulated emission [4]. See text for details

Fig. 1.12 A mechanical analog of collective Landau damping and stimulated emission. See text for description

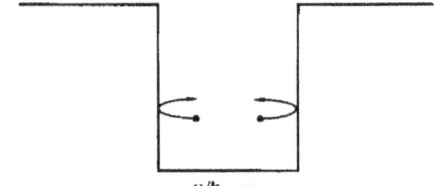

of the wave, the surfer catches the wave and rides on it, exhibiting secular time behavior similar to (1.50). If the surfer has a velocity slightly slower than the phase speed of the wave, he gains energy at the expense of the wave (damping). On the other hand, if the surfer is moving slightly faster than the wave, it pushes on it as he moves uphill transferring energy to the wave (stimulated emission).

While the above analog refers to a single particle interaction, another mechanical analog [7] illustrates the collective nature of Landau damping and stimulated emission. Consider a group of boxes translating along at the phase velocity $V_p = \omega/k$ (Fig. 1.12). Inside the boxes are uniformly distributed particles, some moving slightly slower than ω/k, and some slightly faster. As shown in Fig. 1.12, those particles moving slower than ω/k are overtaken by the wall on their left and gain energy as they are reflected. Similarly, those particles moving faster than ω/k overtake the right wall and lose energy as they are reflected. Considering only times less than a transit time of a particle across the box, the box (representing the wave) will lose energy if more particles are moving slower than ω/k (Landau damping), and will gain energy (stimulated emission) if more particles move faster than ω/k (population inversion).

1.5.5 Tests of the Plasma Dispersion with Damping

Landau damping modifies the dispersion relation for a plasma given by (1.34) to

$$\varepsilon(k, \omega) = 1 - \frac{\omega_e^2 + (3/2)k^2 V_e^2}{\omega^2} + i\frac{\pi \omega_e^3}{2\omega k^2}\left[\frac{\partial F(u)}{\partial u}\right]_{u=\omega/k}. \qquad (1.52)$$

The imaginary part of the dispersion relation in (1.52) represents the Landau damping of the plasma wave.

Figure 1.13 shows the experimental tests of the above dispersion relation 1.52. A modulated electron beam (Fig. 1.8) and a grid with oscillating voltage excited the plasma wave. The wave number k_r and the spatial damping of the wave $1/k_i$ were measured and plotted against the frequency ω. The results were consistent with the theoretical prediction from [10].

Fig. 1.13 Experimental tests of the dispersion relation

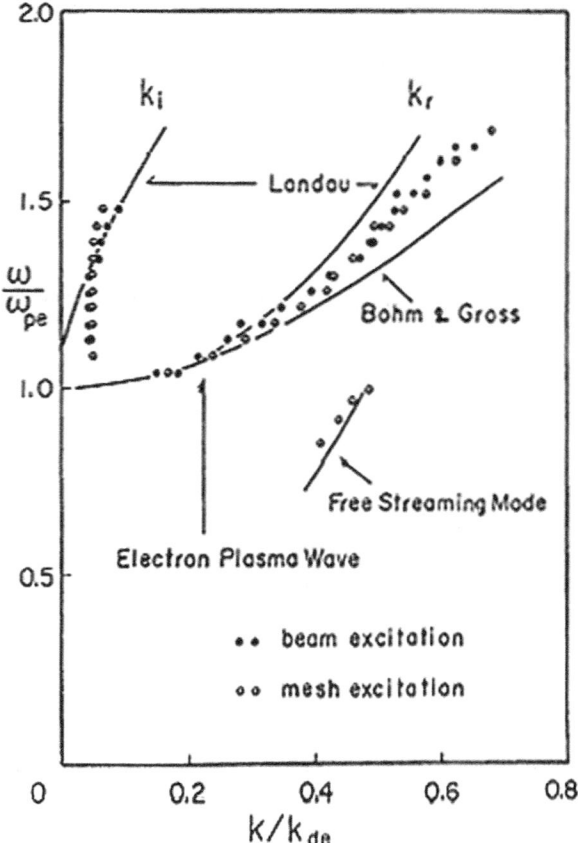

1.5.6 Plasma Waves in Stable Thermal and Non-Thermal Plasmas

In all plasmas there is a finite level of plasma wave activity. The plasma waves are the "normal modes" of the system, and represent degrees of freedom, which according to thermodynamics should be excited in thermal equilibrium. Plasma waves are emitted by discrete particles at the Cerenkov emission rate given by (1.47) and in collisionless plasmas are absorbed at the Landau damping rate given by (1.51). Balancing the emission rate with the absorption rate and integrating over the ω_e resonance, we find that for a Maxwellian plasma with temperature T the energy density of the electric field fluctuations W_k for mode k is given by

$$W_k = \left(\frac{1}{2}\right) \frac{k_B T}{1 + (k\lambda_d)^2}. \tag{1.53}$$

Equation (1.53) shows that weakly damped modes with $k\lambda_d \approx V_e/V_p << 1$ are fully excited to $k_B T/2$ per degree of freedom, while strongly damped modes with

$k\lambda_d \gg 1$ are only weakly excited. Notice that by integrating (1.53) over k we find that the ratio of the total energy density of the electrostatic fluctuations W_L to the thermal energy of the plasma nk_BT is given by

$$\frac{W_L}{nk_BT} \approx \frac{1}{n\lambda_d^3}.$$

The number of particles per Debye sphere is known as the plasma expansion parameter and the formal definition of plasmas is $n\lambda_d^3 \gg 1$ [9].

The level of the plasma waves increases significantly even for stable plasmas under non-equilibrium conditions. Typical examples for such plasmas are ones that contain a long non-Maxwellian tail. Such plasmas abound in space settings. An estimate of the level of the plasma wave energy density can be found by considering a plasma with thermal velocity V_e that includes a high-energy tail modeled as a Maxwellian with thermal velocity $V_E \gg V_e$ containing a few percent of the particles [17]. The ratio of the enhanced level of the energy density to the thermal W_L is given by

$$\frac{W^{NT}}{W_L} = \frac{V_E^2/V_e^2}{\ln[V_E/\beta V_e]}. \tag{1.54}$$

Generalization of this equation for relativistic plasmas can be found in [12]. Notice that the presence of a 10% high-energy tail with $V_E = 20V_e$ results in a factor of 100 enhanced fluctuations. To zero order the effective temperature of the plasma waves corresponds to the temperature of the high-energy tail. Enhanced electron plasma waves due to high-energy tails generate broadband plasma wave spectral with bandwidth $\Delta\omega$, a significant fraction of the plasma frequency. In looking at the plasma wave spectrum shown in Fig. 1.2 we can conclude that it is too narrowband to be caused by high-energy tails.

1.6 Plasma Instabilities

1.6.1 Wave Energy Density – Positive and Negative Energy Waves

Plasmas are notoriously unstable and transform energy contained in non-thermal features such as drifts and beams into very large amplitude plasma waves. Before addressing plasma instabilities per se it is important to introduce a few definitions.

We start with a simple calculation of the wave energy density for electrostatic waves. The presence of a longitudinal wave with average energy density $\varepsilon_0E^2/2$ in plasmas results in electrons quivering with quiver velocity $\tilde{V} = eE/m\omega$. As a result the total average longitudinal energy contained in the wave is given by

$$< W_L > = \frac{1}{2}\varepsilon_o E^2 + \frac{1}{2}nm\tilde{V}^2 = \frac{1}{2}\varepsilon_o E^2\left(1 + \frac{\omega_e^2}{\omega^2}\right) = \frac{1}{2}\varepsilon_o E^2\frac{\partial}{\partial\omega}[\omega\varepsilon(\omega)]. \tag{1.55}$$

In a similar fashion we can show that the wave energy density for an electromagnetic wave is given by

$$< W_T >= \frac{1}{2}\varepsilon_o E^2 \frac{\partial}{\partial\omega}\omega[(\frac{kc}{\omega})^2 - \varepsilon(\omega)]. \qquad (1.56)$$

The Poynting flux is of course given by

$$< S >= V_g < W >,$$

and the appropriate values of W and V_g are used.

These definitions have some peculiar consequences when applied to drifting plasmas. If we consider a cold plasma drifting with velocity V in the laboratory frame, the dispersion relation can be found from (1.33) by simply Doppler shifting ω to $\omega - kV$. As a result the dielectric constant becomes

$$\varepsilon_b(k, \omega; V) = 1 - \frac{\omega_b^2}{(\omega - kV)^2}. \qquad (1.57)$$

Let us look at the modes of (1.57) by taking the right-hand side as zero. We find modes with

$$V_P \equiv \frac{\omega}{k} = V \pm \frac{\omega_b}{k}. \qquad (1.58)$$

The wave moving with phase speed slower than the beam is known as slow wave, while the other as fast wave. Now if we use (1.57) and (1.58) in (1.55) we find that the slow wave has $< W > < 0$ while the fast wave has positive energy. The slow wave is known as **negative energy wave** and plays a critical role in the theory of instabilities. The meaning of negative energy is not that the wave energy is negative, but that in the presence of the wave the average energy of the drift is reduced because the phase of the quivering electrons is such as to reduce it. This occurs only for systems with available free energy. If we take $V = 0$ both waves are positive energy waves.

For single plasma streams the concept of positive and negative energy waves is meaningless, since it is reference frame dependent and thus non-invariant under coordinate transformation. The existence of **negative energy waves** implies a drift or a beam in the plasma. It becomes important when two or more streams are present in which case the type of coupling that occurs among waves of the various streams is invariant and can result in instabilities.

1.6.2 Reactive and Resistive Plasma Response

In addressing the linear and non-linear behavior of plasma instabilities it is important to understand the role of dissipation in the form of classical collisions, Landau damping, or phase mixing in broadband wave instabilities due to thermal velocity spread of particles.

We start by referring to (1.14) that includes dissipation effects on the plasma waves. The value of dissipation v can be due to Coulomb collisions or to an equivalent phase-dependent Landau damping. The dielectric constant of the plasma then becomes

$$\varepsilon(k, \omega; v) = 1 - \frac{\omega_e^2}{\omega(\omega + iv)}. \tag{1.59}$$

If $\omega \gg iv$, (1.59) reduces to (1.23) derived from (1.10). This is a reactive plasma response similar to inductance. It introduces a phase of $\pi/2$ between the current and the electric field without dissipation. In the opposite limit $\omega \ll iv$ the plasma behaves like a conductor or resistor and its dielectric constant is dominated by an imaginary part of the form $i\omega_e^2/\omega v$. As a conductor of course it cannot support any waves. The condition for reactive response is then $\omega \gg v$. At this point we can generalize the result by considering collisionless plasmas. In the presence of Landau damping we simply replace v with the value of the Landau damping at the phase velocity of the relevant mode. This condition becomes more stringent when we consider the dielectric response of a beam streaming through the plasma. Following the same analysis as above for a beam or drift we find instead of (1.57)

$$\varepsilon_b(\omega, k; V, v) = 1 - \frac{\omega_b^2}{(\omega - kV)(\omega - kV + iv)}. \tag{1.60}$$

Again v can represent collisional or collisionless (Landau type) dissipation. It is clear that wave activity can be supported only when

$$\omega - kV \gg v, \tag{1.61}$$

in which case the beam response will be reactive. In the opposite case the beam response will be resistive. Before proceeding we note that when v represents damping of the beam, it is convenient to use an equivalent simplified formula for damping at the main part of the distribution in which case (1.61) is replaced by

$$\omega - kV \gg k\Delta V_T, \tag{1.62}$$

where ΔV_T is the thermal velocity spread of the beam.

1.6.3 Reactive Instabilities – Beam-Plasma Instability

Reactive instabilities in beam-plasma systems result from reactive responses of each one and are due to coupling of a positive to a negative energy wave. A classic example is a uniform cold electron plasma with density n and an electron beam with

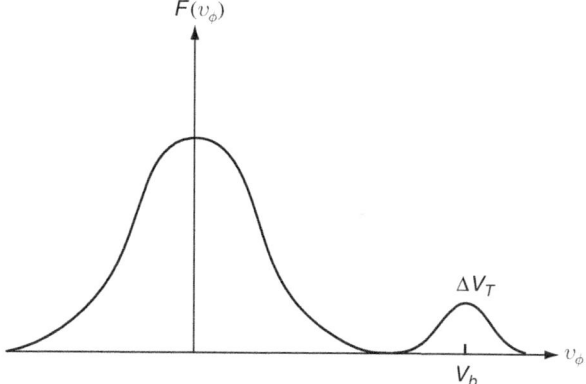

Fig. 1.14 Velocity configuration for beam-plasma instabilities. The beam has a velocity V_b with respect to the background plasma

density n_b streaming through the plasma at velocity V_b (Fig. 1.14). Furthermore, we assume immobile ions forming a neutralizing background and a one-dimensional interaction. The dispersion relation of the system is

$$\frac{\omega_e^2}{\omega^2} + \frac{\omega_b^2}{(\omega - k V_b)^2} = 1. \tag{1.63}$$

In (1.59) ω_b is the plasma frequency of the beam based on n_b. For phase velocities between zero and V_b the plasma waves are positive energy waves while the beam wave is a negative energy wave (slow wave). At any of these phase velocities the plasma wave grows at the expense of the beam energy. For a weak beam in the sense $n_b/n \ll 1$, the analysis [6] predicts maximum growth with the following characteristics

$$
\begin{aligned}
\omega &\approx \omega_e, \\
\Delta\omega &\approx (n_b/n)^{1/3}\omega_e, \\
\gamma_{\max} &\approx (n_b/n)^{1/3}\omega_e, \\
V_p &\approx V_b[1 - (n_b/n)^{1/3}].
\end{aligned}
\tag{1.64}
$$

One can see that the instability excites narrowband (almost monochromatic) plasma waves at the plasma frequency. It is a strong instability and the phase width over which it operates is of the order $(n_b/n)^{1/3}V_b$. For strong beams it affects essentially the entire region of velocities below V_b. This instability is a classic example of a reactive instability. Many instabilities in both magnetized and unmagnetized plasmas in the linear and non-linear regime can be studied by analogy to the cold beam-plasma instability. The instability requires that the beam be cold in the sense given by the inequality (1.62). Using (1.65) in the inequality (1.62) implies that a beam is cold for

$$\Delta V_T / V_b \leq (n_b/n)^{1/3}. \tag{1.65}$$

If inequality (1.65) is satisfied the instability grows creating large amplitude plasma waves that in their turn affect the beam distribution. Approximately half the wave energy is transferred to beam electrons increasing their velocity spread and leading to saturation. The saturation level can be found approximately by setting one-half of the level of the wave energy density W_L as equal to the final thermal energy of the beam required by the equality of (1.65). This gives

$$W_L = n_b m \Delta V_T^2 = n_b m V_b^2 (n_b/n)^{1/3}. \tag{1.66}$$

The reactive beam-plasma instability saturates when approximately $(n_b/n)^{1/3}$ of the beam energy has been transferred to electron plasma waves.

Figure 1.15 show the results of a computer simulation of the beam-plasma instability (courtesy of H. Karimabadi and N. Omidi). In this particular simulation an electric field is acting to accelerate the electrons. Ambient and beam electrons are treated as particles. The phase space shows that the ambient electrons are simply quivering in the presence of the growing plasma wave. The beam electrons on the

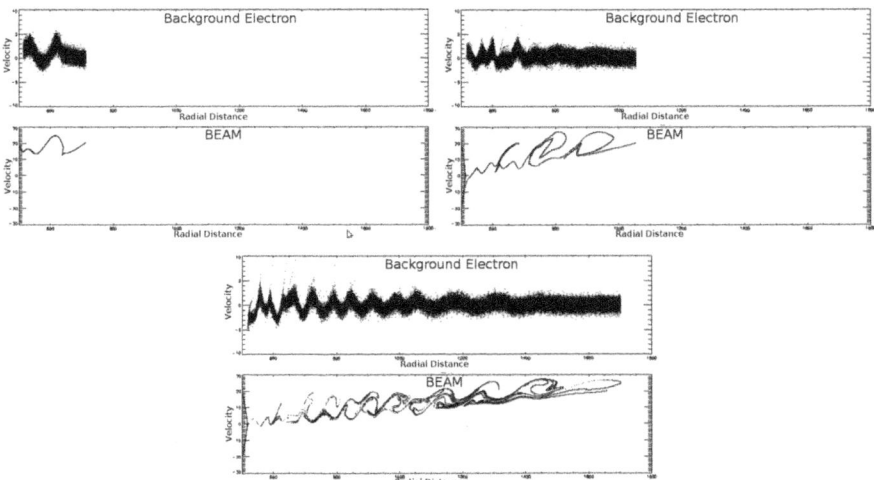

Fig. 1.15 Phase space of background plasma and beam for a reactive beam-plasma instability in the presence of an accelerating electric field at three times. Notice that the ambient electrons simply quiver without any irreversible energy gain. The beam spreads in velocity due to trapping. (Simulation courtesy of H. Karimabadi and N. Omidi)

other hand are trapped by the wave since its phase velocity is close to their velocity. The trapped beam electrons eventually thermalize at the level predicted by (1.66).

1.6.4 Resistive Instabilities – Bump-on-Tail Instability

Many instabilities have a reactive and a resistive form. In reactive instabilities only the real part of the dielectric of the interacting species enters the calculation of the growth rate. For resistive instabilities the imaginary part of the component that drives the instability is important. Resistive instabilities are the result of coupling of a positive energy wave with negative dissipation. For example, consider the coupling of a plasma wave with a dissipative beam given by (1.60). The dispersion relation of the system reads

$$1 - \frac{\omega_e^2}{\omega^2} - \frac{\omega_b^2}{\omega'^2}(1 - i\frac{\nu}{\omega'}) = 0. \tag{1.67}$$

In (1.67) $\omega' = \omega - kv$ and we assumed that $\omega' \gg \nu$. Assuming that $\omega_e \gg \omega_b$ (1.67) becomes

$$1 - \frac{\omega_e^2}{\omega^2} + i\left(\frac{\omega_b^2}{\omega'^2}\right)\left(\frac{\nu}{\omega'}\right) = 0. \tag{1.68}$$

For $\omega \approx \omega_0 + i\gamma$ with $\gamma \ll \omega_0$ we find that

$$\omega_0 \approx \omega_e,$$
$$\gamma \approx -\frac{1}{2}(\frac{\omega_b^2}{\omega'^2}\frac{\nu}{\omega'})\omega_e. \tag{1.69}$$

Instability develops at the plasma frequency if $\nu/\omega' < 0$ (negative absorption). A similar analysis can be followed for the so-called **bump-on-tail** configuration shown in Fig. 1.16. In this case the waves grow because negative Landau damping occurs at phase speeds corresponding to the range of waves that the slope of the bump

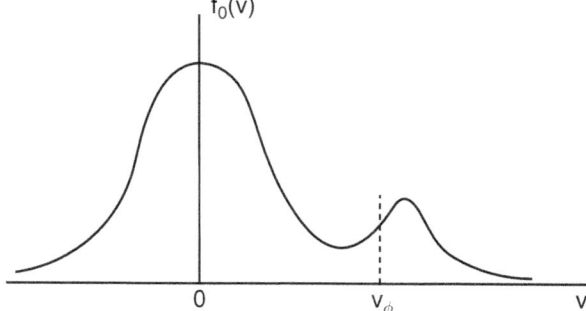

Fig. 1.16 Velocity configuration for bump-on-tail instability

is positive. Following the same analysis as above but replacing the third term of (1.68) with $(-i(\omega_b^2/\omega^2)(k\Delta V_T/\omega))$, we find instability at the plasma frequency with growth rate given by

$$\gamma \approx \frac{1}{2}\frac{\omega_b^2}{\omega_e^2}\frac{k\Delta V_T}{\omega_e}\omega_e = \frac{1}{2}\frac{n_b}{n}\frac{\Delta V_T}{V_b}\omega_e. \tag{1.70}$$

Notice that the growth rate is much weaker than the reactive beam-plasma instability – (n_b/n) vs $(n_b/n)^{1/3}$ – and has much larger bandwidth $\Delta\omega/\omega_e \approx \Delta V_T/V_b$.

This instability has been the prototype in the development of the so-called quasi-linear theory that follows the evolution of the bum-on-tail by a diffusion equation in velocity space along with the development of the wave spectrum in k-space. Figure 1.17 shows the evolution of the bump-on-tail instability [5] resulting from the solution of the quasi-linear equations that can be found in the original publication and many textbooks. The associated wave spectrum as a function of the phase speed $V_p = \omega_e/k$ is given by

$$W(V_p) \approx n_b m \frac{V_p^2}{V_b}. \tag{1.71}$$

In the asymptotic state two-thirds of the free energy of the bump goes into plasma waves while the rest remains in the flat tail of the bump particles. In quasi-linear theory the ambient plasma is adiabatic and the effect of plasma waves is reversible velocity sloshing.

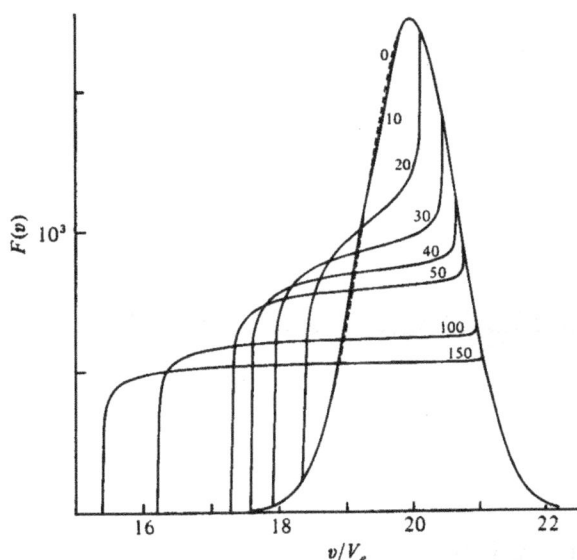

Fig. 1.17 The quasi-linear evolution of bump-on-tail instability. The distribution function of the "bump" particles are shown at a number of different times (in units of inverse initial growth time)

1.7 Electromagnetic Emission from Isotropic Plasmas

Since the early 1960s there have been many observations of electromagnetic emission detected on the ground or space indicative of preferential emission at the plasma frequency and its harmonic (Figs. 1.4 and 1.5). Understanding the physics underlying the emission process has been a major challenge and led to several plasma physics discoveries. From the previous section we see that the excitation of electrostatic plasma waves at the plasma frequency is ubiquitous in non-thermal and unstable plasmas. However, all these processes are related with the phase interactions between waves and plasma particles. Since the phase velocity of the electrostatic plasma waves is relatively small, of the order of a few times the thermal velocity of the plasma, super-thermal particles and electron beams can be in phase with them and generate plasma waves by Cerenkov emission or instabilities. This, however, is not true for electromagnetic plasma waves since in weakly magnetized plasmas, such as those of the interplanetary space and upper corona, they have phase velocities larger than the speed of light and cannot be in resonance with particles. Furthermore under collisionless conditions there is no bremsstrahlung radiation.

As seen in Sects. 3.2 and 3.3 in stationary, homogeneous and isotropic plasmas, the electrostatic and the electromagnetic plasma modes are not coupled and therefore there is no conversion of the electron plasma waves into electromagnetic plasma waves unless one of the above assumptions is violated. This can be present as an apparent dilemma, since the presence of plasma waves implies quivering motions of electrons and therefore time varying currents at the plasma frequency. Why aren't these currents radiating electromagnetic waves at the plasma frequency? From the mathematical point of view generation of a magnetic field requires current vorticity ($\nabla \times \mathbf{J} \neq 0$) that cannot be achieved by currents driven by isotropic plasma waves. However, the physical understanding of the effect provides an important intuitive introduction to the area of scattering from plasmas and to the so-called weak turbulence theories of mode conversion of electrostatic to electromagnetic waves [3].

1.7.1 Scattering of Electromagnetic Waves from Thermal Plasmas – Incoherent Scatter Radars

Consider, for example, a very thin slab of electrons oriented perpendicular to the vector $\mathbf{K}$ defined as $\mathbf{K} = \mathbf{k}_i - \mathbf{k}_s$, where $\mathbf{k}_i$ and $\mathbf{k}_s$ are the wave vectors of incident and scattered electromagnetic radiation with frequency much higher than the local plasma frequency. Take the slab to be so thin that all electrons contribute in phase to the radiation emitted in the direction $\mathbf{k}_s$. Next, choose a second slab parallel to the original one, and separated from it by a distance π/K. Thus the signals from the two slabs arrive exactly out of phase with one another. Since the medium is perfectly homogeneous, the two slabs contain the same number of electrons, and the two signals cancel exactly. This is true for any such pair of plasma slabs. There is no scattered radiation! In fact the same is true for scattering from neutral gases, e.g., Raleigh scattering. However, scattering occurs. How is the paradox resolved?

We got ourselves into this apparent dilemma by considering the relative phases of the radiation. Destructive interference gave the null result. The fact that scattering occurs is entirely due to **density fluctuations** that produce an imbalance in the number of scatterers in the respective slabs. In thermal equilibrium statistical mechanics the root mean square density fluctuation is $\sim \sqrt{N/V}$, where N is the total number of particles and V the total volume. Thus if V is the total volume of each slab the difference in the electron population is $\sim \sqrt{N/V}$ and therefore the net amplitude of the scattered signal $\sim \sqrt{N/V}$ and of the scattered power to the average density n. This is a well-known result that the scattering cross section per unit volume is the product of the Thomson cross section ($\sigma_T = 6.65 \times 10^{-29} \text{m}^2$) times the average density, i.e.,

$$< P >_{scattered} = n\sigma_T < S >_{incident} . \tag{1.72}$$

Notice, however, that the reason for this result is quite subtle, and relies on phase cancelation arguments. Furthermore, we should notice another important consequence of the physical picture introduced above that led to an extremely important result due to the inconsistency of theory and observations. Fluctuations imply that scatterers are in motion; their random thermal velocity causes the scattered light to be Doppler-shifted. Therefore, the scattered signal of a monochromatic wave with frequency $\omega \gg \omega_e$ is received with a spread of frequencies $\Delta\omega = k_i V_e$. As a result, a measurement of the frequency spread of the radiation scattered from plasmas provides a diagnostic measurement of their thermal velocity, or equivalently their temperature.

Electromagnetic wave scattering was first used in the late 1950s in an attempt to measure the electron density of the ionosphere with a device known as an **Incoherent Scatter Radar** (ISR). A radar signal with typical frequency between 50 and 400 MHz, much higher than the plasma frequency of the ionospheric plasma (< 10 MHz), is sent upward toward the ionosphere and the scattered signal recorded as function of time. For backscattering (1.72) predicts a scattering cross section of $1/2n\sigma_T$. A surprising result and of carefully conducted laboratory experiments indicated that

- the scattering cross section was $1/2$ of the expected;
- the measured frequency spread was almost two orders of magnitude smaller than the one expected from the thermal velocity of the electrons and much closer to $k_i V_i$, where V_i is the thermal velocity of the ions.

The second observation indicates that ions are the scattering agents. However, it is unthinkable that high-frequency radiation should scatter from massive and almost immobile ions rather than the much lighter electrons. The resolution of this problem by [19] provided a major insight into the rich collective behavior of plasmas. To assume that scattering is always the sum of single-free-particle scatterings, similar to

Fig. 1.18 A dressed test ion (*left*) and dressed test electron (*right*)

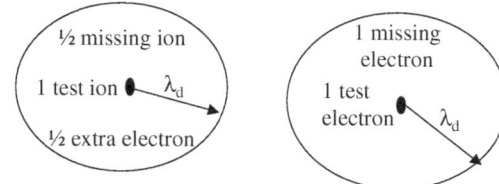

Raleigh scattering from neutral gases oversimplifies the problem. It is the collective behavior that distinguishes the plasma from an assembly of independent electrons. As we show above it is the deviation from homogeneity that allows scattering of radiation. It is the discreteness of the particles that breaks down the homogeneity. A moving ion or electron carries with it a polarization cloud of electrons that shields its electric field to distances larger than a Debye length λ_d. Such a particle is known as **dressed test particle**. Without resorting to the mathematical details dressed test ions or electrons seen over distances longer than λ_d have no charge because they are shielded. The shielding is always due to the mobile electrons.

A dressed test ion consists of a superposition of half an ion and half an electron, because in the shielding process on the average the typical test ion attracts one-half of an electron and repels one-half of an ion (Fig. 1.18). In the same sense a test electron repels one electron. Those puzzled by this description should refer to Fig. 1.19 that shows the electron and ion density without (a) and with (b) discreteness effects. If in the description of the plasma we ignore the fact that the electron and the ions are discrete particles and spread the charge uniformly throughout the volume (known as Vlasov or fluid description) the electron and ion densities are exactly equal and as a result of phase cancelations there is no scattering. If we include discreteness effect a test ion located at position x_1 has an associated electron density fluctuation that over a Debye length increases the average electron density by 1/2 of the electron charge, while reduces the average ion density by 1/2 of the electron charge. In the same picture the average electron density that surrounds a test electron located at x_2 is by one electron charge lower. Of course the overall plasma is charge neutral. In addressing scattering we consider separately the contributions from test ions and test electrons. The presence of the incident electromagnetic wave at the location of a test induces negligible quivering on the test ion and on the missing 1/2 ion. However, it

Fig. 1.19 The presence of a discrete ion at x_1 creates an ion density fluctuation (*blue*) that lower charge by 1/2 and an electron density (*red*) fluctuation that increases the electron charge by 1/2 maintaining neutrality. A test electron at x_2 creates a lowering of the electron density by 1 charge

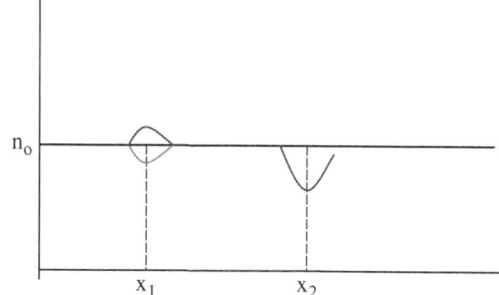

Fig. 1.20 Theoretical (*red*) and experimental (*yellow*) ISR data

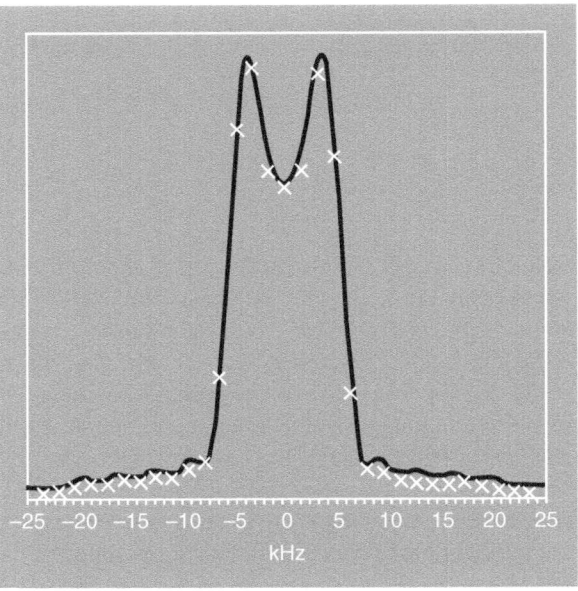

generates a scattered field for the quivering 1/2 surplus electrons. This contributes the measured $1/4n\sigma_T$. On the other hand, the test electron and the electron density with the missing charge cancel each other giving negligible contribution. Since the scattered radiation is associated with the motion of the test ion, the frequency spread characterizes its thermal velocity V_i. A great manifestation of the dressed test particle theory is the operation of ISRs for remote sensing of the parameters of ionospheric plasmas. Figure 1.20 shows the echoes received by the Milstone Hill radar operating at a frequency of 440 MHz. The time of flight indicated that the return was from an altitude of 263.1 km. Incoherent scatter spectra are analyzed by finding the temperatures and velocity that yield a theoretical spectrum based on dressed test particle theory, which most closely matches the measured spectrum. The yellow crosses in Fig. 1.20 are the measured spectrum from 263.1 km. The red curve is a theoretical spectrum corresponding to an ion temperature of 468.6 K, an electron temperature of 2004.0 K, and drift velocity (wind) of -11.0 m/s. More details can be found in http://www.haystack.edu/homepage.html.

1.7.2 Conversion of Electrostatic Plasma Waves into Electromagnetic – Weak Turbulence Theories

Guided by the above analysis of scattering we are led to examine whether similar scattering processes that break the homogeneity of the plasma can lead to conversion of electrostatic plasma waves into electromagnetic waves. These non-linear theories are known as weak turbulence theories because in a fashion similar to Sect. 6.1 they retain terms to the first order in the plasma expansion parameter $(1/n\lambda_d^3)$. In these scattering processes two electrostatic plasma modes $(\omega_1, \mathbf{k}_1)$, $(\omega_2, \mathbf{k}_2)$ scatter from each other and induce a third electromagnetic plasma mode $(\omega_3, \mathbf{k}_3)$. Since energy and momentum must be conserved during the interaction we require that

$$\omega_1 + \omega_2 = \omega_3,$$
$$\mathbf{k}_1 + \mathbf{k}_2 = \mathbf{k}_3. \qquad (1.73)$$

First, since the dominant plasma wave found in plasmas, excited by super-thermal particles or beam instabilities, is the electrostatic plasma wave given by (1.23) and (1.24) and the electromagnetic plasma mode has frequency given by (1.35) that is larger than the frequency of the electrostatic mode, the scattering process must result in frequency upshift. Second, since the wave number of the electromagnetic wave is smaller than c/ω_e and consequently much smaller (of the order of V_e/c or smaller) than the wave number of the electron plasma waves the momentum conservation law of equation requires that to lowest order $\mathbf{k}_1 = -\mathbf{k}_2$ the interacting waves be anti-parallel. Figure 1.21 lists the processes that result in conversion of electrostatic plasma waves to electromagnetic waves. Notice that the main result is electromagnetic radiation at the plasma frequency and its fundamental. These processes are favorable for plasma with isotropic wave spectra so that the requirement of anti-parallel wave numbers can be easily satisfied. This is not the case for beam-excited waves that tend to favor wave numbers along the direction of the beam. These processes are relatively inefficient with conversion to the fundamental of the $(V_p/c)^2$ and of the harmonic of $(V_p/c)^4$, where V_p is the phase velocity of the

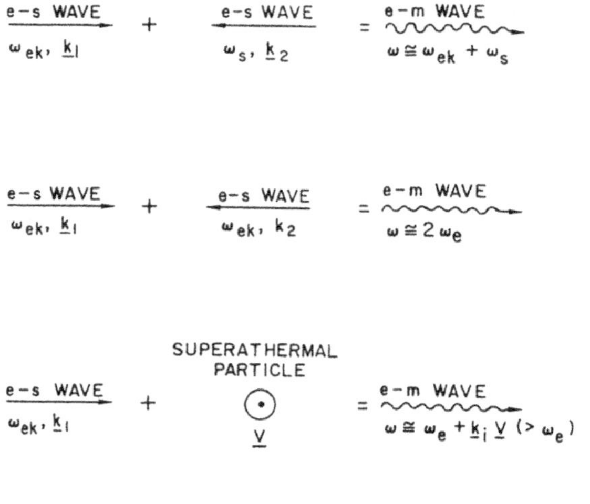

Fig. 1.21 A summary of weak turbulence processes converting es to em radiation [17]. See text for details

electron plasma waves. The mathematical formulae that describe the interactions listed in Fig. 1.21 can be found in [17].

1.8 Type III Radio-Bursts – Puzzles and Resolution – The Triumph of Strong Turbulence Theory

We conclude the discussion of unmagnetized plasma waves and instabilities with a particular example that dominated the literature during the 1960s and 1970s. The subject is the physics of the so-called type III radio-bursts, an example of which was shown in Figs. 1.4 and 1.5. The example clearly demonstrates the application of the concepts discussed above to a real space situation, the puzzles that were created, and the new physics that had to be introduced to resolve them.

Type III radio-bursts are a form of sporadic radio emission originating in the solar corona with continuous generation in the interplanetary region, to heliocentric distances of 1 AU and beyond. They are associated with beams of moderately energetic electrons (10–100 keV) that are accelerated either in solar flares or in active regions and which escape along magnetic field lines that penetrate the corona. Early observations established their frequency as the fundamental plasma frequency (ω_e) and its harmonic ($2\omega_e$). More recent observations (Fig. 1.5) show the presence of a third harmonic. Their frequency drifts at a rate that corresponds to a source of 1/3 to 1/2 of the speed of light, suggestive of a beam of energetic electrons, a fact that has been confirmed by in situ measurements. The observed frequency ranges from few kHz to several hundred MHz. A vast literature of observational studies produced a fairly detailed morphological picture of the radio emission, its scaling, and its decay properties. The observations created major theoretical challenges that ended up with the introduction of a theory that went beyond the weak turbulence considerations discussed above. The new theory is known as **strong turbulence theory**.

As early as 1960 it was recognized that the presence of an electron beam streaming from the corona would drive the instabilities discussed in Sects. 6.3 and 6.4 and will excite electrostatic electron plasma waves. Given the presence of the plasma waves theoretical attempts focused on the processes discussed in Sect. 7.2 that convert the electrostatic waves into electromagnetic waves at the plasma frequency and its first harmonic. In attempting to understand the observations quantitatively along these lines, theorists encountered a large number of puzzles and inconsistencies. We list below some of these puzzles:

1. The first puzzle is known as Sturrock's dilemma after Peter Sturrock who first raised it. Following the theories discussed in Sect. 6.4 the interaction of the beam with the plasma will result in the formation of a plateau such as shown in Fig. 1.17. For type III burst parameters this will occur within the first 50–100 km. After this distance the slope of the beam will be essentially zero and generation of plasma waves should cease. However, in situ observations indicate the presence

of positive slope in the beam as well as the presence of electrostatic waves far into the interplanetary medium.

2. The beam generates electrostatic waves with wave numbers predominantly along its direction. While this does not affect the generation of the fundamental, generation of the harmonic requires counter-streaming wave numbers. This is not expected on the basis of the earlier theories. However, observations indicate that the strength of the fundamental and the harmonic are comparable.

3. Within the context of weak turbulence plasma theories the probability for generation of the third harmonic is totally negligible, contrary to the observational results shown in Fig. 1.5.

1.8.1 The Emergence of Strong Turbulence Theory – The Ponderomotive Force

The inconsistencies between theory and observations led us to question some of the key assumptions of the non-linear theories of Sects. 6.3 and 6.4. A basic assumption was that there is only a high-frequency force that acts on the electrons while the ions simply maintain neutrality. However, there is a very important effect that affects the motion of a single electron when the high-frequency electric field is spatially inhomogeneous, such as is the case for a plasma wave with finite wavelength. An electron in the presence of an electric field $E e^{i\omega t}$ simply quivers with velocity $\tilde{v} = eE/m\omega$, Fig. 1.22 (left). However, if E is a function of x, such as shown in Fig. 1.22 (right), the particle obtains a slow drift because an electron oscillating in the electric field moves further in the half-cycle that moves it from the strong-field region to the weaker vice versa. The motion can be described by introducing a force known as **ponderomotive or Miller force** F_p that for a plasma with density n_o is given by [4],

$$F_p = -\frac{\omega_e^2}{\omega^2} \nabla \left[\frac{< \varepsilon_o E^2 >}{2} \right]. \tag{1.74}$$

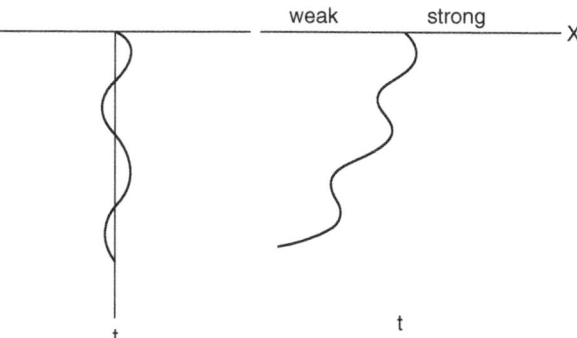

Fig. 1.22 Motion of charged particles in homogeneous (*right*) and inhomogeneous (*left*) electric field

Although the ponderomotive force acts mainly on the electrons it is transmitted to the ions through quasi-neutrality. As a result the ponderomotive force is a low-frequency force that acts on the plasma in the presence of an electron plasma wave.

1.8.2 The Zakharov Equations – The Non-linear Schroedinger Equation

The introduction of the ponderomotive force has profound implications to the physics of plasma waves. First, the presence of the force on the ions indicates that neglect of the ion motion is not a good assumption. Introducing the ion response in the form of density perturbation n into the problem gives [10]

$$\frac{\partial^2 n}{\partial t^2} - c_s^2 \nabla^2 n = \frac{\varepsilon_0 \omega_e^2}{4\omega^2} \nabla^2 E^2. \tag{1.75}$$

where c_s is the ion acoustic speed. In the absence of the ponderomotive force (1.75) describes ion acoustic waves with negligible damping, such as observed in the back of the shock wave of Fig. 1.2. Notice that the presence of the electron plasma wave acts as a driver of acoustic waves. Second, the effect of the density modification n should be included in Eq. (1.19) that describes the plasma waves. Namely, the $\omega_e^2 E$ term is replaced by $\omega_e^2(1 + \frac{n}{n_0})E$ so that (1.19) becomes

$$\frac{\partial^2 E}{\partial t^2} - \alpha V_e^2 \nabla^2 E + \omega_e^2 E = -\omega_e^2(\frac{n}{n_0})E. \tag{1.76}$$

It is convenient to simplify (1.75) and (1.76) assuming one-dimensional interaction by noting that there are two timescales: a fast that involves the plasma frequency and a slower that involves the part $\alpha k^2 \lambda_d^2 \omega_e$ of the frequency dependence. We thus write

$$E(x, t) = \tilde{E}(x, t) \exp(-i\omega_e t). \tag{1.77}$$

From (1.75–1.77) we find the so-called **Zakharov equations**

$$\frac{\partial^2 n}{\partial t^2} - c_s^2 \frac{\partial^2 n}{\partial x^2} = \frac{\varepsilon_0}{4} \frac{\partial^2}{\partial x^2} |\tilde{E}|^2, \tag{1.78}$$

$$i \frac{\partial \tilde{E}}{\partial t} + \alpha \frac{V_e^2}{\omega_e} \frac{\partial^2 \tilde{E}}{\partial x^2} = \frac{\omega_e}{2} \frac{n}{n_0} \tilde{E}. \tag{1.79}$$

Assuming a very slow time variation for the density in Eq. (1.78) we can replace it by

$$n = -\frac{\varepsilon_o}{4c_s^2}\left|\tilde{E}\right|^2 .$$ (1.80)

From (1.79) and (1.80) we find a single equation that describes the envelope of the electron plasma oscillation as

$$i\frac{\partial \tilde{E}}{\partial t} + \alpha \frac{V_e^2}{\omega_e}\frac{\partial^2 \tilde{E}}{\partial x^2} + \frac{\varepsilon_o}{4n_oc_s^2}\left|\tilde{E}\right|^2 \tilde{E} = 0.$$ (1.81)

Equation (1.81) is of the form of the non-linear Schroedinger equation and has stationary solutions of the form of solitons, often called Langmuir solitons.

One of the most profound effects of (1.78) and (1.79) is that once the energy of electrostatic plasma oscillations with wave number k_0 exceeds a certain threshold given by

$$\frac{W_L}{n_0 T} > k_0^2\lambda_d^2,$$ (1.82)

where $n_0 T$ is the ambient plasma, the weak turbulence theory becomes invalid and the homogeneous plasma state becomes strongly inhomogeneous and appears in the form of coupled plasma wave solitons trapped inside density cavities. Figure 1.23 [18] shows the final plasma state for a system driven by a long wavelength pump near the plasma frequency. This state is often called spiky turbulence. The introduction of strong turbulence had profound implications in the resolution of the type III radio-burst puzzles and several mysteries associated with beam-plasma interaction physics.

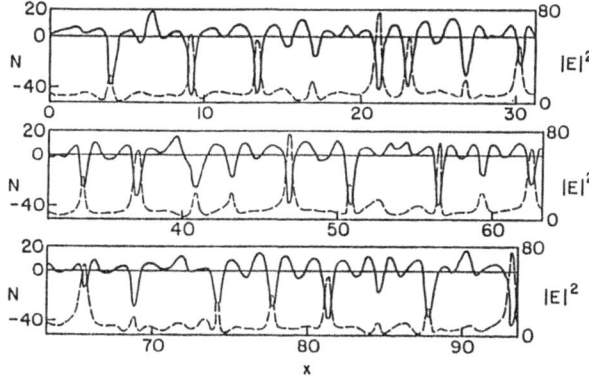

Fig. 1.23 Final state of turbulence (soliton–caviton pairs) in a system driven by a long wavelength electrostatic pump at the plasma frequency (spiky turbulence). The *solid* (*dashed*) *lines* are the density (electric field magnitude)

1.8.3 Non-linear Stabilization of Beam-Plasma Interactions – Resolving Sturrock's Dilemma

In a series of papers [14, 16, 13, 20, 21], was proposed that once the level of electron plasma waves reaches threshold for formation of coupled caviton–soliton pairs the spatial inhomogeneity decouples from the plasma and allows the beam to propagate in a stable fashion, while simply supplementing any energy lost from the waves to particles. In this way it only suffers a weak friction-like force and propagates stably despite the presence of a positive slope. It was also noted that this effect is essentially a one-dimensional effect. However, for type III radio-bursts and other space systems the parameters are such that the presence of the weak interplanetary magnetic field is such as to make the system behave as a one-dimensional system. Figure 1.24 shows the resolution of **Sturrock's dilemma** using the results of a Vlasov simulation [20]. Figure 1.24 (upper part) shows the simulation with infinite mass ions so that the creation of cavities is not allowed. The system forms quickly a quasi-linear plateau as expected from quasi-linear and weak turbulence theories. Figure 1.24 (lower part) shows the same simulation with finite mass ions.

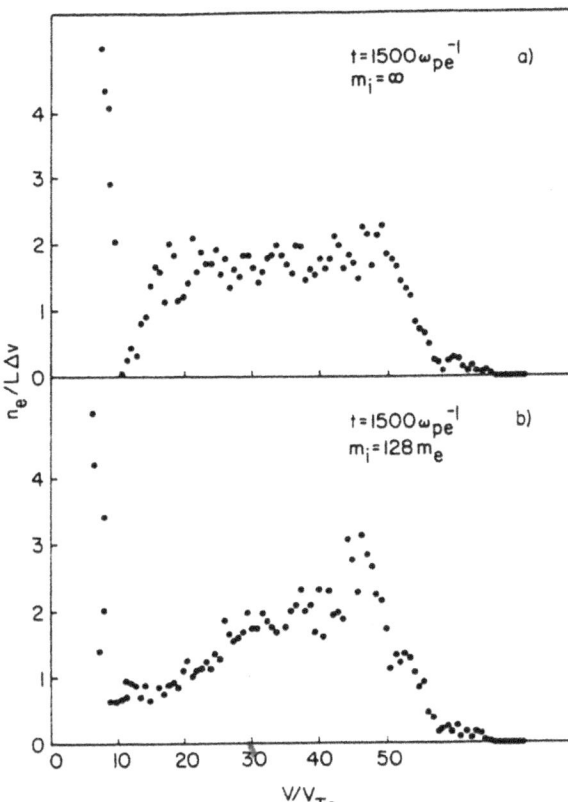

Fig. 1.24 Vlasov simulation showing plateau formation in the absence of ion dynamics and stabilization of the bump-on-tail for finite mass ions

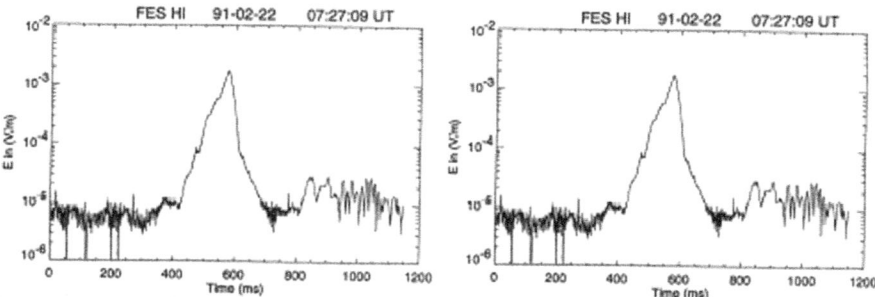

Fig. 1.25 High-resolution Langmuir solitons observed in situ during type III events [23]. One second of electric field data is shown

The instability has been non-linearly stabilized and the beam retains positive slope over very long times. This of course was the resolution of Sturrock's dilemma and marked the first major result of strong turbulence theory. Details of the type III analysis including the strong turbulence effects [13, 21]. We should note that recent in situ measurements [23] have verified the presence of electron plasma oscillation spikes with localization consistent with solitons (Fig. 1.25).

1.8.4 Electromagnetic Emission at the Plasma Frequency and Its Harmonics

In addition to resolving Sturrock's dilemma the introduction of strong turbulence converts with high efficiency electrostatic plasma waves to electromagnetic waves at the plasma frequency and its harmonics, thereby resolving the mysteries of the inefficient conversion associated with weak turbulence processes. The physics associated with the conversion process can be best seen by resorting to the results of a set computer simulations [1]. A 2 1/2-dimensional, fully electromagnetic, two fluid code was used to study the evolution of electron plasma waves with a spectrum consistent with the one expected from the resistive beam-plasma instability to electromagnetic waves. The code was initialized by loading a spatially uniform spectrum of electrostatic plasma waves of the type expected by the resistive beam instability discussed in Sect. 6.4. Figure 1.26a shows the formation of the soliton–caviton pairs similar to Fig. 1.23 but in two dimensions. The beginning of the spatial structuring of the plasma is accompanied by transformation of a significant part of the electrostatic energy of the plasma waves into electromagnetic waves (Fig. 1.26b) at the plasma frequency and its first two harmonics (Fig. 1.26c). Figure 1.26d shows the spectrum of the electrostatic and electromagnetic waves at late times showing excitation of waves at the first and second harmonic in addition to the fundamental plasma frequency.

In exploring the physics of the radiation processes in strongly turbulent plasmas it is important to note that creation of a magnetic field requires generation of local

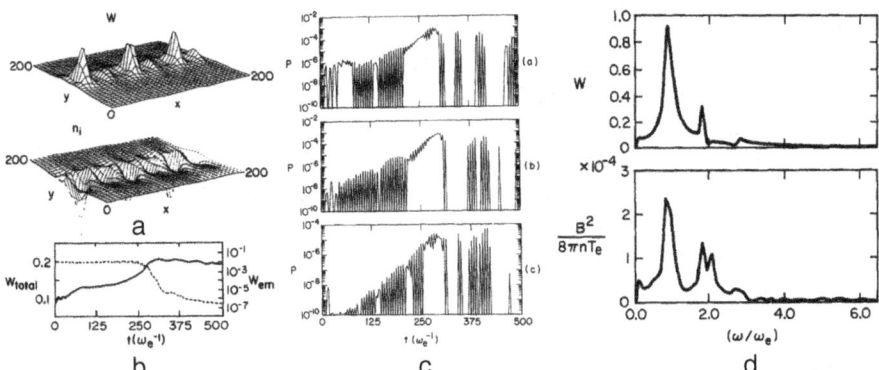

Fig. 1.26 (**a**) Soliton–caviton pair formation; (**b**) Temporal evolution of electrostatic and electromagnetic energy; (**c**) Temporal evolution of the electromagnetic radiation at ω_e, $2\omega_e$, and $3\omega_e$; (**d**) Electrostatic and electromagnetic spectra showing enhanced levels at the plasma frequency and the first two harmonics

current vorticity ($\nabla \times \mathbf{J}$). The current vorticity in the above simulations can be seen by isolating the middle soliton of Fig. 1.26a and plotting the contours of the total electrostatic energy (Fig. 1.27a) and of the resultant radiation source $\nabla \times \mathbf{J}$ (Fig. 1.27b).

It can be seen that the source of radiation is a mixture of dipole and quadrupole radiation related to emission at the fundamental and first harmonic, correspondingly. While the electrostatic energy maximizes at the center of the soliton the electromagnetic emission maximizes at the outer edges where the field curvature is maximal. The final Fig. 1.28 shows the temporal evolution of the radiation as compared to the initial energy density.

Notice that the results show comparable emission at ω_e and $2\omega_e$ and weaker emission at $3\omega_e$. Weak turbulence theory will predict overall radiation level several orders weaker than observed in the simulations. Furthermore, it predicts radiation level at ω_e, the $2\omega_e$ by more than a factor of 100 weaker than at ω_e, and no radiation

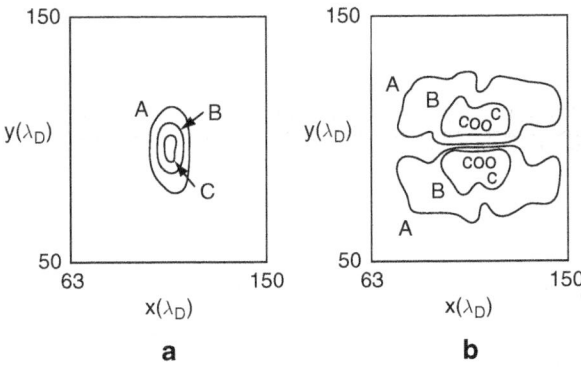

Fig. 1.27 (**a**) Contours of electrostatic energy of single soliton; (**b**) radiation term $\nabla \times \mathbf{J}$

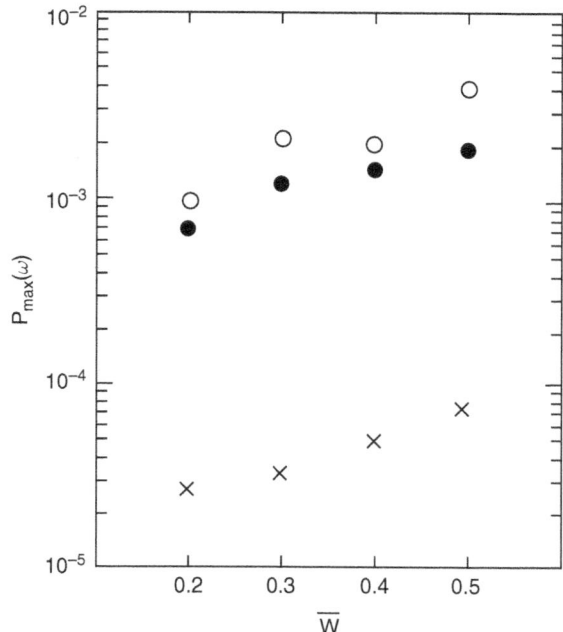

Fig. 1.28 Ratio of radiation power density to initials electrostatic turbulence vs electrostatic turbulence for ω_e, $2\omega_e$, and $3\omega_e$ (*open circles, closed circles*, and *x* respectively)

at $3\omega_e$. The resolution of the type III puzzles is a textbook example of the parallel and interacting paths of theory/simulations and observations in modern plasma physics.

1.9 Epilogue

The objective of this introductory lecture was to present a clear, physically motivated treatment of plasma waves with emphasis on examples from space plasma physics. Since the audience was composed of students and scientists of different specialties and backgrounds very little knowledge of plasma physics was assumed. However, knowledge of electromagnetic theory was a prerequisite. I attempted to start from the most elementary level of plasma physics, electrostatic plasma oscillations and build it up to the level of the most advanced strong turbulence plasma theories. In order to keep the level of presentation and the mathematical complexity to a minimum I restricted the analysis to weakly magnetized plasmas. Furthermore, when possible I attempted to insert and explain advanced concepts of kinetic theory, such as the role of discrete particles and of the plasma parameter that are not covered in recent plasma curricula. In structuring the lecture I felt that there are a number of terms that every physicist involved in plasma, space, and astrophysics should know and tried to at least mention them with an example or a brief picture. These terms are the jargon of our discipline. For the sake of review I list them in the approximate order of presentation.

Generalized Ohm's law, plasma model (cold, thermal, Vlasov, kinetic), plasma response function, electron plasma oscillations, electron plasma waves, reactive plasma response, resistive plasma response, electromagnetic plasma waves, dispersion relations, dispersion diagrams, phase velocity, group velocity, dielectric constant, index of refraction, cutoff and resonance, ionosonde, positive and negative energy waves, Cerenkov emission, Landau damping, stimulated emission, thermal and non-thermal excitation level of plasma waves, reactive instabilities, beam-plasma instability, resistive instabilities, bump-on-tail instability, trapping, quasi-linear theory, scattering, dressed test particle theory, plasma or discreteness parameter, incoherent scatter radar, weak plasma turbulence theory of es to em conversion, ponderomotive force, Sturrock's dilemma, type II and III radio-bursts, strong turbulence theories, non-linear stabilization, Zakharov equations, non-linear Schroedinger's equation, solitons, cavitons, spiky turbulence, collapse, and strong turbulence theory of es to em emission at the plasma frequency and its harmonics.

These terms and their definitions and implications must be in the armory of any space physicist and astrophysicist. Although discussed here only in the context of unmagnetized plasmas their extensions to magnetized plasmas are critical to understanding collective phenomena in space physics and astrophysics.

Acknowledgments In conclusion I would like to thank my dear friends and colleagues Peter Cargill and Loukas Vlahos for their kind invitation. The conference brought back fond memories of exciting scientific discussions and arguments during the years they spend at the University of Maryland. The ideas and concepts presented here would have not been developed without the help, inspiration and participation of the numerous students, postdoctorals and research associates of all levels that over the last 25 years formed the Space Plasma Physics group at the University of Maryland.

References

1. K. Akimoto, H.L. Rowland, K. Papadopoulos: Phys. Fluids, **31**, 2185 (1988)
2. W.P. Allis, S.J. Buchsbaum, A. Bers et al: *Waves in Anisotropic Plasma*. M.I.T. Press, Cambridge, Massachusetts (1963)
3. G. Bekefi: *Radiation Process in Plasmas*. John Wiley and Sons, New York (1966)
4. F.F. Chen: *Introduction to plasma physics and controlled fusion*, Vol.1: Plasma physics. Plenus Press, New York (1990)
5. R.J.-M Grognard: Aust. J. Phys., **28**, 731 (1975)
6. N.A. Krall, A.W. Trievelpiece: *Principles of Plasma Physics*. McGraw-Hill Book Company, New York (1972)
7. Kruer, W.L.: *The Physics of Laser Plasma Interactions*. Frontiers in Physics, Addison-Wesley Publishing Company, New York (1988)
8. D.B. Melrose: *Instabilities in Space and Laboratory Plasmas*, Cambridge University Press, New York (1986)
9. D. Montgomery, D.A. Tidman: *Plasma kinetic theory*. McGraw-Hill Book Co, New York (1964)
10. D.R. Nicholson: *Introduction to plasma theory*, John Wiley and Sons, New York (1983)
11. T. Ohnuma: *Radiation phenomena in plasmas*. World Scientific Singapore (1994)
12. K. Papadopoulos: Phys. Fluids, **12**, 10 2185 (1969)
13. K. Papadopoulos, M. Goldstein, R. Smith: Astrophys. J., **190**, 175 (1974)
14. K. Papadopoulos: Phys. Fluids, **18**, 1769 (1975)
15. K. Papadopoulos: Rev. Geophys. Space Phys., **17**, 624 1979

16. K. Papadopoulos: On the Physics of Strong Turbulence for Plasma Waves, in Diagnostics for Fusion Experiments. In: E. Sindoni and C. Wharton (eds.) Pergamon Press (1979), p. 38
17. K. Papadopoulos, H.P. Freund: Space Sci. Rev., **24**, 511 (1979)
18. N.R. Pereira, R.N. Sudan, J. Denavit: Phys. of Fluids, **20**, 271 (1977)
19. N. Rostoker: Nucl. Fusion, **1**, 101 (1961)
20. H.L. Rowland, K. Papadopoulos: Phys. Rev. Letts., **39**, 1276 (1977)
21. R.A. Smith, M.L. Goldstein, K. Papadopoulos: Solar Phys., **46**, 415 (1976)
22. T.H. Stix:*Waves in Plasmas*. AIP, New York (1962)
23. G. Thejappa, M.L. Goldstein, R.J. MacDowall et al: J. Geophys. Res., **104**, 28279 (1999)

Chapter 2
Solar MHD: An Introduction

C. Chiuderi and M. Velli

2.1 Introduction

The Universe is filled by plasmas. In fact, it can be reasonably estimated that more than 95% of (standard) cosmic matter is found in the plasma state. Given this basic fact, it is apparent that plasma physics is the basic tool to understand the mechanisms that are at work in the astrophysical context and to interpret the observations. The Earth represents a notable exception as far as the presence of natural plasmas is concerned, a lucky circumstance for living beings. In practice, terrestrial plasmas are almost exclusively produced during electrical discharges, such as lightning. If plasmas are almost absent on Earth and in the low-altitude atmosphere, they start to be the dominant state of matter immediately beyond the ionosphere, the magnetosphere, and the whole heliosphere that includes the entire solar system made up of plasmas. The Sun, like the other stars, is made up of ionized gas almost everywhere.

Not only are natural plasmas on Earth rare, but artificial plasmas are also hard to produce and to maintain in the laboratory, as demonstrated by 50 years of (so far unsuccessful) efforts to produce conditions that mimic those in stars' interiors. At present, we try to understand the behavior of plasmas theoretically to advance our ability to produce and confine plasmas in the laboratory and to perform numerical simulations as proxies for real experiments when the latter cannot be done. In any case we shall never be able to reproduce in the lab the conditions prevailing in most astrophysical plasmas, mainly because of the scales involved. The closest natural plasma "laboratory" we have access to is, in fact, the heliosphere, where we can perform both in situ measurements and remote observations to a high degree of detail, due to the relative closeness of the systems under study. The development

C. Chiuderi (✉)
Dipartimento di Astronomia, Università di Firenze, Largo Enrico Fermi, 5 50125 Firenze, Italy,
chiuderi@arcetri.astro.it

M. Velli
Jet Propulsion Laboratory, California Institute of Technology, Pasadena, CA, USA,
velli@arcetri.astro.it

Chiuderi, C., Velli, M.: *Solar MHD: An Introduction*. Lect. Notes Phys. **778**, 45–69 (2009)
DOI 10.1007/978-3-642-00210-6_2   © Springer-Verlag Berlin Heidelberg 2009

of theoretical and numerical techniques to advance our understanding of the plasma mechanisms and the constant comparison between theory, simulations and observations in the heliosphere is therefore of primary importance as the basis for the application to wider systems.

The purpose of this chapter is to briefly illustrate one theoretical approach to plasma dynamics, namely magnetohydrodynamic (MHD) theory, as applied to solar physics. It thus provides the basis for material presented elsewhere in this volume based on the MHD approach. We begin by reviewing the basics of plasma physics and MHD as they apply to the Sun, and continue by providing an introduction to one problem of central importance in space plasma physics, namely coronal heating, subsequently discussed in greater detail in Chap. 4 (Hansteen and Carlsson).

2.1.1 Solar and Heliospheric MHD Texts

We do not pretend to provide an exhaustive list of texts on plasma physics and MHD here, but rather give a brief summary of some useful textbooks for a graduate or postgraduate student entering the field. These texts refer to the material in Sects. 2.2, 2.3, and 2.4 of this chapter: more specific references are provided in Sect. 2.5.

There are a number of books on MHD and plasma physics focused on the solar and heliospheric physical context. Priest's textbook [29], *Solar Magnetohydrodynamics*, has become the classic introduction to the basic large-scale magneto-fluid dynamics of the outer solar atmosphere. The solar – stellar connection is comprehensively explored in Schrijver and Zwaan's text on *Solar and Stellar Magnetic Activity* [32]. Gurnett and Bhattacharjee's [16] text focuses more on plasma physical processes in the space and laboratory. The books by Sturrock [33] and Kulsrud [20] include the immensely interesting insights from two of the founders of the field. More general texts are those of Boyd and Sanderson [6] and Goedbled and Poedts [13]. Finally, on the specific topics of magnetic reconnection and MHD turbulence the two texts by Biskamp [4, 5] give a state-of-the-art account of the problem of small-scale formation and rapid energy conversion in magnetized plasmas.

2.2 Plasma Physics and Magnetohydrodynamics

Although the concept of a plasma can be made more general, a good working definition is the following:
A plasma is a system of charged particles whose dynamics is dominated by collective effects: each particle feels the average electromagnetic fields generated by the distribution and the motions of the other particles of the system. Of course, a particle in the plasma does not feel **only** the effect of electromagnetic fields. It also interacts with the other particles through collisions. If the plasma is sufficiently

dilute, as often happens in astrophysics, inter-particle collisions can be neglected. When the density increases, the effect of collisions has to be taken into account.

So far we have implicitly considered the **particles** as different entities from the **fields**, as usually done in classical physics. In a quantum field-theoretical description, however, the fields may also be described in terms of particles, or better quasi-particles, at least in the linear or weakly nonlinear regimes. In such cases the plasma is defined as the ensemble of the "real" particles and quasi-particles representing the quanta of the electromagnetic fields. This picture suggests that, even in the classical limit, we should consider the plasma to be made up of particles and fields. Although the adoption of a field-theoretical point of view would introduce unnecessary complications (in the quasi-totality of cases classical physics is amply sufficient), it nevertheless provides good insight when dealing with particular processes. An example is given by "collisionless" Landau damping, a process that is much easier to understand if we realize that though collisions between real particles are negligible, those between real particles and quasi-particles are not. The plasma therefore is not totally collisionless, but one type of collisions, that is not usually considered, dominates over the others.

The theoretical description of a plasma poses formidable difficulties, mainly due to the self-consistency requirement: the distribution and motions of the particles generate fields that, in turn, are responsible for those distributions and motions. In principle, we have all we need: the equations of motion for the particles, coupled with the Maxwell equations for the fields. However, the number of particles in a plasma is extremely large, typically of the order of the Avogadro Number, $N_A \simeq 10^{23}$, and there is no hope of solving exactly a set of N_A equations of motion, coupled with Maxwell's equations, and even if these were possible we would be totally unable to use the results.

Thus, we have to strike some sort of compromise between rigor, mathematical tractability, and common sense. It is not surprising, therefore, that a number of theoretical schemes exist, forming a hierarchy in increasing order of simplicity and decreasing order of completeness and "resolving power." At each step information is lost, since we are progressively neglecting certain properties of the system considered to be of lesser interest, and this implies some sort of averaging process. The basic theoretical schemes are given in the following sections.

2.2.1 Orbit Theory

The direct study of particle motions is possible only in **given** fields, i.e., externally imposed fields. Since there is no connection between the particle's dynamics and the fields, we are losing completely the feedback effect and, in fact, we are not even dealing with a plasma, in view of the definition given above. Nevertheless, orbit theory is very useful to get a grasp on particle motions and may be useful in understanding the dynamics of very dilute plasmas.

2.2.2 Kinetic Theory

It is probably the best possible trade-off between completeness of information and mathematical simplicity. In this scheme, all the relevant information is contained in *the particle distribution function (PDF)*, $f(\mathbf{r}, \mathbf{v}, t)$, i.e., the particle density in phase space. Kinetic equations describe the evolution of the particle distribution function in space and time. Kinetic models basically differ by the term that represents the effect of collisions. If collisions can be neglected, the kinetic equation, called in this case the Vlasov equation, assumes its simplest form. Examples of the Vlasov approach can be found in Chap. 1 (Papadopoulos) and such an approach is, of course, essential for any discussion of particle acceleration in solar plasmas (Chap. 5: Vlahos et al.).

When collisions cannot be neglected, some suitable physical model has to be adopted to describe them. Different models give rise to different kinetic equations. We then have the Boltzmann equation (elastic binary collisions), the Fokker–Planck–Landau equation (Coulomb collisions), the Lenard–Balescu equation, and so on.

Whatever the kinetic equation used, it has to be coupled with Maxwell's equations, giving rise to a coupled nonlinear integro-differential system. Needless to say, exact solutions of such a system can be found only in very particular (and not particularly interesting) cases. Even assuming that such a solution is available, a direct comparison with the relevant observations is, in general, not feasible. Particle distribution functions are always very hard to measure, the only notable exception being the case of the solar wind, a plasma sufficiently dilute where in situ measurements are possible. Some relevant measurement techniques for solar wind plasmas are discussed in Chap. 6 (Issautier). The main drawbacks of kinetic theories are, therefore the following

- the theory is still too complicated to deal with realistic cases,
- direct comparison with the observations is difficult,
- the amount of information often exceeds the real needs.

In connection with the last two points we observe that a number of physically relevant parameters, directly comparable with observations, can be obtained by taking the **moments** of the *PDF*, namely,

$$\Phi_{ij..l}(\mathbf{r}, t) = \int v_i v_j ... v_l f(\mathbf{r}, \mathbf{v}, t) d\mathbf{v},$$

the **order** of the moment being defined as the number of velocity components entering the integral above. It is easily seen that the moment of order zero is proportional to the number density of particles in geometrical space, the first-order moment is related to the average speed of particles, and so on. It is therefore natural to try to find a new scheme where the basic quantities are the moments themselves.

2.2.3 Fluid Theories

Fluid theories provide the equations that determine the dynamics of the moments of the *PDF* without making reference to their basic definition in terms of the *PDF* itself. Since there are an infinite number of moments, we are in principle led from a single kinetic equation to an infinite system of coupled equations, a highly impractical situation. We must therefore find ways to truncate the system to a small number of internally consistent equations. This constitutes the **closure** problem that underlies every fluid model and that does not have a unique solution. The standard closure procedure (valid for distribution functions close to Maxwellian) produces the well-known Euler equations (to zeroth order) and the Navier–Stokes equations (to first order). Of course, in fluid theories all information on velocity distributions is lost, but the quantities appearing in those theories have a direct physical meaning in configuration space and are therefore directly comparable with measurable physical parameters.

– Multifluid models. Since there are at least two different species of particles in a plasma (positively charged ions and electrons), there are correspondingly at least two sets of fluid equations. We are therefore faced with the problem of solving $s \times 15$ nonlinear equations (here s is the number of species), the 15 unknowns being n_α, P_α, q_α, $\mathbf{u}_\alpha$, $\mathbf{j}_\alpha$, $\mathbf{E}$, $\mathbf{B}$, respectively, the number density, the partial pressure, the charge density, the drift speed, the current density of each species α plus the electric and magnetic fields. Multifluid models are to be used whenever the interaction among different species is weak, so that each species evolves almost independently of the others, with coupling through the EM fields and collisions, see Chap. 1 (Papadopoulos) for some examples. An example is given by the solar wind, where the electrons and protons are observed to have different values of the temperature. This is due to the different values of the collision times for particles of the same species, and of different species, the latter being much longer. Therefore electrons and protons reach rapidly well-defined values of the temperature, but the establishing of a common temperature between different species takes much longer.
– Single-fluid models. These replace the entire set of equations for the different fluids with just one set of equations for an *equivalent* fluid. In other words the real fluids, one for each species, are replaced by a single *fictitious* fluid, with properties that are intermediate between those of the constituents fluids. For instance, the drift speed of the single fluid that represents an electron–proton plasma is defined as the center-of-mass speed of the composite fluid:

$$\mathbf{U} = \frac{(m_p \, n_p \mathbf{u}_p + m_e \, n_e \, \mathbf{u}_e)}{m_p \, n_p + m_e \, n_e}.$$

– Magnetohydrodynamics. *MHD* is a particular *regime* of single fluid models. In terms of U, L, T, the velocity, length and time scales, respectively, the *MHD* regime is defined by the conditions:

$$L/T \approx U \ll c.$$

MHD is therefore a non-relativistic, low-frequency theory, and it is the **most widely used model in plasma astrophysics**.

The *MHD* scheme (when applicable) achieves a considerable simplification of the single-fluid equations. In fact, a simple dimensional analysis of those equations shows that in the *MHD* regime:

$$E/cB \simeq U/c \ll 1 \quad ; \quad \frac{1}{c^2}\frac{\partial E}{\partial t} \ll |\nabla \times \mathbf{B}|.$$

As a result, it is possible to obtain a closed set of eight equations in the eight primary variables ρ, P, $\mathbf{U}$, $\mathbf{B}$. To recover the other variables one must use the auxiliary equations:

$$\mu_0 \mathbf{j} = \nabla \times \mathbf{B},$$

$$\mathbf{E} = \frac{\mathbf{j}}{\sigma} - \mathbf{U} \times \mathbf{B},$$

$$\frac{q}{\epsilon_0} = \nabla \cdot \mathbf{E},$$

where $\mathbf{U}$, $\mathbf{B}$, $\mathbf{E}$, and $\mathbf{j}$ are the velocity, magnetic field, electric field, and current density respectively, and σ is the electrical conductivity. Gauss' law is included for completeness, though in reality all solar and interplanetary plasmas are quasi-neutral to a very high degree. The complete set of MHD equations turns out to be:

$$\frac{\partial \rho}{\partial t} + \nabla \cdot (\rho \, \mathbf{U}) = 0, \tag{2.1}$$

$$\rho \left(\frac{\partial \mathbf{U}}{\partial t} + (\mathbf{U} \cdot \nabla) \mathbf{U} \right) =$$

$$- \nabla P + \frac{1}{\mu_0} (\nabla \times \mathbf{B}) \times \mathbf{B} + \rho \mathbf{g} + \left[\rho \nu \left(\nabla^2 \mathbf{U} + \frac{1}{3} \nabla (\nabla \cdot \mathbf{U}) \right) \right], \tag{2.2}$$

$$\rho c_p \frac{dT}{dt} - \frac{dp}{dt} = [\mathcal{G} - \mathcal{L}] = \left[\frac{j^2}{\sigma} + \mathcal{H} - \nabla \cdot Q - \mathcal{L}_{rad} \right], \tag{2.3}$$

$$\frac{\partial \mathbf{B}}{\partial t} = \nabla \times (\mathbf{U} \times \mathbf{B}) - [\nabla \times \eta (\nabla \times \mathbf{B})], \tag{2.4}$$

where ρ, P, and T are the mass density, pressure, and temperature respectively, and d/dt represents a convective derivative. Equations (2.1, 2.2, 2.3) express the conservation of mass, momentum, and energy, respectively. Equations (2.4) determines the dynamical evolution of the magnetic field. In the above equations $\eta = 1/(\mu_0 \sigma)$ is the resistivity, ν is the kinematic viscosity, and $\mathbf{Q}$ is the heat flux vector. $\mathcal{G}$ and $\mathcal{L}$ are the generalized gain and loss functions, while $\mathcal{H}$ and $\mathcal{L}_{rad}$ are the heat source term and the radiation loss term. The only external force considered is gravity. The terms in square brackets are **non-ideal** terms, i.e., terms connected with dissipative effects, that are absent when the plasma is considered to be ideal. Closure is accomplished through an equation of state: $P = 2nkT$, where n is the number density of the plasma.

2.3 A Quick Tour of the Sun

The Sun is the only star that allows detailed observations of its visible surface, or photosphere. Remembering that one arcsecond on the surface corresponds to ≈ 700 km and that modern instruments on a spacecraft such as Hinode have a resolving power of less than a few tenths of an arcsecond, we conclude that we are presently able to see details on the surface down to scales of the order 100 km. In the radial direction, the choice of suitable spectral lines allows determination of the physical characteristics of all the layers forming the solar atmosphere, down to what is called the visible solar surface, with the caveat that some of the layers, especially the regions where strong transitions in plasma properties are present, have strong spatial and temporal variability, so that "average" physical characteristics may be misleading.

Below the photosphere the opacity of the solar material inhibits a direct observation: the available information is obtained indirectly, from helioseismology or from theory. Traditionally, the Sun is divided into several regions:

– The **interior** includes

  – the *core*, where the energy is generated by the thermonuclear reactions, that extends from the Sun's center ($T_c \simeq 15 \times 10^6\,K$, $\rho_c \simeq 150\,\mathrm{g\,cm}^{-3}$ up to $\approx 0.25 R_\odot$);
  – the *radiative zone*, where the energy generated in the center diffuses out radiatively, that extends up to the base of the convection zone $\approx 0.7 R_\odot$;
  – The *convection zone*, where the main energy transport process switches from radiation to convection, that extends up to the base of the photosphere at $1\,R_\odot$;

– The **atmosphere** includes

  – the *photosphere*, where the temperature reaches the minimum value $\approx 4300\,\mathrm{K}$ at about 500 km above the solar limb;
  – the *chromosphere*, where the horizontal structuring of the atmosphere starts to show up very clearly (chromospheric network) and where the temperature rises from its minimum value at the photospheric boundary to about $2\,10^4$ K. The average thickness of the chromosphere is $\approx 2500\,\mathrm{km}$;
  – the *transition region*, where the temperature shows an abrupt increase from chromospheric values up to about $2\,10^5$ K over a distance of only few tens of

km. From here, the temperature increases more slowly to reach the million
degree range of the corona in about 2000 km;
– the *corona*, is the outermost and hottest layer of the solar atmosphere, with
  temperatures in the order of 1–$2\ 10^6$ K. It does not have an upper boundary as
  it expands dynamically to fill the interplanetary space, because of the presence
  of the *solar wind*.

The solar material, composed mainly of *H* with about 8% of *He* and much
smaller concentration of heavier elements, is almost completely ionized at $T \gtrsim$
10^4 K, i.e., everywhere except in the lower solar atmosphere. A gas, however, can
be considered a plasma even if the ionization is not complete, so that it is justified to always treat the solar matter as a plasma. The effects of the presence of
a magnetic field start to be felt from the bottom of the convection zone upward
and become particularly evident as we rise in the solar atmosphere, where the field
appears to be the main agent of the observed horizontal structuring. MHD thus appears to be a particularly useful way to describe the physics of the convection zone
and indeed of the whole solar atmosphere, where the plasma is everywhere sufficiently dense to justify the assumption, on which all fluid models are based, that
the *PDF* are almost Maxwellian due to the effect of collisions. The use of MHD
in the solar wind context needs some care, as the limited degree of collisionality
of this system requires rather the use of two-fluid models, or even kinetic models. Similar remarks apply when one is concerned with the production of energetic
particles.

2.3.1 What Are the Major Unanswered Questions?

It is convenient to split these up by region as follows:
In the **interior**:

– How is the magnetic field generated? Is it amplified from a starting "seed"? Is
  there any remnant of the primordial solar nebula field? What is the structure of the
  magnetic field in the Sun's interior? The ensemble of these questions constitute
  the **dynamo problem** and its solution is generally thought to be related to the
  interaction between rotation and convection. The solution of the dynamo problem
  has greatly advanced in recent years, mainly due to the possibility of performing
  realistic numerical simulations of the so-called magneto-convection.
– The explanation of the 11–22 year **cycle of solar activity**, as well as the variation
  of the level of activity of different cycles (e.g., the Maunder minimum) is not
  yet completely satisfactory. Here again the possibility of performing advanced
  numerical simulation of the nonlinear evolution of the fields appears to be a
  promising approach. [The interior is outwith the coverage of this volume, but
  is included here for completeness.]

In the **photosphere**:

– The structure and evolution of **sunspots**. How are the sunspots formed? What is their thermal structure? We also do not know the general structure of the magnetic field outside of the active region, in particular, if and how the field concentrates in **intense flux tubes**. The dynamical evolution of the **solar granulation** has also been a long-standing problem, but modern numerical simulations have been able to reproduce the main observed features. A number of these issues are discussed in Chap. 4 (Hansteen and Carlsson) and Chap. 5 (Vlahos et al.).

In the **chromosphere**:

– How to explain the observed cell network structure? What exactly are the **spicules** and what is their contribution to mass and energy transport? How are **prominences** formed and how are they supported against gravity? What causes prominence **eruptions**? How do the eruptions relate to Coronal Mass Ejection (CME) events observed in the extended solar corona? MHD has proven particularly effective in modeling this class of phenomena. Chapter 4 (Hansteen and Carlsson) addresses the chromosphere further.

In the **transition region**:

– The thermal and magnetic structure of this crucial layer of the solar atmosphere is the main problem. What is the degree of expansion of the photospheric magnetic flux tubes? Is there a **magnetic canopy**? Both the chromosphere and transition regions are time dependent and highly structured both vertically and transversely: this variability may be an intrinsic property, as yet not completely understood, of these layers which separate the lower regions of the solar atmosphere. Here the energetically dominant processes are determined by thermodynamics and plasma motions, and also by the coronal layer, where magnetic fields and currents determine most of the structure.

In the **corona**:

– What are the equilibrium configurations of the coronal structures, such as **loops**, **coronal holes**, and **streamers**? Do we understand the evolutionary timescales of these structures? And, above all, what are the causes of the high coronal temperatures, the celebrated **coronal heating** problem? This, and the associated transition region, is discussed further in Chap. 4 (Hansteen and Chiuderi), Chap. 5 (Vlahos et al.), and Chap. 8 (Guedel).

In the **solar wind**:

– MHD is applicable only at small heliocentric distances. The main problems concern the mechanisms of **acceleration of the solar wind** and of the general structure of the heliospheric magnetic field, in particular, the formation of **magnetic sectors**. The general connectivity of the magnetic field when observed in situ to its source regions in the photosphere is fundamental to the question of propagation of energetic particles. The solar wind, and its associated turbulence, is discussed in Chap. 3 (Carbone and Pouquet), Chap. 6 (Issautier), and Chap. 7 (Velli).

2.4 Plasma Instabilities

2.4.1 Equilibrium and Stability

The MHD equations admit **equilibrium** solutions, namely solutions in which the time derivatives of all quantities vanish, $\partial/\partial t = 0$, and $\mathbf{U} = 0$. Even with these conditions, MHD equations are difficult to solve explicitly and analytical results for MHD equilibria are consequently scarce and generally refer to very simplified cases. Equation (2.1) and (2.3) are then satisfied identically and Eq. (2.2) reduces to

$$0 = -\nabla P + \frac{1}{\mu_0}(\nabla \times \mathbf{B}) \times \mathbf{B} + \rho \mathbf{g}, \tag{2.5}$$

to be supplemented by the condition

$$\nabla \cdot \mathbf{B} = 0.$$

In order to solve explicitly Eq. (2.5), some extra assumptions have to be made. For instance, the equilibria of an ideal plasma, embedded in an axially symmetric magnetic field with negligible gravity, are found by solving Eq. (2.5) in cylindrical geometry for the unknowns: $\mathbf{B}(r) = (0, B_\theta(r), B_z(r))$ and $P(r)$. With these assumptions, Eq. (2.5) reads

$$\frac{B_\theta}{r}\frac{d}{dr}(r B_\theta) + B_z\frac{d B_z}{dr} + \mu_0\frac{d P}{dr} = 0, \tag{2.6}$$

with the divergence-free condition being satisfied identically.

To solve Eq. (2.5) explicitly we have still to make some other assumption either about $\mathbf{B}$ or P. An important class of solutions are the constant pressure solutions, $dP/dr = 0$. From Eq. (2.5) we see that the constancy of pressure implies $\nabla \times \mathbf{B} = \alpha\mathbf{B}$, with α an arbitrary function of r. These fields are called **force-free fields**, since, as shown by Eq. (2.5) they do not contribute to the equilibrium of forces. The special case $\alpha = 0$ corresponds to **potential fields**. Force-free fields are considered to be important in problems in plasma astrophysics with no or negligible gravity, since a dimensional analysis of Eq. (2.2) shows that the pressure gradient term can be neglected with respect to the magnetic term whenever

$$\beta = \frac{P}{B^2/2\mu_0} \ll 1,$$

a circumstance often satisfied in cosmic plasmas.

The existence of equilibrium solutions does not guarantee that they can actually be observed. In fact, any system is subject to a number of forces that are not included in the equilibrium equation, because they are considered small and therefore negligible. However, they do exist and can be included in the theory as **perturbations** of the equilibrium. This brings us to the problem of the **stability** of equilibria. When a given equilibrium is perturbed, the system evolves in time, and equilibria can be classified according to whether the amplitude of the perturbations remains small (**stable** case) or increases in time (**unstable** case). In the unstable case, the system moves away from the initial equilibrium, possibly settling into a different (stable) equilibrium. If perturbations grow, so does the energy associated with them. Therefore, in order to be unstable, the system must have access to some amount of **free energy**, that could be converted into the energy associated with the growing perturbations. This line of reasoning also shows that the amplitude of the perturbations must eventually saturate, due to the finiteness of the free energy available.

Generally speaking, a plasma is always unstable. This is due to the fact that a plasma is a system with a huge number of degrees of freedom, of the order of $3N$, where N is the number of particles in the system. A high number of degrees of freedom implies that the plasma has many ways to "get away" from the constrictions that the existence of an equilibrium imposes, so that it is extremely likely that some unstable degree of freedom could be found. If this is the reality, one may ask why do we speak of equilibria at all. The point is that instabilities develop at different rates and what matters are the **timescales** we are interested in. In other words, what we want to know is whether the system will remain close to the "equilibrium" state for a given time interval. Thus, if we consider the case of laboratory plasmas, and we want the plasma configuration to survive for a time τ, then we must suppress all the instabilities which grow on timescales faster than τ. Conversely, if in the case of astrophysical plasmas we observe a system to remain in a relatively stationary state for a period of time τ, this means that all the instabilities that develop on timescales faster than τ must have been stabilized. The study of instabilities has therefore a central role in plasma physics. On one hand, the understanding of the mechanisms underlying the development of the instabilities is fundamental to controlling them: the history of research on fusion reactors is just that of a continuous fight against instabilities. On the other hand, the observation of configurations lasting longer than expected clearly proves that there are in nature ways to control at least the fastest instabilities.

The stability analysis of a given configuration starts from the assumption that the initial amplitude of the perturbations applied to the system is small. The MHD equations are then expanded to the first order in terms of the amplitudes of the perturbations. In the resulting equations, the perturbations therefore appear linearly, with coefficients that are functions of the unperturbed (equilibrium) quantities, which, by definition, are independent of time. The system of linear equations can therefore be

Fourier analyzed in time, which means that every perturbed quantity $f \propto e^{-i\omega t}$. The resulting equations can be combined to give a single eigenvalue equation, the eigenvalues being proportional to ω^2.

Let us first consider the case of an ideal plasma. It can be shown that in this case the linearized system of equations can be reduced to.

$$-\omega^2 \rho_0 \xi(\mathbf{r}_0) = \mathbf{F}[\xi(\mathbf{r}_0)], \qquad (2.7)$$

where ξ is the Lagrangian displacement from the equilibrium position $\mathbf{r}_0$, $\mathbf{r} = \mathbf{r}_0 + \xi$,

$$\xi(\mathbf{r}_0, t) = \xi(\mathbf{r}_0) e^{-i\omega t},$$

and $\mathbf{F}$ is a Hermitian linear operator on ξ, and the index 0 indicates, from now on, unperturbed quantities. Here ω^2 is real, but not necessarily positive. $\omega^2 > 0$ implies a real ω and the system exhibits an oscillatory behavior around the equilibrium position, $\omega^2 < 0$ indicates that ω is purely imaginary and the perturbations grow. The sign of ω^2 can therefore be used to discriminate between stable and unstable states. When $\omega = 0$, the system is said to be in a *marginal* stability state. Marginal stability thus represent the boundary (in parameter space) between stable and unstable regions and, given the fact that it is generally easier to find the conditions for marginal stability, it is easy to understand the relevance of the study of marginal states.

As we have seen, a stable ideal system oscillates around the equilibrium configuration, the eigenvalues ω giving the proper oscillation frequencies of the system. The oscillations can be either *stationary* or *propagating* waves, giving rise to a complex and varied phenomenology. An unstable system moves away from the equilibrium configuration, $|\omega|$ defining the *growth rate* of the instability. After a certain time, the amplitude of the perturbation is no longer small and the whole scheme on which the linear stability analysis is based is no longer valid. We are entering the *nonlinear regime*, that normally requires the use of numerical simulations to be analyzed. We may say that the linear stability analysis gives an indication whether instabilities start or not, but is unable to predict the outcome of the instability, in particular, its saturation limit.

Moving now to non-ideal plasmas, where dissipative processes are at work, and following the same linearization process, we end up with a system of equations that contains ω in addition to ω^2. The eigenvalue equation gives complex values for ω and the solutions contain both an oscillatory part and a damped or amplified part, according to the sign of the imaginary part of ω. In the first case, we are in the presence of damped oscillations, in the latter, sometimes referred to as an *overstable* case, of amplified oscillations. Damped oscillations are interesting in a number of astrophysical situations since propagating waves transport energy away from the region where they are generated and the damping process deposits part of that energy in far away parts of the system. We shall return to this point when dealing with the coronal heating problem.

Ideal instabilities are generally the fastest and thus have to be stabilized first. Once this is done, non-ideal instabilities come into play. They are classified according to the dissipative terms that cause them. A finite thermal conductivity, coupled with the presence of radiative losses, gives rise to *thermal* instabilities, while a finite electrical conductivity is the origin of *resistive* instabilities.

2.4.2 Ideal Instabilities

2.4.2.1 Generalities

As we have seen, the purpose of a linear stability analysis is to determine the set of eigenvalues of Eq. (2.7). If the coefficients of this equation do not depend on certain spatial variables, we can make a further Fourier expansion in terms of those "ignorable" variables, $\xi \propto exp(ik_j x_j)$ where x_j are the ignorable coordinates.

For instance, the solution of Eq. (2.7) for a constant density plasma supported against gravity by a uniform magnetic field gives

$$\omega^2 = -k\,g + \frac{(\mathbf{k} \cdot \mathbf{B}_0)^2}{\mu_0 \rho_0}.$$

We see therefore that an instability sets in when $\mathbf{k} \cdot \mathbf{B}_0 = 0$. If we replace the uniform field with a rotating one, $\mathbf{B}_0(y)$, the instability for a given mode (i.e., a given $\mathbf{k}$) is confined to a layer close to the surface $y = y_s$ where $\mathbf{k} \cdot \mathbf{B}_0(y_s) = 0$.

Although this conclusion has been drawn for a very special case, it has a much more general validity. In fact, surfaces where $\mathbf{k} \cdot \mathbf{B}_0 = 0$, called *resonant surfaces*, have a special role in stability analysis. This is connected to the fact that Eq. (2.7) has singular points located where

$$\mu_0 \rho_0 \omega^2 - (\mathbf{k} \cdot \mathbf{B}_0)^2 = 0.$$

The singularities therefore are present only for stable configurations, $\omega^2 > 0$, but for marginal stability ($\omega = 0$) they occur at the resonant surfaces, and the instabilities tend to be localized there.

Such a behavior is shown for instance by a system with axial symmetry and no z-dependence (such as, an infinite cylinder), for which $\mathbf{B}_0 = [0, B_\theta(r), B_z(r)]$, then we may write:

$$\xi(\mathbf{r}, t) = \xi(r) \exp(im\theta + ikz) \exp(-i\omega t),$$

and the different modes can be classified according to the values of m and k. For marginal stability Eq. (2.7) can be reduced to

$$\frac{d}{dr} \left(\frac{(\mathbf{k} \cdot \mathbf{B}_0)^2}{m^2 + k^2 r^2} \frac{d\xi_r}{dr} \right) + g(r)\xi_r = 0,$$

where $g(r)$ is a regular function of its argument. We see therefore that the the resonant surface,

$$(\mathbf{k} \cdot \mathbf{B}_0) = \left(\frac{m B_\theta(r_s)}{r_s} + k B_z(r_s) \right) = 0$$

is a singular surface. For future reference, we mention here that the presence of resistivity removes the singularity and gives rise to a new kind of instability, the *resistive tearing-mode*.

Returning to ideal instabilities, it is possible to find the eigenvalues of Eq. (2.7) for each value of m. The first mode $m = 0$, the *sausage* mode, retains the axial symmetry, while the successive ones do not. In particular it is found that the $m = 1$ mode, the *kink* mode, is the only one that displaces the axis of the cylinder. The kink mode is the most dangerous, as it grows faster than the others and has therefore to be stabilized first.

In an axially symmetric plasma on each cylindrical surface all the lines of force have the same inclination, but this may vary with the radius. From the definition of a line of force, we have

$$\frac{B_\theta}{B_z} = \frac{r \, d\theta}{dz}.$$

The amount of twist of a line of force over a distance L in the z-direction, sometimes referred to as the *pitch* of the field, is therefore given by

$$\Phi = \int d\phi = \frac{1}{r} \int_0^L (B_\theta / B_z) \, dz = \frac{L B_\theta}{r B_z}.$$

It is possible to find a *necessary* criterion for the stability of a cylindrical plasma, in terms of the *safety factor*:

$$q = \frac{2\pi}{\Phi},$$

in the following form:

$$B_z^2 \left(\frac{q'}{q} \right)^2 + \frac{\mu_0 P'}{2r} > 0,$$

where the primes indicate radial derivatives (Suydam criterion).

2.4.3 Application to Solar Coronal Loops and Coronal Arcades

Infinite cylinders are not very good representations of real magnetic structures, both in the lab and on the Sun, where we are typically dealing with toruses (tokamaks)

Fig. 2.1 Coronal loops as seen by TRACE. The emission is from a highly ionized state of iron at a temperature of roughly 10^6 K

or half-toruses (coronal loops: Fig. 2.1). If the aspect ratio of the torus, i.e., the ratio between the large and the small radius of the torus, is sufficiently large, so that the curvature effects can be neglected, a cylinder with periodic conditions in the z-direction can mimic a torus to a reasonable degree of approximation. The above analysis is then still valid, k being now a discrete variable as dictated by the periodicity length.

Coronal loops are the basic building blocks of the active corona, and are discussed much more extensively in Chap. 4 (Hansteen and Carlsson) and Chap. 8 (Guedel). They are observed from space in several spectral lines, covering the temperature range from the transition region to the corona. The most spectacular images of loops refer to observations made onboard the satellite TRACE, like the one shown in Fig. 2.1.

The tantalizing complexity of the structures appearing in the above picture makes it clear that in order to make any progress in explaining what we see we must replace the real thing with some kind of highly simplified theoretical object, hopefully without loosing too much of the physics. As a first, rough, approximation, loops are replaced with cylinders of length $2L$ and cross-sectional area A, gravity is neglected and the boundary conditions at the bases of the cylinder are assumed to represent the photospheric (or chromospheric) plasma.

However, the representation of loops as cylinders of finite length, even if periodic conditions are imposed, ignores one important aspect of the problem, namely, the fact that magnetic field lines are tied in the dense photospheric plasma. Coronal loops are only half-toruses and the stabilizing effect of line tying must be taken into account. Even in this case it is possible to derive a necessary stability criterion, the so-called *ballooning* criterion, analogous to the Suydam criterion for the infinite cylinder:

$$a\left(\frac{qB}{2r}\right)^2 + b\frac{B_z^2}{4}\left(\frac{q'}{q}\right)^2 + c\frac{2\mu_0 P}{r} + d\frac{\mu_0\gamma P B_\theta^2}{r^2(B^2 + 4\pi\gamma P)} > 0,$$

where a, b, c, and d are constants. The problem of the stability of line-tied loops has been the subject of intense investigation in recent times: the results are generally presented in terms of a *critical twist*, Φ_{crit}. Although the results change according to the different assumptions made, it is typically found:

$$\Phi_{crit} \approx 2.5\pi.$$

The influence of line-tying extends also to the case of resistive plasmas. Without line-tying, singularities develop, resulting physically in current sheet formation, and tearing-mode instability sets in. With line-tying, singularities are removed, the linear tearing-mode is stabilized, but the possibility of nonlinear reconnection remains.

Observations show that magnetic loops are often organized into arcades. These configurations are of great interest since they have been shown to provide magnetic support for material heavier than the ambient and are therefore considered as the prime candidates for sustaining the cold and dense structures seen at coronal levels, known as filaments when they are observed on the disc and as prominences when seen at the limb. The footpoints of the magnetic arcades are themselves tied to the photospheric plasma and are subject therefore to shearing motions. When the amount of shearing reaches a critical level, the system may become unstable and the prominence may erupt, a phenomenon well known to solar observers. The consequences of the eruption of prominences are felt also on heliospheric scale being correlated with the so-called Coronal Mass Ejections (CME).

2.5 Coronal Heating, Waves, and Turbulence

The solar corona is kept at a temperature above 10^6 K by a poorly understood mechanism which must non-radiatively transfer the abundant mechanical energy in the turbulent photospheric velocity field thousands of kilometers above, and then dissipate this energy within 1–2 solar radii [35]. The estimated energy flux required to balance radiative and conductive losses is $\epsilon \simeq 10^7$ erg/cm^2/s for active regions, $\epsilon \simeq 8\,10^5 - 10^6$ erg/cm^2/s for the quiet sun, and $\epsilon = 5\,10^5 - 8\,10^5$ erg/cm^2/s, including solar wind losses, for a coronal hole [38, 1].

Because of the strong magnetic field, the energy flux propagates upward as a Poynting flux $\mathbf{S} = 1/\mu_0\,(\mathbf{E} \times \mathbf{B})$ where the electric field is induced by the photospheric motions ($\delta\mathbf{v}$) perpendicular to the magnetic field (if we consider the photosphere to be a perfect conductor), so that $\mathbf{E} = -\delta\mathbf{v} \times \mathbf{B}$. The boundary motions of magnetic footpoints are due essentially to the solar granulation, with characteristic speeds $\delta v \simeq 0.25 - 2$ km/s, sizes l_c of order $l_c \simeq 10^3$ km, and lifetimes τ_c of order $\tau_c \simeq 300$ s, as well as the supergranulation, with characteristic speeds $\delta v \simeq 0.3$ km/s, sizes $l_c \simeq 3 \ 10^4$ km, and lifetimes $\tau_c \simeq 10^5$ s. In fact, photospheric motions are distributed on all scales in between because the convection is turbulent. In any case it is clear that there is more than enough energy present in convection to supply the total coronal losses. The solar coronal heating problem is thus more a question of how the energy is transmitted upward and dissipated in the right place, within a few solar radii of the surface. Because of the enormous kinetic and magnetic Reynolds numbers ($\sim 10^{12}$), and the coincidence of magnetic and thermal structures, one is led to the conclusion that the magnetic fields must play a crucial role, and the mechanism must involve the formation of very small scales, reconnection, and/or turbulence.

The rest of this section introduces a number of popular theories for coronal heating. A much more detailed discussion of one model, heating due to nanoflares, as well as a discussion of computational approaches, can be found in Chap. 4 (Hansteen and Carlsson), which also discusses in some detail how coronal emission is generated, and what can be learned from measurements.

2.5.1 Wave-Based Heating Mechanisms

How the corona responds to perturbations on the timescale τ depends on the ratio τ/τ_a, with $\tau_a = l/V_a$ the timescale for typical (Alfvén) wave propagation along a coronal loop. For $B = 100$ Gauss, $n = 10^9$ cm^{-3}, and a loop length $L = 10^9$ cm this amounts to 1.5–2 s. Roughly speaking, if $\tau/\tau_a \leq 1$ perturbations will propagate as waves, while if the inequality is reversed, the corona will respond quasi-statically. Coronal heating theories based on waves must answer two basic questions: (i) Which waves are likely to be generated with an adequate energy flux? and (ii) Is this energy deposited at the right heights at the appropriate rate?

Three types of MHD waves exist: (a) slow waves, which are essentially sound waves constrained to propagate along the magnetic field; these may be ruled out as a significant source of heating because of the small observed flux. (b) Fast magnetoacoustic waves, which, however, are totally reflected at transition region heights, unless their wave vectors are strictly aligned with the magnetic field, in which case they are indistinguishable from (c) Alfvén waves. These appear to be the most promising candidates for coronal heating since because of their anisotropic dispersion relation, they do not suffer from reflection in the geometrical optics approximation. Also, Alfvén waves propagating away from the Sun are observed to be a dominant component of the turbulence observed in the solar wind [2], and are discussed extensively in Chap. 3 (Carbone and Pouquet) and Chap. 7 (Velli). These waves might be the remnant of a wave flux with sufficient energy for coronal heating, though there are counter-arguments: Alfvén waves of periods greater than a few minutes no longer propagate according to geometrical optics and are therefore strongly reflected in the chromosphere and transition [50].

It must be pointed out also that energy stored by slow motions in the photosphere can be released in bursts in the corona (next section), where as a consequence large amplitude Alfvén waves may be generated.

Because of the extremely small dissipation coefficients, large gradients must develop for damping to become significant. Since there is very little energy directly available at the dissipative scales, the generation of such scales must rely directly on the linear or nonlinear evolution of the waves. Hence, we must invoke either a nonlinear cascade (leading to shock formation and/or turbulence) or the interaction of waves with the nonuniformities of the underlying magnetic field.

For waves propagating along magnetic fields which are inhomogeneous in a transverse direction, two different effects have been shown to be important. The first one, known as phase mixing [18], is due to the frequency (or wavelength) detuning between neighboring oscillating magnetic field lines due to the Alfvén velocity gradients. As a result, the oscillations become rapidly out of phase as they propagate (or evolve in time) and the wavefronts are rapidly deformed and corrugated transversely as the perturbation moves upward.

The second process, known as resonant absorption ([22, 25] and references therein), is due to the pressure gradients associated with the wave, which have the tendency to concentrate the energy in vicinity of the point where the frequency of the wave is equal to the local Alfvén frequency, i.e., where $\omega = k_{\parallel} V_a$. Corresponding to such processes there exist two different types of normal modes [10], with a dissipation rate which is very much enhanced with respect to that of the corresponding uniform case: in fact, it is independent of the magnetic Lundquist number S, defined here as $S = (l V_a m u_0)/(\eta)$.

Propagation in more complex geometries leads to generalizations of the previous results: for example, propagation of Alfvén waves in a magnetic configuration containing x-points [7] leads to normal modes with a dissipation rate (normalized to the real frequency) $\gamma \tau_a \sim -\ln S$, and a similar damping rate is obtained if the coronal magnetic field is chaotic. In this case it is the exponential separation of neighboring field lines which causes the rapid deformation of the wavefronts. Alfvén waves can also mode-convert into (fast) shocks at x-points [21] which might provide additional sources of heating and particle acceleration. In conclusion to establish the relevance of wave heating requires a strong numerical simulation effort with realistic coronal magnetic fields, and high enough resolution.

2.5.2 DC Heating Mechanisms

Gold [12] first put forward the idea that coronal heating is due to the dissipation of field-aligned electric currents. Convective flows below the solar surface cause a random footpoint shuffling of magnetic field lines, which in regions of closed magnetic topology cause a secular increase in the stresses within the coronal magnetic field.

Parker [26, 27] conjectured that such motions must lead to singularities (current sheets or filaments) appearing in the coronal field configuration. The relaxation of such currents would then result in coronal heating. Although Parker's conjecture has not been proved for continuous footpoint motions and fields lacking separatrices or x-points, it is certainly verified for the random motion of finite-sized flux tubes (discontinuities appear when different flux tubes come into contact) [28, 34, 36, 3]. Numerical simulations [24, 17] confirmed that the transverse magnetic field cascades to small scales, leading to an exponential growth of the coronal current.

Most of the above referenced papers obtained a heating function due to the processes considered of the form

$$F_H = q \, B^2/\mu_0 \, \delta v, \qquad (2.8)$$

where B is the normal (vertical) component of the magnetic field, δv the rms photospheric velocity field, and q an efficiency factor, which, depending on the model, may vary. The efficiency factor should generally also depend on the dimensionless variables in the problem, for closed loops typically the parameter $\alpha = lB/\sqrt{\mu_0 \rho L \delta v}$, where l/L is the aspect ratio of the loop as defined by the ratio of loop length to typical photospheric velocity correlation length.

2.5.3 Twisting and Braiding

Consider for example Sturrock and Uchida [34], who studied the random twisting of an isolated flux tube. The mean square winding angle $\theta^2(t)$ will depend on the characteristics of the turbulent photospheric velocity field. Considering for simplicity the flow to lie within the plane of the photosphere, the vorticity ω will contribute to the rotation in the form

$$\frac{d\theta^2}{dt} = \tau_c < \omega^2 >$$

with τ_c the correlation time of the flow. A comparable contribution to the twist comes from shearing the flux tube footpoints (as long as the tube is not perfectly circular). In terms of the correlation length λ_c and the rms velocity δv, $< \omega^2 >$ may be written as $4\delta v^2/\lambda_c^2$. For motions on the granular scale $\tau_c \simeq 800$ s, $\lambda_c = 10^5$ cm, $\delta v = 10^3$ cm/s, the root mean square twist reaches about four turns in 24 h.

How much energy is injected in this way? Consider a flux tube of length L, radius R, and axial field strength B, the energy of the non-potential magnetic field in such a tube is $\delta E = \pi R^4 B^2 < \theta^2 > /4L\mu_0$ and the input power per unit area becomes

$$P = \frac{3R^2 B^2 \tau_c \delta v^2}{4\mu_0 \lambda_c^2 L},$$

where the factor 3 arises from geometrical considerations and so $q = 3/4R^2/(\lambda_c L)$. For $L = 10^{10}$ cm, $B = 100$ Gauss, and $R = \lambda_c$, a heating rate of $P = 5 \times 10^5$ ergs/cm^2/s is obtained, an order of magnitude lower than the 10^7 required for active region heating.

Parker considered the braiding of several flux tubes. For simplicity, consider a cartesian geometry in which the photosphere is represented by a pair of conducting planes separated by a distance L. The (x, y) axes will be taken in the planes parallel to the photosphere, while the z-axis is orthogonal to the photosphere. We will consider flux tubes which fill all space in the corona but are pinched into discrete footpoints in the photosphere, and neglect the narrow photosphere/corona transition layer. The photospheric motions will move the fluxtube footpoints around in a random walk, generating a transverse magnetic field $B_\perp$. The total projected arc length of the field line connecting two footpoints on the photosphere, $l(t)$, will be greater than or equal to the projected transverse distance $s(t)$ separating the footpoints. The transverse magnetic field is given by $B_\perp/l = B/L$. If $s(t)$ is much less than the initial distance d_0 between footpoints of separate flux tubes, the field lines will be practically straight, and $l(t) = s(t)$. The input power thus increases linearly with time, as in the Sturrock and Uchida case. With $K = l(t)^2/8t$, the diffusion coefficient of random footpoint motions,

$$P = 2KB^2/\mu_0 L,$$

yields 3×10^5ergs/cm^2/s. If, however, $s(t) \geq d_0$, the footpoints are close enough so that a significant amount of tangling between different flux tubes occurs, so the field lines can no longer straighten out. Hence the transverse field must increase linearly with t, and the stored energy quadratically with t; the input power therefore also increases linearly with t:

$$P = \frac{2\delta v^2 B^2 t}{\mu_0 L}.$$

Parker considers supergranular motions with correlation time $\tau_c = 5.0 \times 10^4$ s and an axial field strength of 100 Gauss ($\delta v = 0.5$ km/s). In this way one finally obtains the required energy flux of 10^7 ergs/cm^2/s. A consequence of this choice is that reconnection should set in once the angle between neighboring field lines exceeds $\theta = 14°$, beyond which current sheet dissipation limits any additional current build up. The magnetic field variation across the current sheet is $\delta B_\perp = 25$ Gauss.

2.5.4 Large-Scale Turbulence Models

Several authors [36, 19, 14] considered the nonlinear cascade driven from boundary motions. Starting from an initially uniform field between two plates representing the photosphere, the slow dragging of magnetic field lines, formation of current sheet, and subsequent cascade was modeled as a turbulent cascade. Reference [36]

assumed Gaussian turbulence, and found the heating rate a factor of ten lower than
required for the active region corona.

The approach in [19] attempted self-consistency: the photosphere drives coronal
turbulence, which is treated as an eddy viscosity term in a large-scale description
of a viscous and resistive coronal equilibrium. The amount of energy driven into
the corona by the photospheric motion depends on the coronal "effective" Reynolds
number. On the other hand, the effective Reynolds number must have just the value
required to dissipate the energy which is input. A detailed calculation is given for the
case of a 1-D incompressible photospheric shear flow. Assuming that the turbulence
endows the coronal medium with an effective viscosity and resistivity, it is easy
to calculate the stationary coronal solution for velocity and magnetic field with the
given boundary motion. For a photospheric shear flow with a Kolmogorov spectrum,
and characteristic size of 10^8 cm (the loop length being 10^9 cm) the Poynting flux
they found was

$$F_H = 1.5710^6 (\delta v / 10^5)^{4/3} (B/100)^{5/3} (n/10^{10})^{1/6} \text{ ergs/cm}^2/\text{s},$$

where δv is the photospheric rms field in cm/s, B the magnetic field in Gauss, and
n the number density. Typical heating rates are then of order 10^6 ergs/cm^2/s.

Another fully turbulent model for coronal heating was presented by Gomez and
Ferro-Fontan [14]: in their computation the goal was not to obtain the correct heating
rate, which is fixed by the photospheric turbulent forcing (a broadband Kolmogorov-
type spectrum), but rather to derive the coronal magnetic and velocity spectrum on
the basis of a given photospheric velocity spectrum. Isotropy and homogeneity in
planes parallel to the photosphere are assumed, and consideration is restricted to
averages along the loop length. Their results imply magnetically dominated coronal
turbulence, with both kinetic and magnetic energy spectra tending to a $k^{-5/3}$ power
law at large wave numbers. The fraction of kinetic to magnetic energy remains at a
very small level (4%).

2.5.5 Numerical Simulations of the Parker Scenario

Thus far, we have discussed time-average heating rates, although implicitly hinting
at the possibly discrete nature of coronal heating through a collection of small energy releases. These small events, often referred to as nanoflares because they are
believed to involve energies nine order of magnitude smaller than the largest flare,
are now the subject of widespread study from both theoretical and observational
viewpoints. [One key question beyond our scope is the power-law distribution of
observed events at optical, ultraviolet, and X-ray wavelengths.] Chapter 4 discusses
nanoflares more fully, but we here present some examples of how a computational
approach can be used to address such bursty coronal heating.

Early works by Mikic et al. and Hendrix and Van Hoven [24, 17] involved simulations using a 3D straightened out loop and imposing photospheric shearing given
by alternate direction flow patterns. They showed that a complex coronal magnetic

field results from the photospheric field line random walk, and though the field does not, strictly speaking, evolve through a sequence of static force-free equilibrium states, magnetic energy nonetheless tends to dominate kinetic energy in the system, which may be thought of as evolving in a special regime of MHD turbulence. In this limit the field is structured by current sheets elongated along the axial direction, separating quasi-2D flux tubes which constantly move around and interact. Longcope and Sudan [23] carried out low-resolution 3D simulations using the reduced MHD approximation to study the current sheet formation process.

Subsequently, to carry out the lengthy simulations necessary to define the statistics of heating events, Einandi et al. [9] first carried out 2D numerical simulations of incompressible magnetohydrodynamic (MHD) turbulence using a random large-scale magnetic forcing function to mimick the forcing exerted in three dimensions by the photosphere. Georgoulis et al. [11] extended these simulations to longer times, and were able to show how the magnetically dominated turbulence in the 2D system displays bursts in the dissipation, corresponding to the formation and dissipation of current sheets separating the longer living magnetic eddies, which follow a power-law behavior in total energy, peak dissipation, and duration indexes not far from those determined observationally in X-rays. This was the first direct numerical proof of a possible relationship between reconnection driven by the random photospheric motions and the observed statistics of coronal activity.

Studies on the magnetic forcing have been carried out to further complexity by Gudiksen and Nordlund [15], who presented simulations of full 3D sections of the solar corona with a realistic geometry. While this approach has advantages when investigating the coronal loop dynamics within its neighboring coronal region, modeling a larger part of the solar corona drastically reduces the number of points occupied by the coronal loops, making interpretation potentially more difficult (Chap. 4).

Finally, Rappazzo et al. [30] have carried out a comprehensive series of simulations allowing for a full solution to the Parker problem in cartesian geometry. The simulations showed that:

(a) coronal loops unamabiguously develop small scales following a turbulent cascade, with an overall kinetic energy much smaller than the magnetic energy;

(b) the heating rate and energy spectra depend on two parameters: the loop aspect ratio and the ratio of coronal Alfvén speed to the exciting photospheric velocity field magnitude;

(c) as these parameters vary, the turbulent spectra correspondingly span the various regimes of MHD turbulence, from weak to strong: spectral slopes of magnetic energy are steeper for strong axial magnetic fields and short loops, while they are flatter for weak fields and long loops. As a consequence the scaling of the heating rate with axial magnetic field intensity B, which depends on the spectral index of magnetic energy for given loop parameters, varies from $B^{3/2}$ for weak fields to B^2 for strong fields at a given aspect ratio;

(d) the heating rate for coronal loops obtained with the photospheric velocities and axial magnetic fields estimated at present are on the lower side of the coronal heating requirements.

This study has therefore provided a fairly complete understanding of the very idealized problem of field line-tangling. The generalization to the realistic solar corona however is still far away. Outstanding issues include: the inclusion of other energizing process such as emerging flux, a full understanding of the the the thermodynamic and radiative response in the presence of a jungle of field lines of differing heights and lengths in the corona, and the understanding of the coupling of the sub-photospheric MHD to the corona above. These exciting challenges await further experimental and theoretical investigation.

2.5.6 Coronal Plasma Properties from Heating Mechanisms

Finally, we show how a given (time-averaged) magnetic energy dissipation rate can be used to estimate a "steady state" coronal temperature and density. In turn, these derived quantities can then be used to create "observables" that can then be compared with data. If s is the coordinate measured along the loop axis and we assume that the loop is in a stationary state, we may build a model based on the energy balance among heating, radiation losses, and thermal conduction. The basic equation for these models reads as

$$\frac{d}{ds}\left(\kappa_0 T^{5/2}\frac{dT}{ds}\right) = \mathcal{L} - \mathcal{H} = \chi n^2 T^\alpha - \mathcal{H},$$

where the temperature dependence of radiation loss term, $\mathcal{L}$, has been parametrized as a power law, with the exponent α assuming different values in different temperature ranges and the heating term, $\mathcal{H}$, is considered to be a constant. Assuming that the temperature reaches its maximum T_{max} at the top of the loop $(dT/ds)_{s=L} = 0$ and that $T_{max} \gg T_0$, where T_0 is the temperature at the base and furthermore that

$$P = 2nkT = const. \qquad (dT/ds)_{s=0} = 0,$$

the above equation can be integrated to give the following results:

$$\mathcal{H} = \frac{7P^2}{8k_B^2}T_{max}^{-5/2},$$

and

$$PL = const.T_{max}^3, \tag{2.9}$$

where we have taken $\alpha = -\frac{1}{2}$ as an appropriate value for T_{max} in the million-degree range. Equation (8.33) is the well-known scaling law of Rosner et al. [31], the simplest of the many possible scaling laws obtainable with more sophisticated approaches. The virtue of the scaling law (8.33) is that it involves quantities that can be measured or estimated, like the pressure, the top temperature of the loop, and its semi-length.

The fact that a structure whose footpoints are kept at the same temperature T_0 and in the absence of heat flux, $(dT/ds)_{s=0} = 0$ would develop a nonuniform temperature profile rather than an isothermal one may look surprising at first sight. But it can be shown that a constant temperature loop is unstable to thermal instability, driven by the different temperature dependence of the heating and radiation terms, so that an initially isothermal loop eventually settles in a stable equilibrium with temperature increasing from T_0 to $T_{max} \gg T_0$ as we go from the base to the apex of the loop.

Finally, we note that dynamic models for the evolution of the coronal temperature and density in response to, for example, nanoflares, involve the solution of Eq. (2.3). Such models lead to a wide range of coronal temperatures and densities [8] in a given large-scale structure and would appear to be better able to account for the range of observed coronal properties than steady-state models.

References

1. M.J. Aschwanden, A. Winebarger, D. Tsiklauri, H. Peter: Astrophys. J., **659**, 1673 (2007)
2. J.W. Belcher, L. Davis: J. Geophys. Res., **76**, 3534 (1971)
3. M.A. Berger: Astron. Astrophys., **252**, 369 (1991)
4. D. Biskamp: *Magnetic Reconnection in Plasmas*. Cambridge University Press, Cambridge (2000)
5. D. Biskamp: *Magnetohydrodynamic Turbulence*. Cambridge University Press, Cambridge (2003)
6. T.J.M. Boyd, J.J. Sanderson: *The Physics of Plasmas*. Cambridge University Press, Cambridge (2003)
7. S.V. Bulanov, S.G. Shasharina, F. Pegoraro: Plasma Phys. and Contr. Fusion, **34**, 33 (1992)
8. P.J. Cargill, J.A. Klimchuk: Astrophys. J., **605**, 911 (2004)
9. G. Einaudi, M. Velli, H. Politano, A. Pouquet: Astrophys. J., **457**, L113 (1996)
10. G. Einaudi, C. Chiuderi, C.F. Califano: Adv. Space Res., **13**, 85 (1993)
11. M. Georgoulis, M. Velli, G. Einaudi: Astrophys. J., **497**, 957 (1998)
12. T. Gold: The Physics of Solar Flares. In: W. Hess (ed.), *NASA SP-50*, p. 389 (1964)
13. H. Goedbloed, S. Poedts, *Principles of Magnetohydrodynamics, with Applications to Laboratory and Astrophysical Plasma*. Cambridge University Press, Cambridge (2004)
14. D.O. Gomez, C.F. Ferro-Fontan: Astrophys. J., **394**, 662 (1992)
15. B.V. Gudiksen, A. Nordlund: Astrophys. J., **618**, 1020 (2005)
16. D. Gurnett, A. Bhattacharjee: *Introduction to Plasma Physics: with Space and Laboratory Applications*. Cambridge University Press, Cambridge (2005)
17. D.L. Hendrix, G. Van Hoven: Astrophys. J., **467**, 887 (1996)
18. J. Heyvaerts, E.R. Priest: Astron. Astrophys., **117**, 220 (1983)
19. J. Heyvaerts, E.R. Priest: Astrophys. J., **390**, 297 (1992)
20. R. Kulsrud: *Plasma Physics for Astrophysics*. Princeton University Press, Princeton (2005)
21. S. Landi, M. Velli, G. Einaudi: Astrophys. J., **624**, 392 (2005)
22. M.A. Lee and B. Roberts: Astrophys. J., **301**, 430 (1986)
23. D.W. Longcope and R.N. Sudan: Astrophys. J., **437**, 491 (1994)
24. Z. Mikic, D.D. Schnack, G. Van Hoven: Astrophys. J., **338**, 1148 (1989)
25. Y. Mok, G. Einaudi: J. Plasma Phys., **33**, 199(1985)
26. E.N. Parker: Astrophys. J., **174**, 499 (1972)

27. E.N. Parker: Astrophys. J., **330**, 474 (1988)
28. E.N. Parker: *Spontaneous Current Sheets in Magnetic Fields*. Oxford University Press, Oxford (1994)
29. E.R. Priest: *Solar Magnetohydrodynamics*. Kluwer Academic Publishers, Dodrecht (1982)
30. A.F. Rappazzo, M. Velli, G. Einaudi, R.B. Dahlburg: Astrophys. J., **657**, L47 (2007)
31. R. Rosner, W.H. Tucker, G.S. Vaiana: Astrophys. J., **220**, 643 (1978)
32. C. Schrijver, C. Zwaan: *Solar and Stellar Magnetic Activity*. Cambridge University Press, Cambridge (2000)
33. P.A. Sturrock: *Plasma Physics: An Introduction to the Theory of Astrophysical, Geophysical, and Laboratory Plasmas*. Cambridge University Press, Cambridge (1994)
34. P.A. Sturrock, Y. Uchida: Astrophys. J., **246**, 331(1981)
35. P. Ulmschneider, E.R. Priest, R. Rosner (eds.): *Mechanisms of Chromospheric and Coronal Heating*. Springer Verlag, Heidelberg, (1991)
36. A.A. Van Ballegooijen: Astrophys. J., **311**, 1001 (1986)
37. A. Verdini, M. Velli: Astrophys. J., **662**, 669 (2007)
38. G. Withbroe, R.W. Noyes: Ann. Rev. Astron. Astrophys., **15**, 363 (1977)

Chapter 3
An Introduction to Fluid and MHD Turbulence for Astrophysical Flows: Theory, Observational and Numerical Data, and Modeling

V. Carbone and A. Pouquet

3.1 Introduction

The turbulent evolution of fluid flows is characterized by randomness in *both* space and time (e.g., [11, 27]). Turbulence is a nonlinear phenomenon, ubiquitous in Nature, where chaotic dynamics and power law statistics co-exist. Turbulence sets in when the Reynolds number $Re = L_0 U / \nu$ becomes greater than a threshold value. Here U and L_0 are the characteristic velocity and length, respectively, and ν the kinematic viscosity, and Re is the ratio of the nonlinear and dissipative terms in the Navier–Stokes equation, which describes the dynamics of fluid flows. The main feature of turbulence is the presence of *specific structures*, for example, vortices (or eddies), at all dynamically interesting scales.

At low Reynolds numbers, structures are present at some typical large scale L_0 (for example, the size of an obstacle in a flow, or the mesh size of a grid, or the distance between the walls in a channel). As the Reynolds number increases, and as the nonlinear terms begin to be non-negligible, random activity of the flow is observed on all scales. Finally as $Re \rightarrow \infty$ the flow is said to be in a *fully developed turbulence* regime. In these conditions, the fields are highly chaotic, with the overlapping of different eddies over an "infinite" range of scales. However, in this regime it is possible to separate the scaling behavior into three well-defined ranges according to the cascade phenomenology worked out by Richardson [80]. The energy-containing scales (L_0) in which the energy is injected into the system due to some external forcing are called *integral scales*. The scales at which dissipation is dominant (ℓ_D) belong to the *dissipative range*. In between these two ranges $\ell_D \ll \ell \ll L_0$ the dominant terms in the equations are the nonlinear terms, and an

V. Carbone (✉)
Dipartimento di Fisica – Universitá della Calabria, and Istituto Nazionale di Fisica della Materia, unitá di Cosenza, Cubo 31C 87036 Rende (CS), Italy
carbone@fis.unical.it

A. Pouquet
NCAR, P.O. Box 3000, Boulder, Co 80304-7000 U.S.A.
pouquet@ucar.edu

Carbone, V., Pouquet, A.: *An Introduction to Fluid and MHD Turbulence for Astrophysical Flows: Theory, Observational and Numerical Data, and Modeling*. Lect. Notes Phys. **778**, 71–128 (2009)
DOI 10.1007/978-3-642-00210-6_3   © Springer-Verlag Berlin Heidelberg 2009

energy redistribution over different scales ℓ dominates the dynamics. This range of scales is called "inertial range".

Even though wind tunnel or atmospheric measurements allow us to gain some insight into turbulent behavior, it is in space that turbulence fully reveals its attractive universality. The interest in investigating turbulence in the astrophysical context is related to the fact that, because of the relatively large collision lengths, a wide inertial range can be observed, up to very small scales where kinetic effects, rather than classical dissipative phenomena, are at work. As an order of magnitude estimate, while Reynolds numbers investigated with numerical simulations are in general relatively modest ($Re \simeq 10^3 \nabla \times 10^5$), Reynolds numbers in geophysical flows such as the atmosphere, the oceans, or in space are of order $Re \simeq 10^8 \nabla \times 10^{12}$.

Flows in space are associated with highly conducting fluids (a plasma), comprised of charged particles, and effects related to the presence of magnetic fields and electrical currents can sometimes prevail. This makes the already complex problem of turbulence yet more complicated since one has now to take into account interactions between different fields on top of interactions between a wide range of scales. However, things can perhaps be more interesting!

A turbulent plasma is characterized by three gross features: (1) spatio-temporal evolutions of the electromagnetic field and/or of the plasma without any apparent order; (2) a superposition of very different scales where the flow is active; (3) the flow cannot be predicted in details, but only on average. Solitary structures, which are currently observed in space plasmas, are not necessarily independent phenomena since they may often be considered as elementary entities within a turbulent flow. In general, turbulence plays an important role in the exchanges of energy and momentum in a collisionless plasma and is also involved in the stochastic acceleration of particles.

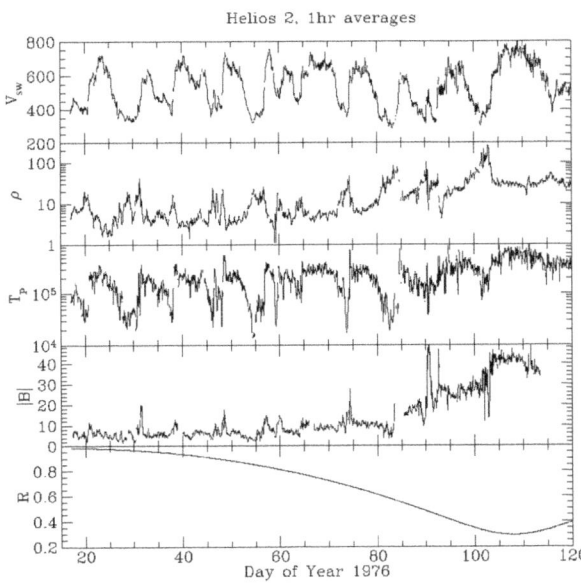

Fig. 3.1 One hour averages of fields measured by the Helios 2 satellite in the inner heliosphere, from 1 to 0.3 AU. The various panels, from the top to the bottom, refer to time evolution of bulk wind speed measured in km s^{-1}, mass density measured in number of particles per cm^{-3}, kinetic proton temperature measured in K, magnetic intensity measured in mT, and distance from the Earth measured in units of AU

The solar wind, a high Reynolds number plasma flow of solar origin that fills all the heliosphere (e.g. [97], Chaps. 5 and 7 (Issautier and Velli)), is the most accessible medium to study collisionless magneto-hydrodynamic (MHD) turbulence by in situ measurements from spacecraft flying in interplanetary space. This is a topic of fundamental importance for both plasma physics and astrophysics, and can be regarded also as a fundamental topic in the study of turbulence itself. Many examples of the occurrence of turbulent flows in space plasmas can be given. For example, in interplanetary space turbulent fluctuations are observed from scales $\ell \sim 180 \times 10^6$ km, down to $\ell \sim 100$ km. In Fig. 3.1 we show the time evolution of a series of fluctuations obtained from the Helios 2 spacecraft. Chapter 7 (Velli) discusses solar wind turbulence in more detail.

A look at images from the Voyager spacecraft reveals the turbulent characteristics of the Jovian atmosphere on planetary scales (see Fig. 3.2). Turbulence in the solar environment is currently investigated with the aid of satellite images or telescopes on the Earth. The solar photosphere is a beautiful example of turbulent convection, convective cells being of the order of 10^3 km (actually the instrumental resolution) up to the 10^5 km for supergranulation [90]. The solar corona is a turbulent medium where small scales dissipative isolated events play a crucial role in determining the heating process. On larger scales, compressible turbulent effects within the interstellar medium are invoked as key ingredients to explain the

Fig. 3.2 Observations of Jupiter's atmosphere from the Voyager satellite. Structures at all scales, within a turbulent environment, are evident

formation of stars within galaxies [55]. Finally on cosmological scales accretion disks around neutron stars or black holes are turbulent. For references on turbulence in astrophysical context see, for example, the proceedings of Ref. [7] and references therein.

3.2 Fundamental Equations

3.2.1 Hydrodynamics

The Navier–Stokes (NS) equations are simply Newton's law expressed as

$$\partial_t \mathbf{u} + \mathbf{u} \cdot \nabla \mathbf{u} = -\nabla P + \nu \nabla^2 \mathbf{u} + \mathbf{F}, \tag{3.1}$$

where $\mathbf{u}$ is the velocity, P the pressure, and $\mathbf{F}$ stands for any other force that can be present, such as gravity, rotation, or radiative processes. In order to focus on the problem of turbulence, that of coupling scales through nonlinear terms, we restrict the analysis to the incompressible case for which $\nabla \cdot \mathbf{u} = 0$. We are then able to absorb the (constant) mass density which usually appears on the left-hand side into the other variables without loss of generality. These equations, in the absence of dissipation ($\nu \equiv 0$), conserve the total kinetic energy $E^V = \langle u^2 \rangle / 2$ and the total kinetic helicity $H^V = \mathbf{u} \cdot \boldsymbol{\omega}$, with $\boldsymbol{\omega} = \nabla \times \mathbf{u}$ being the vorticity. It is straightforward to see the latter when rewriting the advection term as $\mathbf{u} \cdot \nabla \mathbf{u} = -\nabla u^2 / 2 + \mathbf{u} \times \boldsymbol{\omega}$, with the first term being absorbed in the pressure. An important parameter arises when non-dimensionalizing the above equations, namely the Reynolds number $Re = U L_0 / \nu$ which measures the relative strength of the nonlinear term to the dissipative one. It is also convenient to define, from a dynamical point of view, the Taylor Reynolds number $R_\lambda = U \lambda / \nu$ with $\lambda^2 = \langle u^2 \rangle / \langle \omega^2 \rangle$ being the Taylor scale.

3.2.2 The Kolmogorov Phenomenology of Turbulence

A fundamental step in understanding the basic phenomenological properties of turbulence, as obtained from the NS equations, is due to A.N. Kolmogorov [48]: hereafter K41. Its "legacy" for turbulence research has been discussed in a large number of papers and textbooks (among others, see [27] and references therein). The main idea is that at very large Reynolds numbers the injection scale L_0 and the dissipative scale ℓ_D are completely separated. In a stationary situation, the energy injection rate must be balanced by the energy dissipation rate and must also be the same as the energy transfer rate ε measured at any scale ℓ within the inertial range $\ell_D \ll \ell \ll L_0$.

The NS equations possess scaling properties [27], that is, there exists a class of solutions which are invariant under scaling transformations. In fact, introducing a length scale ℓ, it is straightforward to verify that the scaling transformation

$\ell \rightarrow \lambda\ell'$ and $\mathbf{u} \rightarrow \lambda^h\mathbf{u}'$ (λ is a scaling factor) leaves invariate the inviscid NS equations for any scaling exponent h. In real space, the usual tool to investigate statistical properties of turbulence is by way of the velocity field increments $\delta u_\ell(\mathbf{r}) = [\mathbf{u}(\mathbf{r} + \ell) - \mathbf{u}(\mathbf{r})] \cdot \mathbf{e}$ ($\mathbf{e}$ being the longitudinal direction along the velocity field). These stochastic quantities represent fluctuations across eddies at scale ℓ. The scaling invariance of the NS equations from a phenomenological point of view implies that we expect solutions where $\delta u_\ell \sim \ell^h$. All the statistical properties of the field depend only on the scale ℓ, the mean energy dissipation rate ε, and the viscosity ν. Also, ε is supposed to be the common value of injection, transfer, and dissipation rates. Moreover, the dependence on the viscosity only arises at small scales, near the bottom of the inertial range. Under these assumptions the typical energy dissipation rate per unit mass scales as $\varepsilon \sim \delta u_\ell^2/t_\ell$. The time t_ℓ associated with the scale ℓ is the typical time needed for the energy to be transferred to a smaller scale, say the eddy turnover time $t_\ell \sim \ell/\delta u_\ell$, so that a scaling law for the field increments can be obtained (K41):

$$\delta u_\ell \sim \varepsilon^{1/3}\ell^{1/3} . \tag{3.2}$$

Note that, since from dimensional considerations the scaling of the energy transfer rate should be $\varepsilon \rightarrow \lambda^{1-3h}\varepsilon'$, $h = 1/3$ is the choice to guarantee the absence of scaling for ε.

In real space, turbulence properties can be described using either the *probability density function* (PDF) of increments or the *longitudinal structure functions*, which represent nothing but the higher order moments of the field, namely $S_\ell^{(p)} = \langle \delta u_\ell^p \rangle$. In the inertial range these quantities behave as a power law $S_\ell^{(p)} \sim \ell^{\xi_p}$, so that what is interesting is to compute the set of scaling exponent ξ_p. In this approach a very important result in turbulence theory, due to Kolmogorov [48], is the so-called "4/5 law". This is an exact result stemming from the NS equations under usual hypotheses, i.e., isotropy, homogeneity, stationarity, incompressibility, and high Reynolds number (see Ref. [27]). In the inertial range the third-order longitudinal velocity structure function behaves linearly with ℓ:

$$S_\ell^{(3)} = -\frac{4}{5}\varepsilon\ell. \tag{3.3}$$

Using, from a phenomenological point of view, the scaling for field increments (3.2), it is straightforward to compute the scaling laws $S_\ell^{(p)} \sim \ell^{p/3}$ and $\xi_p = p/3$.

For a Gaussian distribution of fields, a particular role is played by the second-order moment, because all moments can be computed from $S_\ell^{(2)}$. It is straightforward to translate the dimensional analysis results to Fourier spectra. The spectral property of the field can be recovered from $S_\ell^{(2)}$ which is proportional to $S_\ell^{(2)} \simeq kE(k)$ (where $k \sim 1/\ell$ is the wave number), so that in the inertial range

$$E(k) \sim \varepsilon^{2/3}k^{-5/3} . \tag{3.4}$$

This *Kolmogorov spectrum* (3.4) is largely observed in all experimental investigations of turbulence, and is considered as the main result of the Kolmogorov's phenomenology of turbulence. However, the spectral analysis does not provide a complete description of the statistical properties of the field, unless it has Gaussian properties.

3.2.3 Coupling to a Magnetic Field in the MHD Approximation

3.2.3.1 Introduction

Magnetic fields are ubiquitous in the Universe, and are often dynamically important. If at small scales kinetic (plasma) effects may be dominant, large scales can be modeled using the MHD approximation. Furthermore, dissipative phenomena can be all but neglected at large scales although their effects will be felt because of nonlocality of nonlinear interactions.

MHD turbulence occurs, for example, in the laboratory (fusion plasmas), in the generation and dynamical evolution of magnetic fields of planets, stars and galaxies, in the interstellar medium (ISM), and in the solar wind (Chap. 7). Even though most of these celestial bodies can be looked at as compressible fluids, with Mach numbers based on the *r.m.s.* velocity sometimes greater than unity, as a simplifying assumption, only the incompressible case will be examined.

In the presence of a magnetic field, the Lorentz force $\mathcal{L} = \mathbf{j} \times \mathbf{b}$ must be added to the momentum Eq. (3.1), with $\mathbf{j} = \nabla \times \mathbf{b}$ being the electric current density. The MHD equations stem from the Maxwell equations in which the displacement current is neglected in Ampere's law under the assumption that the velocity of the plasma is much smaller than the speed of light. This is a common occurrence in astrophysical flows; for example, in both the solar photosphere and the interstellar medium, characteristic speeds are of the order of 1 km s^{-1}, and in the solar wind, the advection speed is between 400 and 800 km s^{-1}.

3.2.3.2 The Equations

With the further hypothesis that the velocity field is incompressible, the MHD equations take the form

$$\partial_t \mathbf{u} + \mathbf{u} \cdot \nabla \mathbf{u} = -\nabla P + \nu \nabla^2 \mathbf{u} + \mathbf{j} \times \mathbf{b}, \tag{3.5}$$

$$\partial_t \mathbf{b} = \nabla \times (\mathbf{u} \times \mathbf{b}) + \eta \nabla^2 \mathbf{b} , \tag{3.6}$$

together with the constraint $\nabla \cdot \mathbf{u} = 0$ stemming from mass conservation, with $\nabla \cdot \mathbf{b} = 0$ (no magnetic monopoles) and with η the magnetic diffusivity. One must add the fact that $\mathbf{b}$ is an axial vector, like the vorticity $\omega = \nabla \times \mathbf{u}$, since it stems from the vector potential, viz. $\mathbf{b} = \nabla \times \mathbf{a}$ for divergence-free fields, so that $\mathbf{j} = -\nabla^2 \mathbf{a}$.

Similar to the traditional Reynolds number, a magnetic Reynolds number R_M can be defined, namely

$$R_M = \frac{U_0 L_0}{\eta} \ .$$

This number in most circumstances in astrophysics is very large, but the ratio of the two Reynolds numbers or in other words the magnetic Prandtl number

$$P_M = \frac{\nu}{\eta}$$

can differ widely from unity. In the core of the Earth (and for liquid metals in the laboratory), this number is very small, whereas in the interstellar medium it is large.

The change of variable due to Elsässer [25], viz. $\mathbf{z}^{\pm} = \mathbf{u} \pm \mathbf{b}$, leads to the more symmetrical form of the MHD equations in the incompressible case:

$$(\partial_t + \mathbf{z}^{\mp} \cdot \nabla)\mathbf{z}^{\pm} = -\nabla \mathbf{P}_* + \nu^{\pm}\nabla^2\mathbf{z}^{\pm} + \nu^{\mp}\nabla^2\mathbf{z}^{\mp} + \mathbf{F}^{\pm} \ , \qquad (3.7)$$

where $\mathbf{P}_*$ is the total pressure, including the magnetic pressure stemming from the Lorentz force decomposed into $\mathcal{L} = -\nabla b^2/2 + \mathbf{b} \cdot \nabla \mathbf{b}$; we have $2\nu^{\pm} = \nu \pm \eta$, and $\mathbf{F}^{\pm}$ are forcing terms. The relations $\nabla \cdot \mathbf{z}^{\pm} = \mathbf{0}$ complete the equations. Since $\nabla \cdot \mathbf{z}^{\pm} = 0$, one sees easily that it is the gradient of the $z_i^+ \, z_j^-$ tensor that enters Eq. (3.7). [Note that the permeability constant μ_0 that usually appears in MKS versions of the MHD equations has been absorbed into the other variables in the non-dimensionalization process as has ρ.]

Note that the Elsässer variablesare a bit special: when changing from a right-handed to a left-handed system of reference, the velocity is unchanged and the magnetic field, an axial vector, changes sign; hence the Elsässer variablesare neither vector nor axial vector; under such a change of system, they are interchanged $(\mathbf{u} + \mathbf{b}) \rightarrow (\mathbf{u} - \mathbf{b})$. The Elsässer form of Eq. (3.7) reveals that there are no self-interactions for $\mathbf{z}^+$ waves or $\mathbf{z}^-$ waves, but that it is only the $z^+ \, z^-$ interactions that carry the energy to small scales. This fact leads to possible profound changes in the dynamics of MHD fluids, compared to neutral ones, as we shall see in the following in a phenomenological way.

[We note in passing that to obtain (3.7), one rewrites the induction Eq. (3.6) using the identity

$$\nabla \times (\mathbf{u} \times \mathbf{b}) = -\mathbf{u} \cdot \nabla \mathbf{b} + \mathbf{b} \cdot \nabla \mathbf{u}$$

for divergence-free vectors $\mathbf{u}$ and $\mathbf{b}$. In other words, the flux-conserving r.h.s. of the induction equation transforms itself into an advection term and a stretching term of the magnetic field by velocity gradients. Similarly, one can transform the Lorentz force into a pressure term $\nabla b^2/2$ and a curvature term $\mathbf{b} \cdot \nabla \mathbf{b}$, as we saw before.]

Let us examine briefly these equations. Assuming that the Elsässer variablesbe-have in a similar manner, i.e., $z^+ \sim z^- \sim z$, then we recover the Navier–Stokes equation for z; in such a case, the corresponding MHD turbulence should behave like K41. On the other hand, forgetting for a moment the vector-like nature of $\mathbf{z}^{\pm}$, then the Elsässer equation is the advection of a (passive?) quantity by the other Elsässer field. As such, if it were to behave like a passive scalar, it should be very intermittent, with localized fronts.

Finally, remembering that the equations can be linearized around a uniform magnetic field $\mathbf{B_0}$, it was remarked independently by Iroshnikov and by Kraichnan [49, 44] in the mid-1960s (hereafter IK) that the fact that the only nonlinear interactions in MHD take place between z^+ and z^- fields (i.e., there are no self-interactions in that formulation) renders nonlinear transfer to small scales less efficient than for the Navier–Stokes equations, leading to a different behavior from that of Kolmogorov, namely with an (isotropic) energy spectrum $\sim k^{-3/2}$.

What actually happens is greatly debated today.

3.2.4 Scaling Invariance of MHD Equations

The scaling invariance properties of the MHD equations generalize those of the NS equations. Let us consider the MHD equations neglecting dissipative terms, and let us introduce the scaling transformations $\ell \to \lambda \ell'$ and $\mathbf{u} \to \lambda^h \mathbf{u}'$. Moreover let us introduce scaling transformations for all fields with different scaling exponents $\mathbf{B} \to \lambda^\beta \mathbf{B}'$, $p \to \lambda^\nu p'$, and $\rho \to \lambda^\mu \rho'$. Inserting these relations into the MHD equations, we can see that they are invariants (the same scaling factor is found for the whole equation) when $\alpha = 1 - h$, $\mu = 2(\beta - h)$, and $\nu = 2\beta$, for each value of h and β. In the incompressible case, $\rho = const$, the scaling exponent is the same for velocity and magnetic field $\beta = h$, and the scaling structure of MHD is similar to the NS equations.

3.2.4.1 Invariants and Cascades

Let us try to build our knowledge and understanding of MHD turbulence from first principles. We shall begin by examining the dissipationless case ($\nu = 0$, $\eta = 0$). In that case, there are three quadratic invariants, the total energy $E^T = \langle u^2 \rangle/2 + \langle b^2 \rangle/2$, the cross-correlation $E^C = \langle \mathbf{u} \cdot \mathbf{b} \rangle$, and the magnetic helicity in three space dimensions, $H^M = \langle \mathbf{a} \cdot \mathbf{b} \rangle$. The latter two are pseudo-scalars, like the kinetic helicity $H^V = \langle \mathbf{u} \cdot \boldsymbol{\omega} \rangle$. The conservation of magnetic helicity was first demonstrated by Woltjer [98]; in fact, it was this paper that prompted fluid dynamicists to examine the conservation of kinetic helicity. The conservation of H_M is easily shown by writing the equation for the magnetic potential and noting that $d_t H_M = 2 \int \mathbf{b} \cdot d_t \mathbf{a}$, assuming that boundary terms are zero. Note that the magnetic helicity is not positive definite; this fact has interesting consequences in the context of the dynamo problem[1] (see below, and [12]).

---

[1] Note that in the 2D case, with fields depending on x and y only (say), and with $u_z \equiv 0$ and $b_z \equiv 0$, $H_M \equiv 0$; however, there is another invariant in that case, the squared magnetic potential $\langle a^2 \rangle$.

Such invariants, according to the phenomenology of K41, cascade nonlinearly, to either small or large scales (i.e., in a direct or inverse cascade). Finally, note that the two energy invariants E^T and E^C can be combined in terms of the (pseudo) energies of the Elsässer variables, viz. $E^\pm \equiv \langle |z^\pm|^2 \rangle = E^T \pm E^C$.

One way to decide whether the cascade of an invariant is direct or inverse is to examine the physical dimension of the invariants with respect to one another, simplifying an argument due to Heyvaerts and Priest [41] using characteristic timescales. Indeed, the dimension of $\langle a^2 \rangle$ is larger than that of H^M which itself is also larger than that of the energies; by larger, what is meant here is that larger scales are involved in the definition of $\langle a^2 \rangle$, compared to the magnetic helicity, since the former involves lower derivatives. It is then argued that the cascade to larger scales will occur for the invariant having the larger dimension, having also slower characteristic times.

Another way to decide what direction the cascade will take is to follow Onsager [66] and build the statistical mechanics of such a truncated system: when more than one invariant occurs the equilibrium solution to the system with a finite number of modes is no longer that of equipartition on the hyper surface of constant energy in phase space; the Maxwellian distribution needs as many temperatures as there are invariants that survive truncation, and these ideal spectra can in some cases peak at large scale. This phenomenon is general, and not limited to hydrodynamics or MHD.

3.2.4.2 Small-Scale Dynamics

In order to stress the small-scale dynamics, it is useful to write down small-scale MHD equations for the current and vorticity; this is mainly an exercise in vector calculus better left to the reader; the end result is, omitting dissipation and forcing:

$$\frac{\partial \boldsymbol{\omega}}{\partial t} + \mathbf{v} \cdot \nabla \boldsymbol{\omega} = \boldsymbol{\omega} \cdot \nabla \mathbf{v} - \mathbf{j} \cdot \nabla \mathbf{b} + \mathbf{b} \cdot \nabla \mathbf{j}, \tag{3.8}$$

$$\frac{\partial \mathbf{j}}{\partial t} + \mathbf{v} \cdot \nabla \mathbf{j} = \mathbf{j} \cdot \nabla \mathbf{v} - \boldsymbol{\omega} \cdot \nabla \mathbf{b} + \mathbf{b} \cdot \nabla \boldsymbol{\omega} - 2 \sum_m \nabla v_m \times \nabla b_m , \tag{3.9}$$

where the following identity is of use:

$$\nabla \times (\mathbf{u_1} \cdot \nabla \mathbf{u_2}) = \mathbf{u_1} \cdot \nabla \boldsymbol{\omega}_2 - \boldsymbol{\omega}_2 \cdot \nabla \mathbf{u_1} + \sum_m \nabla u_{1_m} \times \nabla u_{2_m} ,$$

with $\boldsymbol{\omega}_2 = \nabla \times \mathbf{u_2}$, and where $\nabla \cdot \mathbf{u_1} = 0$ has been assumed. Note that the last term on the r.h.s. of (3.9) is new, compared to the Navier–Stokes equivalent of the stretching of vorticity (and current) by velocity (and magnetic field) gradients, and compared to the stretching of magnetic fields by gradients of the vorticity and the current.

The induction Eq. (3.6) in its dissipationless version ($\eta = 0$) is of a similar flux-conserving form as the vorticity equation with $\nu = 0$, leading to the batchelor analogy between vorticity and magnetic field. One can then deduce immediately the equivalent form of Kelvin's theorem, but instead for the induction; it is called the Alfvén theorem for (magnetic) flux conservation: magnetic field lines, in the absence of magnetic resistivity, are frozen into the fluid and move with it conserving their topology. Furthermore, one expects also from this analogy that the temporal development of the vorticity and of the magnetic induction be similar, at least for small times, because of the stretching of field lines by velocity gradients. This can be seen by using the following standard vector identity:

$$\nabla[\mathbf{v} \times \mathbf{c}] = -\mathbf{v} \cdot \nabla \mathbf{c} + \mathbf{c} \cdot \nabla \mathbf{v} \ ;$$

in other words, the flux-conserving term decomposes into advection and stretching.

Finally, Ohm's law can be written in more general forms than has been assumed here: for example, a term $\mathcal{H}$ corresponding to the Hall current can be added to (3.6), in the form

$$\mathcal{H} = -\frac{1}{\xi}\nabla \times \left[\frac{1}{\rho_0}(\nabla \times \mathbf{b}) \times \mathbf{b}\right],$$

where ξ, in front of the dispersive Hall term, is the ionization fraction of the gas (provided it is not too low, otherwise the ambipolar drift term would become predominant); in general, for ionized plasmas, ξ is assumed to be equal to unity. The linearization of the induction equation with a Hall term leads to the presence of whistler waves.

3.2.5 The 1D Case

3.2.5.1 An Exact Law in One Space Dimension

The 1D advection–diffusion equation, also called Burgers equation is,

$$\partial_t u + u \partial_x u = \nu \partial_{xx}^2 u \ ,$$

and occurs in a variety of phenomena (see, e.g., [99]). In the absence of viscosity, it conserves the total energy $\langle u^2 \rangle / 2$.

The solution of the advection–diffusion equation can be found by using the Hopf–Cole transformation, such that the velocity is derived from a potential $u(x, t) = \partial_x s(x, t)$, and choosing $s(x, t) = -2\nu ln \Psi(x, t)$. We obtain

$$\Psi_t = \nu \Psi_{xx} - \Psi f(t)/(2\nu) \ .$$

We can eliminate the last term in the above equation by using the gauge freedom since the fields $\tilde{s}(x, t) = -2\nu\tilde{\Psi}(x, t)$ with $\tilde{\Psi}(x, t) = \Psi(x, t)G(t)$ and $s(x, t)$ lead to the same velocity field. The heat equation is obtained with the choice of gauge

$\dot{G}(t)/G(t) = -f(t)/(2\nu)$. But it should be noted that inverting the solution of Burgers equation (i.e., going from Ψ to u) leads to numerical difficulties when the viscosity is small; this remark becomes relevant when one tries to test a numerical code, such as an adaptive code, against exact solutions.

The Burgers equation can be taken, with some precautions, as the archetype of fluid turbulence: it develops structures in space that are highly localized, or intermittent, with a corresponding power law Fourier spectrum, non-Gaussian PDFs of velocity gradients, and a self-similar decay of energy in time. However, two essential features of turbulence are missing, namely the presence of a pressure gradient, and vorticity, i.e., the ability of the flow not only to stretch and bend but also to rotate and form more complex, involuted features. Indeed, in a Burgers flow, sharp shocks develop locally, with linear ramps between them, corresponding to the self-similar solution $u(x, t) = x/t$; the shocks have a thickness proportional to $\nu^{1/2}$. Such shocks lead to a pseudo-singularity in the first derivative, with an ensuing k^{-2} spectrum.

It is easy to derive an exact law for third-order structure functions on Burgers equation. Take two independent points x and $x' = x + r$, with $u' = u(x')$ and $\partial' = \partial/\partial x'$; one can thus write

$$u'\partial_t u = -\frac{1}{2}\partial_x(u^2 u')$$

and similarly

$$u\partial_t u' = -\frac{1}{2}\partial'(uu'^2) \ .$$

Since $\partial' = \partial_r = -\partial_x$ assuming homogeneity, the equation for the correlation function of the velocity follows immediately:

$$2\partial_t \langle uu' \rangle = -\partial_r \langle uu'^2 - u^2 u' \rangle \ .$$

Introducing now the velocity difference

$$\delta u(r) = u' - u \ ,$$

developing in terms of correlation functions, viz. $\langle \delta u^2 \rangle = 2\langle u^2 \rangle - 2\langle uu' \rangle$ and $\langle \delta u^3 \rangle = 3\langle u'u^2 \rangle - 3\langle uu'^2 \rangle$, and recalling that $-2\partial_t \langle u^2 \rangle \equiv \epsilon$ leads to, assuming stationarity:

$$\langle \delta u^3(r) \rangle = -12\,\epsilon\,r \ .$$

Thus, third-order structure functions have linear scaling in r; in fact, it can be shown easily that all structure functions $\langle \delta u^p(r) \rangle \sim r^{\xi_p}$ of order p are linear in their argument ($\xi_p = 1$) for $p \geq 3$, whereas $\xi_p = p$ for $p < 3$ as expected from a

Taylor expansion. Thus, the Burgers case is said to be bi-fractal, corresponding to the two physical structures present in such simplified flows, namely the ramps and the shocks.

3.2.5.2 An Exact Law for 1D MHD

Let us guess at the case for magneto-hydrodynamics by taking, following Thomas [95] the simple configuration of a velocity field $(u, 0, 0)$ and a magnetic induction $(0, b, 0)$ and assuming that derivatives are non-zero only in the x-direction. These model equations for MHD in 1D can be written by assuming conservation of total energy $E_T = \langle u^2 + b^2 \rangle / 2$ and cross-correlation E_C, defined here as $\langle u\,b \rangle$, with $\mathbf{b}$ in the y-direction as stated before in order to ensure incompressibility of the induction field. With these assumptions, the dissipationless MHD equations in 1D read

$$\partial_t u = -\partial_x [(u^2 + b^2)/2],$$
$$\partial_t b = -\partial_x (ub).$$

The change of variables $y_s = u + sb$, with $s = \pm 1$ leads to two (uncoupled) Burgers equations for y_s; the exact law for this 1D model of MHD is then immediately deduced from the result of the preceding section:

$$\langle \delta y_s^3(r) \rangle = -12\,\epsilon_s\,r$$

with $\epsilon_\pm = -\dot{E}_\pm$. Defining now

$$\epsilon^T = \frac{\epsilon_+ + \epsilon_-}{2} \quad , \quad \epsilon^C = \frac{\epsilon_+ - \epsilon_-}{2}$$

as, respectively, the rates of transfer of the total energy and the cross–correlation, the exact law for this 1D model of MHD becomes, in terms of the velocity and magnetic fields,

$$\langle \delta u^3(r) \rangle + 3\langle \delta u(r)\delta b^2(r) \rangle = -12\epsilon^T r$$

and

$$\langle \delta b^3(r) \rangle + 3\langle \delta b(r)\delta u^2(r) \rangle = -12\epsilon^C r \ .$$

In other words, in the more complex case of MHD, *mixed* velocity–magnetic field structure functions appear, showing the dynamical importance of such correlations between the basic physical fields.

3.2.6 Three Laws for MHD Turbulence

3.2.6.1 Energy Spectra in MHD

What could be a good phenomenology for MHD turbulence? A first answer is to say that it does not differ from the fluid case, and hence the energy spectrum is going to follow a Kolmogorov $k^{-5/3}$ law. On the other hand, Iroshnikov and Kraichnan proposed that Alfvén waves have a profound influence on the rate at which energy is being transferred to small scales. Such $z^{\pm}$ waves propagate in opposite directions along the strong magnetic field B_0, assumed to be quasi-uniform, and they interact only when they meet; hence the nonlinear transfer is weaker than in the NS case. Specifically, they considered a slowing-down of such a transfer (compared with the fluid case), in a manner that is proportional to the ratio of the Alfvén time to the eddy turnover time, i.e., to the time it takes waves to interact. Thus, one writes that the time of energy transfer τ_{tr} is reduced in a $\tau_{NL}/\tau_A \gg 1$ ratio, with $\tau_{NL} = \ell/u_\ell$ the eddy turnover time and $\tau_A = \ell/B_0$ the Alfvén time. The expression for transfer becomes

$$\tau_{tr}(k) = \tau_{NL}(k)^2/\tau_A(k) \ . \tag{3.10}$$

The energy spectrum follows directly by assuming that the rate of transfer of energy to the small scales is independent of scale and that it is expressed as the ratio of the total energy $kE_T(k)$ to the time of energy transfer τ_{tr}. Hence, assuming no correlation between the velocity and the magnetic field, one obtains

$$E_T(k) \sim (\epsilon_T B_0)^{1/2} k^{-3/2} \ . \tag{3.11}$$

Note that if we drop the condition of isotropy and recall that the dispersion law of Alfvén waves is $\omega_k = \mathbf{k} \cdot \mathbf{B}_0$, and if we further assume that in the presence of a strong uniform magnetic field, the flow becomes 2D and hence the eddy turnover time in this case becomes $\tau_{NL}^{quasi\text{-}2D} = \ell_\perp/u_{\ell_\perp}$, then the same phenomenological argument as above, basically of a dimensional nature, but now introducing the parallel and perpendicular length-scales, gives for the energy spectrum,

$$E_T^{quasi\text{-}2D}(\mathbf{k}) \sim (\epsilon_T B_0 k_\parallel)^{1/2} k_\perp^{-2} \ . \tag{3.12}$$

Here $k_\perp$ ($k_\parallel$) referring to the direction perpendicular (parallel) to $\mathbf{B}_0$. Indeed, for $k \sim k_\parallel \sim k_\perp$, one recovers (3.11). This spectrum is what is derived analytically from the weak MHD turbulence approach (see below).

In this framework, the scaling law for velocity increments must be changed with respect to the fluid-like case [24, 14], namely

$$\delta u_\ell \sim U_0 \left(\frac{\ell}{L_0}\right)^{1/4} ,$$

so that the pth-order velocity structure function must behaves like $S_\ell^{(p)} \sim \ell^{p/4}$, and the IK spectrum (3.11) can be easily calculated from the second-order structure function.

3.2.6.2 Temporal Decay of the Total Energy in MHD

We proceed in MHD as for the Navier–Stokes case, and the hypotheses are the same, namely, a large-scale spectrum taken as a power law $\sim k^{s_M}$, a self-similar decay of total energy in time $E_T(t) \sim (t - t_*)^{-\alpha_M}$ and a power law increase of the integral scale $\ell \sim (t - t_*)^{\beta_M}$. The only difference is now that the IK spectrum is assumed to be achieved in MHD flows, with a characteristic time of energy transfer given by (3.10). One now finds, for $s_M = 4$ in 3D:

$$\beta_M = \frac{1}{6}, \quad \alpha_M = \frac{5}{6} .$$

This results, as expected, in a slower decay of energy than for a Kolmogorov flow for which (see Sect. 3.1), a $t^{-10/7}$ decay law is expected for the kinetic energy in 3D.

3.2.6.3 An Exact Law for MHD

As said before, one of the very few exact results in turbulence is the so-called "4/5th" law of Kolmogorov for the linear scaling of third-order longitudinal structure functions of the velocity field. In incompressible MHD, the equivalent exact result is obtained [68] using again the Elsässer variables with transfer rates $-\dot{E}^\pm = \epsilon^\pm$ (i.e., $\epsilon^\pm$ are the mean energy transfer rates of the $z^\pm$ variables):

$$\langle \delta z_L^{+2}(\mathbf{r}) \delta z_L^-(\mathbf{r}) \rangle - 2\langle z_L^+(\mathbf{x}) z_L^+(\mathbf{x}) z_L^-(\mathbf{x'}) \rangle = -C_d \epsilon^+ r, \tag{3.13}$$

where $C_d = 2K_d/3$ and $d(d + 2)K_d = 12$ in dimension d (thus, $K_3 = 4/5$). Taking now into account all components of the fields, one also has for the third-order correlators defined below [69], following a similar approach developed by Yaglom [100] for the passive scalar:

$$Y_3^\pm(r) = \langle \delta z_L^\mp(\mathbf{r}) \Sigma_i (\delta z_i^\pm(\mathbf{r}))^2 \rangle = -\frac{4}{d} \epsilon^\pm r , \tag{3.14}$$

where $\delta z_L^\pm(r) = [\mathbf{z}^\pm(\mathbf{x'}) - \mathbf{z}^\pm(\mathbf{x})] \cdot \mathbf{r}$ is the longitudinal increment of the fields $\mathbf{z}^\pm$. Note that the flux functions $Y_3^\pm(r)$ involve all components of the physical fields.

A similar law has been derived for magnetic helicity [71] (see also [31] for the 1D version). It reads, in the coordinate system $L, 2, 3$:

$$\langle [v_L(\mathbf{x}) b_2(\mathbf{x}) - v_2(\mathbf{x}) b_L(\mathbf{x})] \, a_2(\mathbf{x'}) \rangle = -\frac{1}{6} \tilde{\epsilon}^{H_M} r . \tag{3.15}$$

In terms of the electromotive force due to the turbulent motions $\mathcal{E}^t = \mathbf{v} \times \mathbf{b}$, this becomes equivalently

$$\langle \left[\mathcal{E}^t(\mathbf{x}) \times \mathbf{a}(\mathbf{x}') \right]_L \rangle = +\frac{1}{3}\tilde{\epsilon}^{H_M} r . \tag{3.16}$$

3.3 The Problem of Intermittency in Turbulence

3.3.1 Scaling Laws of Structure Functions

The linear behavior of the structure functions scaling exponents with the moment order as deduced from K41-like phenomenology and the Gaussianity of the PDFs of increments, representing the Kolmogorov scaling, are not observed in experimental data except at second order for which the K41 theory works quite well. But when more and more accurate experimental techniques permitted the investigation of higher moments, the need for a different interpretation arose [27]. In Fig. 3.3 we show the normalized scaling exponents $\zeta_p = \xi_p/\xi_3$ for high-order moments of velocity and magnetic field increments, as measured in the solar wind plasma. In Appendix A we report on data coming from different experiments; both in fluid flows on the Earth, in laboratory fusion plasmas and in numerical simulations of 2D MHD. Data in the interplanetary space are obtained from the Helios 2 spacecraft, sampling low-speed streams, when the spacecraft orbited near the Earth at a distance $R = 0.9$ AU. As a comparison, in Fig. 3.3 we show the scaling exponents for

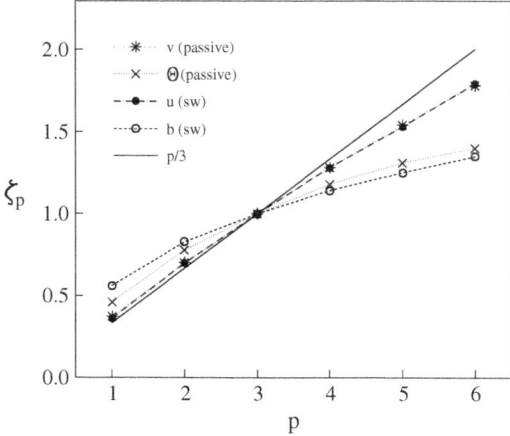

Fig. 3.3 The normalized scaling exponents ζ_p as a function of the moment order p, along with the linear value $p/3$ (*full line*) expected from the K41 theory. Data refer to the bulk velocity (*black circles*) and the magnitude of the magnetic field (*white circles*), as measured by the Helios satellite in the inner heliosphere at 0.9 AU during slow wind streams. Shown for comparison are the normalized scaling exponents for longitudinal velocity field (*stars*) and the temperature field (passive scalar) in fluid flows [82]

velocity and temperature (considered as a passive field) as obtained in a wind tunnel experiment. We also show normalized scaling exponents $\zeta_p = \xi_p/\xi_3$ calculated through the extended self-similarity approach [6]. It can be seen that the departure from the Kolmogorov linear scaling is similar in all experiments, that is, $\zeta_p < p/3$ for $p > 3$ while $\zeta_p > p/3$ for $p < 3$. What is interesting is the fact that both velocity fields display the same degree of intermittency (calculated as the distance between the linear law $p/3$ and the actual values of scaling exponents); there is no difference between scaling of flows on the Earth and in space [17]. This gives us the idea of a kind of universality of turbulence. The magnetic field is much more intermittent than the velocity field, that is, as far as the intermittent properties are considered, the magnetic field behaves more like the passive temperature field [19].

3.3.2 What is Intermittent in Turbulence?

Figures 3.4 and 3.5 show examples of heliospheric turbulence. The panels represent the longitudinal velocity differences and the magnetic field intensity differences, computed from solar wind data for three different values of τ (shown on the figures). The bottom panel is the large-scale case, and the signal looks like Brownian motion. As the scale decreases the signal becomes more and more intermittent. We can observe the emergence of localized regions whose fluctuations are stronger as the scale decreases. The intermittent events at small scales are clearly visible on both figures. We can thus see that they play a key role in the statistics of turbulence.

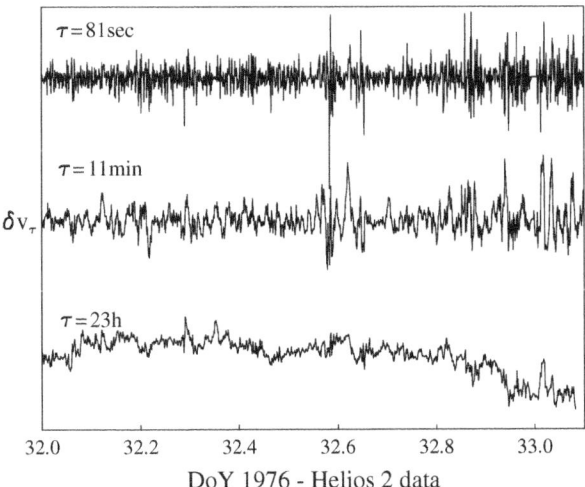

Fig. 3.4 Fluctuations of the bulk velocity field $\delta u_\tau = u(t + \tau) - u(t)$ as a function of time t for three different scales τ. The fluctuations are calculated from a turbulent sample from the Helios 2 spacecraft in the solar wind

Fig. 3.5 Fluctuations of the magnitude of the magnetic field $\delta b_\tau = B(t + \tau) - B(t)$ as a function of time t for three different scales τ. The fluctuations are calculated from a turbulent sample from Helios 2 spacecraft in the solar wind

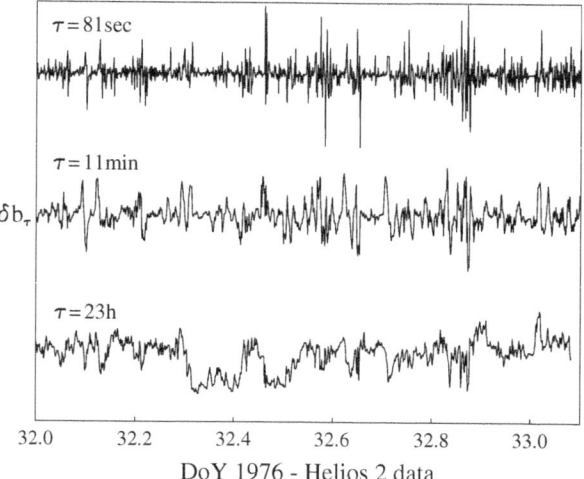

DoY 1976 - Helios 2 data

3.3.3 Probability Density Functions (PDFs) of Fluctuations

The presence of scaling laws for fluctuations in general is a signature of self-similarity. In fact the observable δu_ℓ, which depends on a scaling variable ℓ, is invariant with respect to the scaling relation $\ell \rightarrow \lambda\ell$, when there exists a parameter $\mu(\lambda)$ such that $\delta u_\ell = \mu(\lambda)\delta u_{\lambda\ell}$. The solution of this last relation is a power law $\delta u_\ell \sim \ell^h$ with scaling exponent $h = -\log_\lambda \mu$. Then the ratio of fluctuations at two scales $\delta u_{\lambda\ell}/\delta u_\ell \sim \lambda^h$ depends only on the value of h, that is, we cannot define any characteristic scale. This means that PDFs of scaling variables are related through $P(\delta u_{\lambda\ell}) = P(\lambda^h \delta u_\ell)$. Let us consider the standardized variables $y_\ell = \delta u_\ell/\langle(\delta u_\ell)^2\rangle^{1/2}$. It can be easily shown that when h is unique, say in a pure self-similar situation, PDFs are such that $P(y_\ell) = P(y_{\lambda\ell})$, namely by changing scale the PDFs collapse.

In Fig. 3.6 we show PDFs for the standardized velocity fluctuations, as observed in atmospheric flow, and fluctuations $\Delta b_\ell = \delta b_\ell/\langle\delta b_\ell^2\rangle^2$ at three different scales ℓ, for three different data sets: a set from the solar wind, inside a laboratory plasma and in 2D numerical simulations. It appears evident that the global self-similarity in turbulence is broken. PDFs at different scales do not collapse, their shape seems to be strongly dependent on ℓ. In particular at large scales PDFs look almost Gaussian, but they become more stretched as ℓ decreases. At the smallest scale PDFs are stretched exponentials. This scaling dependence of PDFs is a different way to say that scaling exponents of fluctuations are anomalous, which is a different definition of intermittency. Note that the wings of PDFs are higher than a Gaussian function. This implies that intense fluctuations have a probability of occurrence greater than what they would have if they had a Gaussian distribution. Said differently, intense stochastic fluctuations are less rare than we should expect from the point of view of a Gaussian approach to the statistics of turbulence.

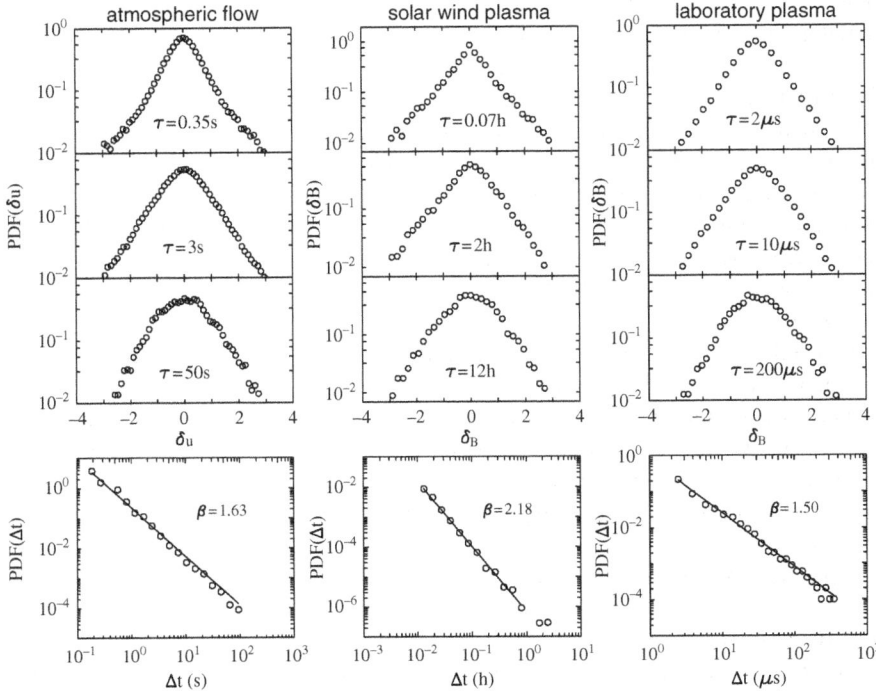

Fig. 3.6 The *first three panels* show PDFs of fluctuations at three different scales τ, for three different experiments: atmospheric flow (fluctuations of velocity), the solar wind, and a laboratory plasma (fluctuations of magnitude of magnetic field). Note the same scaling behavior of the PDFs, even if the scales τ are completely different for each experiment. The *bottom three panels* show the PDFs of waiting times Δt between structures at the smallest scale for each experiment. The PDFs of waiting times behave like $P(\Delta t) \sim \Delta t^\beta$, values of β for each experiment are reported on the figures

3.3.4 Intermittency and Non-Poisson Events

Intermittency generates rare and intense events, which can be seen as coherent structures present on all dynamically interesting scales. The times t_j of the occurrence of the maxima for the time evolution of both $|\delta u_\ell|$ and $|\delta b_\ell|$ can be extracted from a time series by using a threshold [10]. Then for each scale ℓ we can get a set of waiting times $\Delta t = t_{j+1} - t_j$. The distributions $P(\Delta t)$ for both velocity and magnetic variables, at a given shell, are shown in the bottom panels of Fig 3.6. As can be seen, a power law is recovered: $P(\Delta t) \sim \Delta t^{-\beta}$, with different values for the scaling exponents β. The presence of a power law instead of an exponential decay is a signature that the process underlying the formation of rare bursts does not follow Poisson statistics, implying a certain degree of memory.

3.3.5 The Multiplicative Cascade Model

To understand the way the phenomenology must be changed to include intermittency in turbulent fields, the picture of the Richardson cascade can be modified. One of the main points on which the Kolmogorov K41 theory was based is the fact that the actual spatial statistics of the energy dissipation rate ε does not come into play. The idea of universality implied by the model suggests a uniform distribution. However, only the *global* mean value of the energy transfer rate is constant through the cascade, while its local value can be a (stochastic) fluctuating function, with both bursty and quiet zones alternatively. In this framework the presence of strong activity regions must be scale dependent, showing the concentration of active structures at definite positions in space, and such concentrations becoming more and more evident as the scale decreases.

Kraichnan [50] pointed out that in order to obtain phenomenological information about inertial range quantities, the energy dissipation rate which appears in the phenomenology must be replaced by the local energy transfer rate, namely $\delta u_\ell \sim \varepsilon_\ell^{1/3} \ell^{1/3}$. As a consequence the pth-order structure function will take into account the scaling of fluctuations of the energy transfer rate, namely

$$S_p(\ell) \sim \langle \varepsilon_\ell^{p/3} \rangle \ell^{p/3}.$$

Then, by assuming a scaling law for the energy transfer rate $\varepsilon_\ell^q \sim \ell^{\tau_q}$, the correction to the scaling exponents of the structure functions $\xi_p = p/3 + \tau_{p/3}$ due to intermittency comes from the scaling behavior of ε. This opens a "Pandora's box" of possibilities [50] to model the energy transfer rate and to compare the scaling exponents of the model with that from real experiments. The most common way to recover a model is to interpret the energy cascade as a multiplicative process, according to Richardson's picture. In this framework the energy transfer rate at a generic scale $\varepsilon_n = \varepsilon_{\ell_n}$, where $\ell_n = 2^{-n} L_0$, is viewed as a stochastic variable, computed as the result of a multiplicative process

$$\varepsilon_n = \varepsilon_0 \prod_{i=1}^{n} \beta_i \; .$$

The statistics of the scaling exponents ξ_p depends on the statistics of the multipliers β_i. Assuming that all multiplicative factors are derived from the same process, we get

$$\xi_p = p/3 - \log_2 \langle \beta^{p/3} \rangle.$$

Then, given a model for the cascade we can try to work out an expression for the statistics of β, and then obtain a model for ξ_p. The model can be fitted to the various data sets in order to derive the values for the parameters of the model. The most common models encountered in the literature are reported in Ref. [27].

3.3.6 The Multifractal Approach

An elegant way to describe the occurrence of intermittency in turbulence has been worked out by Frisch and Parisi [27], who introduced the multifractal model for intermittency. The multifractal idea is to consider a continuous spectrum of possible values of the scaling exponent h. That is, space is described as being made of an infinite number of subsets $S(h)$, each one of fractal dimension $D(h)$, and on each of which the scaling is described by the exponent $h(\mathbf{x})$ [27]. Summing together all the subspaces contribution to the scaling of the fields gives

$$\delta v_\ell(x) \sim U_0 \left(\frac{\ell}{L_0} \right)^h , \quad x \in S_h.$$

The structure functions can now be considered as the superposition of infinite power laws, each one representing the set where its exponent is valid. Since $(\ell/L_0) \ll 1$, the leading exponent for each value of the order p is the minimum, so that

$$S_p(\ell) \sim U_0^p \int_{h_{min}}^{h_{max}} \left(\frac{\ell_n}{L_0} \right)^{ph+3-D(h)} d\mu(h),$$

$\mu(h)$ being the measure representing the probability distribution function of the exponents h. The integral can be solved using the saddle point method and leads to

$$\xi_p = \inf_h [ph + 3 - D(h)].$$

This expression can be inverted for a given value of p. Using the model, at a fixed value of p, we select singularities of order h within a set of fractal dimension $D(h)$, where the scaling $\delta u_\ell \sim \ell^h$ holds.

3.3.7 Scaling Behavior of PDFs

Several models can be introduced to capture the scaling behavior of PDFs of the increments of fields. The fundamental relation linking such scaling is

$$P(\delta\psi_0) = P(\delta\psi) \left| \frac{\delta\psi}{\delta\psi_0} \right|, \tag{3.17}$$

which needs some further assumptions about the energy cascade in order to get a model for PDFs scaling.

Using the multifractal framework, a quantitative analysis of the continuous scaling departure of PDFs from a Gaussian can be performed. In order to describe the PDFs at a given scale ℓ, two ingredients are needed: the parent distribution at the large scale ℓ_0 and the distribution of the energy transfer rate. In fact a correspondence exists between the energy transfer rate ϵ and the variance σ of the conditioned PDFs. The dependence on scales of the PDFs can be eliminated by looking at the PDFs *conditioned* to a given value of the energy transfer rate at the scale ℓ, so that for each scale ℓ the field can be decomposed into a set of Gaussian curves, each one corresponding to a given value (or bin of values) of the energy transfer rate. The energy transfer rate distribution can thus be represented using a distribution for the variances of the conditioned PDFs. The PDF of the field increments is then seen as a superposition of curves (the conditioned PDFs) whose standard deviations are distributed according to a given phenomenological law [88].

This can be done by computing the convolution

$$P_\lambda(\delta\psi_\ell) = \int_0^\infty \mathcal{L}_\lambda(\sigma_\ell) P_0(\delta\psi_\ell, \sigma_\ell)\, d\ln\sigma_\ell \ . \tag{3.18}$$

Both the large-scale PDF P_0 and the distribution of the variances $\mathcal{L}_\lambda(\sigma)$ of the above general relation could in principle be determined experimentally. In a turbulent flow it is even observed that the parent distribution is clearly Gaussian:

$$P_0(\delta\psi_\ell, \sigma_\ell) = \frac{1}{\sigma_\ell\sqrt{2\pi}} \exp\left(-\frac{\delta\psi_\ell^2}{2\sigma_\ell^2}\right) . \tag{3.19}$$

Moreover, it is well known that large-scale PDFs of turbulent increments are Gaussians, both in fluids [27, 20] and in plasmas [97, 89].

The determination of the function $\mathcal{L}_\lambda(\sigma_\ell)$ could be determined by computing the variances of the conditioned PDFs, and then observing the PDF of such variances. Unfortunately, a very large amount of data would be necessary in order to have enough values of variances for their PDF to be computed. It is then useful to approach the problem using models for the energy transfer rate. In the paper by Castaing et al. [20] (see also Ref. [89]) a log-normal ansatz has been tried

$$\mathcal{L}_{\lambda_\ell}(\sigma_\ell) = \frac{1}{\sqrt{2\pi}\lambda_\ell} \exp\left(-\frac{\ln^2 \sigma_\ell/\sigma_{0,\ell}}{2\lambda_\ell^2}\right) \tag{3.20}$$

(but different possibilities could be easily investigated). In Eq. (3.20), λ^2 is the variance of $\ln\sigma_\ell$ distribution and $\sigma_{0,\ell}$ is the most probable value of σ at scale ℓ. When the model is used on actual turbulent data set [20, 89], a power law is observed for the scaling behavior of λ, namely $\lambda^2 \sim r^\beta$. Perhaps this solves the difficulties encountered with the log-normal model when used to simulate the energy transfer rate [27].

3.3.8 Laws and Structures

How does one make the bridge between the laws, exact or phenomenological, that are derived for turbulent flows and the structures that develop in such flows? A good example is that of the Burgers equation. It is known that shocks develop that can be approximated by a localized jump in the velocity, such jumps being connected through linear ramps. In such a case, the intermittency exponents can easily be shown to be proportional to p for $p \leq 1$ and to be equal to unity thereafter, leading to a bi-fractal behavior. The connection between statistical properties of turbulent flows and the structures such as vortex filaments is not so straightforward when going to higher dimensions with more complex flows.

3.4 Modeling Turbulence

Fully developed turbulence involves a hierarchical process, in which many scales of motion are involved. To look at this phenomenon it is often useful to investigate the behavior of Fourier coefficients of fields. Assuming periodic boundary conditions the αth component of Elsässer variables can be Fourier decomposed as

$$z_\alpha^\pm(\mathbf{r}, t) = \sum_{\mathbf{k}} z_\alpha^\pm(\mathbf{k}, t) \exp(i\mathbf{k} \cdot \mathbf{r}) \ . \tag{3.21}$$

When used in Eqs. (3.7) the nonlinear term becomes a convolution sum

$$\frac{\partial z_\alpha^\pm(\mathbf{k}, t)}{\partial t} = M_{\alpha\beta\gamma}(\mathbf{k}) \sum_{\mathbf{q}} z_\gamma^\pm(\mathbf{k} - \mathbf{q}, t) z_\beta^\mp(\mathbf{q}, t), \tag{3.22}$$

where $M_{\alpha\beta\gamma}(\mathbf{k}) = -ik_\beta(\delta_{\alpha\gamma} - k_\alpha k_\beta/k^2)$. The quadratic nonlinearities of the original equations correspond to a convolution term involving wave vectors $\mathbf{k}$, $\mathbf{p}$, and $\mathbf{q}$ related by the triangular relation $\mathbf{p} = \mathbf{k} - \mathbf{q}$. Fourier coefficients locally couple to generate an energy transfer from any pair of modes $\mathbf{p}$ and $\mathbf{q}$ to a mode $\mathbf{k} = \mathbf{p} + \mathbf{q}$.

3.4.1 Why Dynamical Models for Turbulence?

In the limit of fully developed turbulence, when dissipation goes to zero, an infinite range of scales are excited, that is, energy lies on all available wave vectors. Dissipation takes place at a typical dissipation length scale which depends on the Reynolds number R through $\ell_D \sim L R^{-3/4}$ (we used a Kolmogorov spectrum $E(k) \sim k^{-5/3}$). In a 3D numerical simulation the minimum number of grid points which are necessary to obtain information on the fields at these scales is given by $N \sim (L/\ell_D)^3 \sim R^{9/4}$. This rough estimate shows that a considerable amount of memory is required when we want to perform numerical simulations with high R. Typical values of Reynolds numbers at present reached in 2D and 3D numerical simulations are, respectively, of the order of 10^4 and 10^3. At these values the inertial range spans approximately one decade or little more.

Because of the situation described above, the question of the best description of the dynamics which result from the original equations, using only a small amount of degree of freedom, becomes an important issue. This can be achieved by introducing turbulence models which are investigated using tools of dynamical system theory [11]. Dynamical models then represent minimal sets of ordinary differential equations that can mimic the gross features of the energy cascade in turbulence. These studies are motivated by the famous Lorenz model [54], which, containing only 3 degrees of freedom, simulates the complex chaotic behavior of turbulent atmospheric flows, becoming a paradigm for the study of chaotic systems. Low-order truncations of fluid equations give rise to Lorenz-like dynamical models which are able to simulate various routes to chaos (see Appendix B).

3.4.2 Shell Models of Turbulence Cascade

The shell model approach to turbulence is viewed as a consistent and relevant approach to the energy cascade of turbulence [11]. These models mimic the gross features of the time evolution of spectral Navier–Stokes or MHD equations. The 3D hydrodynamic shell model is usually quoted in literature as the GOY model, and was been introduced by Gledzer [36] and by Ohkitani and Yamada [64] (for a review see Ref. [8]). The MHD shell model, which coincides with the GOY model when the magnetic variables are set to zero, has been introduced independently by Frick and Sokoloff [30] and Giuliani and Carbone [34] (see also Ref. [35]). In the following we will refer to the MHD shell model as the FSGC model. Here we derive and discuss the basic features of the FSGC shell model, the hydrodynamical version can be obtained by imposing zero magnetic variables.

A typical shell model can be built up through four different steps.

(a) Introduce discrete wave vectors

As a first step we divide the wave vector space into a discrete number of shells whose radii grow according to a power $k_n = k_0 \lambda^n$, where $\lambda > 1$ is the inter-shell ratio, k_0 is the fundamental wave vector related to the largest available length scale L, and $n = 1, 2, ..., N$. Of course the maximum number (N) of shells depends on the Reynolds number we want to investigate. The dissipative wave vector is given by $k_D = k_0 \lambda^{n_D} \sim \nu^{3/4}$, so that the number of modes required grows as $\ln R$. To appreciate the reduction of modes, and then of computer time required to numerically solve our equations, that number must be compared to the usual power $R^{9/4}$.

(b) Assign to each shell discrete scalar variables

Each shell is assigned two or more complex scalar variables $u_n(t)$ and $b_n(t)$, or Elsässer variables $Z_n^{\pm}(t) = u_n \pm b_n$. These variables describe the chaotic dynamics of modes in the shell of wave vectors between k_n and k_{n+1}. It is worth noting that the discrete variables, mimicking the average behavior of Fourier modes within each shell, represent characteristic fluctuations across eddies at the scale $\ell_n \sim k_n^{-1}$. That is, the fields have the same scalings as field differences, for example, $Z_n^\pm \sim |Z^\pm(x + \ell_n) - Z^\pm(x)| \sim \ell_n^h$ in fully developed turbulence. In this way we ruled out the possibility of describing spatial behavior within the model. We can only get, from a dynamical shell model, time series for shell variables at a given k_n, and we lose the fact that turbulence is a typical temporal *and* spatial complex phenomena.

(c) Introduce a dynamical model which describes nonlinear evolution

Looking at Eq. (3.22) a model must have quadratic nonlinearities among opposite variables $Z_n^\pm(t)$ and $Z_n^\mp(t)$, and must couple different shells, that is, in general

$$\frac{dZ_n^\pm}{dt} = ik_n \sum_\sigma \sum_{j,m} A_{j,m}^\sigma Z_{n+j}^{\sigma*} Z_{n+m}^{-\sigma*}, \tag{3.23}$$

where $\sigma = \pm$, $A_{j,m}^\pm$ represent arbitrary real coupling coefficients (the symbol $*$ being complex conjugation). We exclude from the sum in (3.23) terms where $j = m$ and terms where $j = 0$ or $m = 0$. The first constraint comes from the fact that in the original equations the coupling coefficient involving wave vectors where $\mathbf{p} = \mathbf{q}$ is zero. The second constraint comes from the Liouville theorem, which states that the nonlinear term in the primitive equation must conserve volume in phase space. This means that $(\partial/\partial Z_n^\sigma)(dZ_n^\sigma/dt) = 0$. We introduce the shell model where couplings happen among next and next-nearest shells, that is, both $j = \pm 1, \pm 2$ and $m = \pm 1, \pm 2$.

(d) Fix as much as possible the coupling coefficients

This last step is not standard. A numerical investigation of the model might require the scanning of the properties of the system when all coefficients are varied. Coupling coefficients can be fixed by requiring that system (3.23) satisfies the quadratic conservation laws of the original equations, namely the pseudo-energies $E^\pm(t)$

$$E^\pm(t) = \frac{1}{4} \sum_n |Z_n^\pm|^2 .$$

Then, by requiring that Eq. (3.23) must satisfy $dE^\pm/dt = 0$, we get

$$\frac{dZ_n^\pm}{dt} = ik_n \Phi_n^{\pm*}, \tag{3.24}$$

where

$$
\Phi_n^\pm = \left(\frac{2-a-c}{2}\right) Z_{n+2}^\pm Z_{n+1}^\mp + \left(\frac{a+c}{2}\right) Z_{n+1}^\pm Z_{n+2}^\mp +
$$

$$
+ \left(\frac{c-a}{2\lambda}\right) Z_{n-1}^\pm Z_{n+1}^\mp - \left(\frac{a+c}{2\lambda}\right) Z_{n-1}^\mp Z_{n+1}^\pm +
$$

$$
- \left(\frac{c-a}{2\lambda^2}\right) Z_{n-2}^\mp Z_{n-1}^\pm - \left(\frac{2-a-c}{2\lambda^2}\right) Z_{n-1}^\mp Z_{n-2}^\pm . \qquad (3.25)
$$

In terms of velocity and magnetic shell variables $u_n(t)$ and $b_n(t)$, from (3.25) we can write down immediately a set of equations as

$$
\frac{du_n}{dt} = ik_n \Big[(u_{n+1}u_{n+2} - b_{n+1}b_{n+2}) +
$$

$$
- \frac{a}{\lambda}(u_{n-1}u_{n+1} - b_{n-1}b_{n+1}) - \frac{1-a}{\lambda^2}(u_{n-2}u_{n-1} - b_{n-2}b_{n-1}) \Big]^* , \quad (3.26)
$$

$$
\frac{db_n}{dt} = ik_n \Big[(1-a-c)(u_{n+1}b_{n+2} - b_{n+1}u_{n+2}) +
$$

$$
+ \frac{c}{\lambda}(u_{n-1}b_{n+1} - b_{n-1}u_{n+1}) + \frac{1-c}{\lambda^2}(u_{n-2}b_{n-1} - b_{n-2}u_{n-1}) \Big]^* . \quad (3.27)
$$

Conservation of pseudo-energies $E^\pm(t)$ implies that these equations conserve equivalently both total energy $E(t)$ and cross-helicity $H_c(t)$, say

$$
E(t) = \frac{1}{2} \sum_n |u_n|^2 + |b_n|^2 , \quad H_c(t) = \sum_n 2Re\left(u_n b_n^*\right) .
$$

The equations we recover describe the nonlinear evolution of the shell model; on the right-hand side of (3.24) in the following we will add the dissipative terms and forcing terms to restore turbulence.

3.4.3 2D and 3D Shell Models

As we said before shell models cannot describe spatial geometry of nonlinear interactions in turbulence, so that we lose the possibility of distinguishing between 2D and 3D turbulent behavior. The distinction is, however, of primary importance, for example, as far as the dynamo effect is concerned in MHD. The shell model (3.24) contains two free parameters which can be fixed by introducing a third ideal invariant, $H(t)$, which can be later identified as a surrogate of the magnetic helicity or of the squared vector potential.

As can be easily verified from (3.24), the invariant which can be conserved takes the form

$$H(t) = \sum_n \frac{|b_n|^2}{k_n^\alpha}. \tag{3.28}$$

By imposing that Eq. (3.27) must satisfy $dH/dt = 0$, we get two classes of models: The first class identifies a shell model where the third invariant (3.28) is positive definite. When we choose $\alpha = 2$, $H(t)$ can be dimensionally identified with the squared magnetic potential, so that this model mimics a kind of 2D MHD turbulence. The second class of models, obtained by requiring that α be complex, identifies a shell model where the third invariant is not positive, as defined by

$$H(t) = \sum_n (-1)^n \frac{|b_n|^2}{k_n^{\alpha_R}}. \tag{3.29}$$

When the real part of α, namely α_R is equal to unity, the invariant $H(t)$ can be dimensionally identified as the magnetic helicity and the shell model mimics a kind of 3D MHD turbulence. Finally the most common choice $\lambda = 2$ for the inter-shell ratio [11] fixes the free parameters of the MHD shell model to the values $a = 5/4$ and $c = -1/3$ for the 2D case, $a = 1/2$ and $c = 1/3$ for the 3D case.

The MHD shell model evolves in a phase space built up by considering (u_n, b_n) as coordinates. When $b_n = 0$, the phase space of the system reduces to a subspace described by the GOY hydrodynamical shell model [36, 64].

3.4.4 Basic Properties of Shell Models

Taking into account both the dissipative and forcing terms, the FSGC model can be written as

$$\frac{dZ_n^\pm}{dt} = ik_n \Phi_n^{\pm*} + \frac{\nu \pm \mu}{2} k_n^2 Z_n^+ + \frac{\nu \mp \mu}{2} k_n^2 Z_n^- + F_n^\pm. \tag{3.30}$$

In the following we will consider only the case where the dissipative coefficients are the same, so $\nu = \mu$.

3.4.4.1 The Inertial Range of the Energy Cascade

The existence of a cascade toward small scales is expressed by an exact relation which is equivalent to the 4/5 law for MHD turbulence. Using Eqs. (3.30) the scale-by-scale pseudo-energy budget is given by

$$\frac{d}{dt} \sum_n |Z_n^\pm|^2 = k_n Im \left[T_n^\pm \right] - \sum_n 2\nu k_n^2 |Z_n^\pm|^2 + \sum_n 2Re \left[Z_n^\pm F_n^{\pm*} \right].$$

The second and third terms on the right-hand side represent respectively, the rate of pseudo-energy dissipation and the rate of pseudo-energy injection. The first term represents the flux of pseudo-energy along the wave vectors, responsible for the redistribution of pseudo-energies on the wave vectors, and is given by

$$
T_n^\pm = (a + c) Z_n^\pm Z_{n+1}^\pm Z_{n+2}^\mp + \left(\frac{2 - a - c}{\lambda} \right) Z_{n-1}^\pm Z_{n+1}^\pm Z_n^\mp +
$$

$$
+ (2 - a - c) Z_n^\pm Z_{n+2}^\pm Z_{n+1}^\mp + \left(\frac{c - a}{\lambda} \right) Z_n^\pm Z_{n+1}^\pm Z_{n-1}^\mp. \qquad (3.31)
$$

Using three classical assumptions, namely (i) the forcing terms act only on the largest scales; (ii) the system can reach a statistically stationary state; (iii) in the limit of fully developed turbulence, $\nu \to 0$, the mean pseudo-energy dissipation rates tend to finite positive limits $\epsilon^\pm$, it can be found that

$$
\langle T_n^\pm \rangle = -\epsilon^\pm k_n^{-1}. \qquad (3.32)
$$

This is an exact relation which is valid in the inertial range of turbulence and it can be used as an operative definition of the inertial range. Thus the inertial range of the energy cascade in the shell model is defined as the range of scales k_n where law (3.32) is verified.

It is worth pointing out that in the case of the hydrodynamical model, apart for kinetic energy, an exact relationship exists also for the flux of kinetic helicity [8]. No similar result exists for the magnetic helicity in the MHD shell model.

3.4.4.2 Fixed Points of the Shell Model

The shell models contain some interesting fixed points, defined as solutions of the nonlinear term, $\Phi_n^\pm = 0$. From Eqs. (3.26) and (3.27), the main fixed point can be cast as a scaling law for the Elsässer variables $Z_n^\pm \sim k_n^{-h^\pm}$. In fact by using this scaling law in Eq. (3.24) it can be found that scaling exponents must be related by $2h^\pm + h^\mp = 1$. In this case Elsässer variables have the same scaling $h^+ = h^- = h$ and this reduces to the Kolmogorov's scaling $h = 1/3$, which is in fact the only fixed point of the GOY hydrodynamical shell model.

As far as the MHD FSGC shell model is concerned, a new interesting fixed point appears. In fact a trivial solution of Eqs. (3.26) and (3.27) is that $u_n(t) = \pm b_n(t)$ for each shell. This corresponds to having $Z_n^\sigma \neq 0$ and $Z_n^{-\sigma} = 0$ (or vice versa). This solution, even if trivial, is particularly interesting because it shows that an Alfvénic fluctuation is an exact nonlinear solution of the MHD equations.

3.4.5 Numerical Simulations

Here we investigate the basic behavior of the energy cascade described by shell models. We use numerical simulations of the equations, carried out through a fourth-order Runge–Kutta integrator. When the dissipation coefficients are non-zero and we want to investigate the smallest scales of the turbulent cascade, the total number of shells N must be carefully chosen according to the condition $k_N > k_D$, where $k_D \sim \nu^{-3/4}$ is the wave number at which dissipative effects start to be effective. Here we present results for $N = 18$ shells, the kinematic viscosity and magnetic diffusivity are set to $\nu = \mu = 0.5 \times 10^{-7}$, and the system is run for about 2×10^4 large-scale turnover times. We use the FSGC model which critically depends on the kind of forcing term we use.

3.4.5.1 Dynamical Properties of FSGC Shell Models

The FSGC has remarkable properties which closely resemble those typical of MHD phenomena. One of these is the magnetic dynamo action, that is, the amplification of a seed of magnetic field and its maintenance against losses of dissipation. Starting from a well-developed turbulent velocity field, a seed of magnetic field is injected and the growth of the magnetic spectra monitored in time. The energy is injected at the shell $n = 4$ with $k_0 = 1$ through a forcing $F_4^+ = F_4^- = (1 + i) \times 10^{-3}$, which corresponds to injecting only kinetic energy at large scales. We used both the 2D and the 3D models[2].

As shown in Ref. [34] a kind of dynamo effect is visible in the MHD shell model. The magnetic energy grows rapidly in time and forms a spectrum which on average is of the same order as that for the kinetic energy, the spectral index being closed to the Kolmogorov value. This kind of dynamo effect is absent in the 2D version of the model [34]. Since b_n is small, its back reaction on the velocity field in this case is negligible, thus the kinematic part of the model evolves independently from the magnetic one and the scaling $|u_n|^2 \sim k_n^{-4/3}$ clearly emerges. This scaling immediately follows from a cascade process of the quantity $\sum_n k_n |u_n|^2$, which is the 2D hydrodynamic invariant conserved by the kinematic part. Let us stress that the sign of the third ideal invariant is essential as far as the growth of a seed of magnetic energy is concerned.

The presence of a constant forcing term in the shell model induces a dynamical alignment. In fact it is found that unless the model is forced appropriately, it evolves invariably toward a state in which velocity and magnetic fields are strongly correlated, such that $Z_n^{\pm} \neq 0$ and $Z_n^{\mp} = 0$. In this state the nonlinear terms vanish, the energy cascade is stopped, and magnetic and kinetic spectra become steeper. We can say that the Alfvénic state is a strong attractor for the system. When we want to investigate statistical properties of turbulence described by MHD shell models, this should be avoided. In fact the Kolmogorov transient and the aligned states can become mixed during the averaging procedure, thus leading to unreliable results for scaling laws. It is possible however, to replace the constant forcing term by an

---

[2] Since the 2D shell model belongs to a family of shell model which do not present energy cascade, a large-scale viscosity ν' has been used to remove energy injected by the forcing. The term $-\nu' k_n^{-2} |u_n|^2$ has been added to the equation for the velocity field.

exponentially time-correlated Gaussian random forcing which is able to destabilize the Alfvénic fixed point, thus assuring the energy cascade. The forcing is obtained by solving the following Langevin equation:

$$\frac{dF_n}{dt} = -\frac{F_n}{\tau} + \mu(t), \tag{3.33}$$

where $\mu(t)$ is a Gaussian stochastic process δ-correlated in time $\langle \mu(t)\mu(t') \rangle = 2D\delta(t'-t)$. In this case the fixed point is destabilized, the system spends some large-scale turnover times around one of the Alfénic attractor, jumping from one to the other at irregular periods. This kind of forcing will be used to investigate statistical properties.

3.4.5.2 Statistical Properties of FSGC Shell Models

In this section we investigate the statistical properties of MHD shell models by using the forcing obtained in the previous section in order to get a statistically stationary state. This stationary state is reached by the system, as shown in Ref. [34], with a well-defined inertial range, a region where relation (3.32) is verified. In Figs. 3.7 and 3.8 we show the spectra for both the velocity $|u_n(t)|^2$ and magnetic $|b_n(t)|^2$ variables, as a function of k_n obtained in the stationary state. Fluctuations are averaged over time. The Kolmogorov spectrum is also reported as a solid line.

The Intermittency Correction

Intermittency in the shell model is due to the time behavior of shell variables. It has been shown [65] that the evolution of GOY model consists of short bursts traveling through the shells and long period of oscillations before the next burst arises. In Figs. 3.9 and 3.10 we report the time evolution of the real part of both velocity variables $u_n(t)$ and magnetic variables $b_n(t)$ at three different shells. It can be seen that,

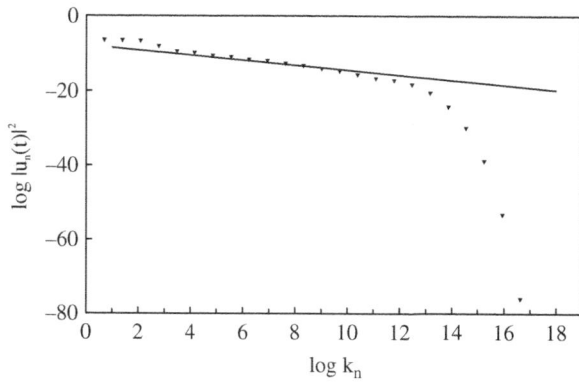

Fig. 3.7 The kinetic energy spectrum $|u_n(t)|^2$ as a function of $\log k_n$ for the MHD shell model. The *solid line* refers to the Kolmogorov spectrum $k_n^{-2/3}$

Fig. 3.8 The magnetic
energy spectrum $|b_n(t)|^2$ as a
function of $\log k_n$ for the
MHD shell model. The *solid
line* refers to the Kolmogorov
spectrum $k_n^{-2/3}$

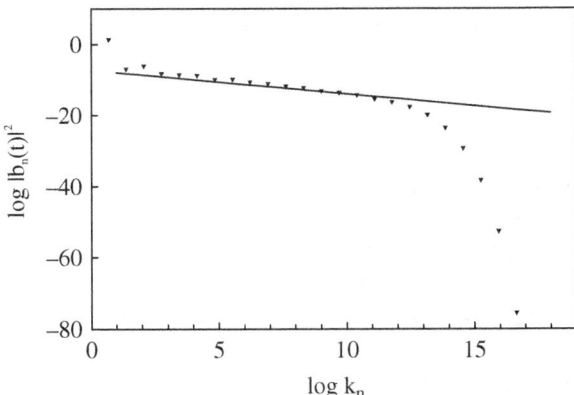

while at smaller k_n variables seems to be Gaussian, at larger k_n, variables present
very sharp fluctuations in between very low fluctuations.

The temporal behavior of variables at different shells changes the statistics of
fluctuations. In Fig. 3.11 we report the probability density functions $P(\Delta u_n)$ and
$P(\Delta B_n)$ of standardized variables

$$\Delta u_n = \frac{\Re e(u_n)}{\sqrt{\langle |u_n|^2 \rangle}}, \quad \Delta B_n = \frac{\Re e(b_n)}{\sqrt{\langle |b_n|^2 \rangle}}$$

for different shells n. Typically we see that PDFs look different in different shells:
at small k_n fluctuations form roughly a Gaussian distribution, while at large k_n they
tend to become increasingly non-Gaussian, by developing extensive tails. "Rare"
fluctuations have a probability of occurrence larger than for a Gaussian distribution.
This is the typical behavior of intermittency as observed in usual fluid flows and
described in previous sections.

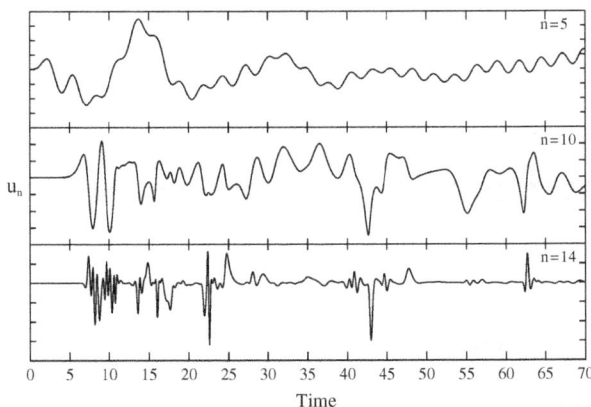

Fig. 3.9 The temporal
behavior of the real part of
velocity variable $u_n(t)$ at
three different shells n
reported on the figures

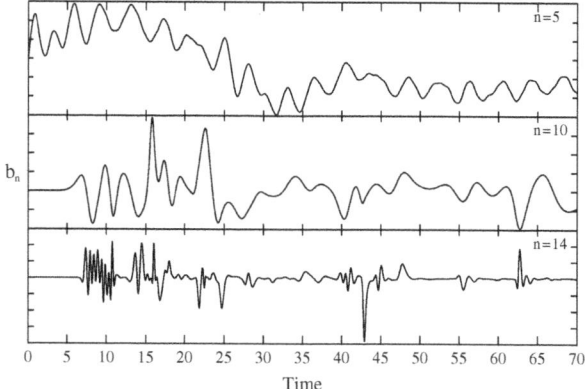

Fig. 3.10 The temporal behavior of the real part of magnetic variable $b_n(t)$ at three different shells n reported on the figures

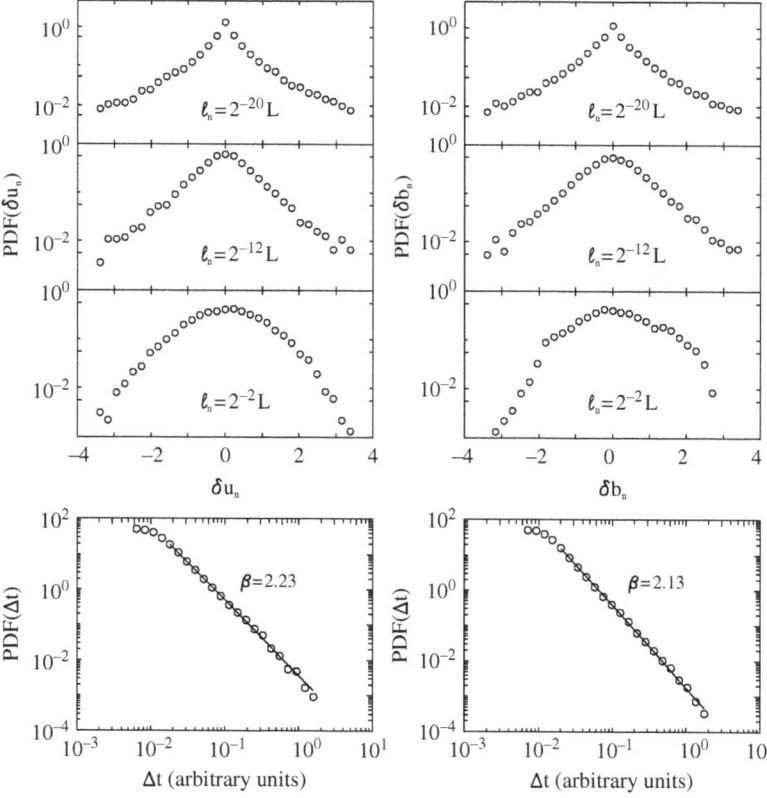

Fig. 3.11 The *first three panels* show PDFs of both velocity (*left-hand panels*) and magnetic (*right-hand panels*) shell variables, at three different shells ℓ_n. The *bottom panels* show PDFs of waiting times between intermittent structures at shell $n = 12$ for the corresponding velocity and magnetic variables

Table 3.1 Scaling exponents for velocity and magnetic variables, Elsässer variables, and fluxes. Errors on $\beta_p^\pm$ are about an order of magnitude smaller than the errors shown

p	ζ_p	η_p	ξ_p^+	ξ_p^-	β_p^+	β_p^-
1	0.36 ± 0.01	0.35 ± 0.01	0.35 ± 0.01	0.36 ± 0.01	0.326	0.318
2	0.71 ± 0.02	0.69 ± 0.03	0.70 ± 0.02	0.70 ± 0.03	0.671	0.666
3	1.03 ± 0.03	1.01 ± 0.04	1.02 ± 0.04	1.02 ± 0.04	1.000	1.000
4	1.31 ± 0.05	1.31 ± 0.06	1.30 ± 0.05	1.32 ± 0.06	1.317	1.323
5	1.57 ± 0.07	1.58 ± 0.08	1.54 ± 0.07	1.60 ± 0.08	1.621	1.635
6	1.80 ± 0.08	1.8 ± 0.10	1.79 ± 0.09	1.87 ± 0.10	1.91	1.94

The same phenomenon gives rise to the departure of scaling laws of structure functions from a Kolmogorov scaling. Within the framework of the shell model the analogies of structure functions are defined as

$$\langle |u_n|^p \rangle \sim k_n^{-\xi_p}, \ \langle |b_n|^p \rangle \sim k_n^{-\eta_p}, \ \langle |Z_n^\pm|^p \rangle \sim k_n^{-\xi_p^\pm}.$$

For MHD turbulence it is also useful to report mixed correlators of the flux variables

$$\langle [T_n^\pm]^{p/3} \rangle \sim k_n^{-\beta_p^\pm}.$$

Scaling exponents have been determined from a least-square fit in the inertial range $3 \leq n \leq 12$. The values of these exponents are reported in Table 3.1. It is interesting to notice that velocity, magnetic, and Elsässer variables are more intermittent than the mixed correlators. This could be due to the cancellation effects among the different terms defining the mixed correlators.

Time intermittency in the shell model generates rare and intense events. These events are the result of the chaotic dynamics in the phasespace typical of the shell model [65]. Such dynamics is characterized by a certain amount of memory, as can be seen through the statistics of waiting times between these events. The distributions $P(\Delta t)$ of waiting times are shown in the bottom panels of Fig. 3.11, at a given shell $n = 12$. The same statistical law is observed for the bursts of total dissipation [10].

3.4.6 Further Shell Models

Let us consider a phase transformation of the Elsässer variables

$$Z_n^\pm = z_n^\pm \exp(i\theta_n^\pm) \tag{3.34}$$

and let us consider the fact that these variables are random. Then assuming that the statistical properties of pair correlations remain invariant under the phase transformation (3.34), we get

$$\langle Z_n^\pm Z_m^{\pm *}\rangle = \langle z_n^\pm z_m^{\pm *}\rangle \exp[i(\theta_n^\pm - \theta_m^\pm)].$$

Since phases might be random this must imply that $\langle Z_n^\pm Z_m^{\pm *}\rangle = 0$ for all $n \neq m$. On the contrary, it can be immediately verified that Eqs. (3.24) are invariant under these transformations providing that the following two relations between phases hold:

$$\theta_{n+1}^\pm + \theta_n^\pm + \theta_{n-1}^\mp = 0, \, mod(2\pi).$$

Owing to the presence of these phase relationships, correlations are present for variables at two different shells. In particular, apart for the pseudo-energies $\langle Z_n^\pm Z_n^{\pm *}\rangle \neq 0$, there is a quadratic form which is different from zero, namely $\langle Z_n^\pm Z_{n+3m}^{\pm *}\rangle \neq 0$. This phase invariance can be exploited by defining a slightly different shell model where the spectrum of possible correlations is reduced. The model reads

$$
\begin{aligned}
\frac{dZ_n^\pm}{dt} = ik_n \Bigg[&\left(\frac{2-a-c}{2}\right) Z_{n+2}^\pm Z_{n+1}^{\mp *} + \left(\frac{a+c}{2}\right) Z_{n+1}^{\pm *} Z_{n+2}^\mp + \\
&+ \left(\frac{c-a}{2\lambda}\right) Z_{n-1}^{\pm *} Z_{n+1}^\mp - \left(\frac{a+c}{2\lambda}\right) Z_{n-1}^{\mp *} Z_{n+1}^\pm + \\
&- \left(\frac{c-a}{2\lambda^2}\right) Z_{n-2}^\mp Z_{n-1}^\pm - \left(\frac{2-a-c}{2\lambda^2}\right) Z_{n-1}^\mp Z_{n-2}^\pm \Bigg].
\end{aligned}
\tag{3.35}
$$

As can be verified immediately, model (3.35) has the same phase invariance, except that in that case the following phase relations hold:

$$\theta_{n+1}^\pm - \theta_n^\pm - \theta_{n-1}^\mp = 0, \, mod(2\pi).$$

Owing to these new relations, the only quadratic forms different from zero are the pseudo-energies. Even in this case, when $b_n = 0$ we recover a further version of the hydrodynamical model [8].

3.4.7 Galerkin Approximations of the 2D MHD Equations

The dynamics of low-frequency plasmas described by MHD is very interesting because it represents the first approach to the study of a wide variety of phenomena in both laboratory and astrophysical plasmas [9]. In some fusion plasmas, or in the solar corona, the plasma β parameter (that is, the ratio between the kinetic and magnetic pressure) is low ($\beta \simeq 10^{-2}$), and MHD equations can be simplified into the so-called reduced MHD [81]. This approximation is valid, for example, for a plasma column with a low aspect ratio $a/L \ll 1$ (a and L being, respectively, the radius and the length of the cylinder), with a strong magnetic field $\mathbf{B}_0$ along the z-direction.

Let us consider a plasma inside a cylinder, of diameter a and length L, whose axis is directed along the z-axis. A constant background magnetic field B_0 is assumed to be in the z-direction. The plasma is described by the velocity field $\mathbf{u}(\mathbf{r}, z, t)$ and the magnetic field $\mathbf{b}(\mathbf{r}, z, t)$ with the usual non-dimensionalization and $\mathbf{r} = (x, y)$ being directed in the plane perpendicular to the z-axis. In the (x, y) plane we Fourier transform the fields in a 2D box of size a as follows:

$$\mathbf{u}(\mathbf{r}, z, t) = \sum_{\mathbf{k}} u(\mathbf{k}, z, t) \mathbf{e}(\mathbf{k}) \exp(i\mathbf{k} \cdot \mathbf{r}), \tag{3.36}$$

$$\mathbf{b}(\mathbf{r}, z, t) = \sum_{\mathbf{k}} b(\mathbf{k}, z, t) \mathbf{e}(\mathbf{k}) \exp(i\mathbf{k} \cdot \mathbf{r}), \tag{3.37}$$

where $\mathbf{e}(\mathbf{k})$ is a unitary vector in the direction of $\mathbf{k} = 2\pi \mathbf{m}/a$, and $\mathbf{m}$ is a pair of integers. After some algebra, it can be shown that, for each value of z, the MHD equations reduce to

$$\frac{\partial u(\mathbf{k})}{\partial t} = B_0 \frac{\partial b(\mathbf{k})}{\partial z} + \sum_{\mathbf{k}=\mathbf{p}+\mathbf{q}} c(k, p, q)(p^2 - q^2) \left[u(\mathbf{p})u(\mathbf{q}) - b(\mathbf{p})b(\mathbf{q}) \right]$$

$$\frac{\partial b(\mathbf{k})}{\partial t} = B_0 \frac{\partial u(\mathbf{k})}{\partial z} + \sum_{\mathbf{k}=\mathbf{p}+\mathbf{q}} c(k, p, q)k^2 \left[b(\mathbf{p})u(\mathbf{q}) - u(\mathbf{p})b(\mathbf{q}) \right] \tag{3.38}$$

(for simplicity we omit the time and z dependence of the Fourier amplitudes), where $c(k, p, q) = (p_x q_y - p_y q_x)/2kpq$ is a geometrical factor. The sum on the r.h.s. of (3.38) is related to the Kronecker symbol

$$\sum_{\mathbf{k}=\mathbf{p}+\mathbf{q}} = \left(\frac{2\pi}{L}\right)^2 \sum_{\mathbf{p}, bq} \delta_{\mathbf{k}, \mathbf{p}+bq},$$

which means that the sum is extended over all wave vectors $\mathbf{p}$ and $\mathbf{q}$ which satisfy the triad-interaction relation $\mathbf{k} = \mathbf{p} + \mathbf{q}$. We will consider a box of size $a = 2\pi$, so that each wave vector $\mathbf{k}$ turns out to be represented by a couple of integers.

Equations (3.38) have quadratic invariants [9], namely the total energy

$$E(t) = \sum_{\mathbf{k}} [|v(\mathbf{k}, t)|^2 + |b(\mathbf{k}, t)|^2]$$

and the cross-helicity

$$H_C(t) = \sum_{\mathbf{k}} Re[v(\mathbf{k}, t)b^*(\mathbf{k}, t)].$$

When the background magnetic field is set to zero ($B_0 = 0$), the mean square of the vector potential

$$A(t) = \sum_{\mathbf{k}} |b(\mathbf{k}, t)|^2 / k^2$$

is also conserved and an infinite number of non-quadratic invariants exist [9]. However, quadratic invariants, also called rugged invariants, play a key role, because they survive *each finite number of triads wave vectors* $(\mathbf{k}, \mathbf{p}, \mathbf{q})$ *which satisfy the triad-interaction relation*. In other words, even if the r.h.s. of (3.38) contains an infinite number of nonlinear interactions, $E(t)$, $H_C(t)$, and $A(t)$ remain invariant for each triad of wave vectors $\mathbf{k} + \mathbf{p} + \mathbf{q} = 0$.

3.4.8 Relaxation Processes in MHD

One of the most fascinating problems in MHD concerns the possibility of predicting the final fate of solutions starting from quite general initial conditions. A crucial role is played by the quadratic invariants. Since during ideal relaxation quadratic invariants remain constant, the problem of predicting the final state reduces to determining the ensemble-averaged equilibrium spectra of invariants, which are predicted by the ensemble-averaged initial values of invariants. Dissipative relaxation processes in 2D MHD are much more complicated, since the values of the ideal invariants change during the evolution. It seems, however, that some long-lived nontrivial states exist toward which dissipative flows are attracted and that these states could be derived by minimizing an energy integral subject to some constraints.

Dissipative relaxations have been investigated in connection with measurements in laboratory plasmas, and observations in the solar wind turbulence. Taylor [93] conjectured that an MHD system relaxes toward a state where the energy tends to a minimum, subject to the constraint that magnetic helicity is conserved. Physically the relaxation represents a *selective decay* between the two invariants. The solution of the problem can be obtained through a variational principle

$$\delta \int (|\mathbf{v}|^2 + |\mathbf{b}|^2)dV - \lambda \delta \int (\mathbf{a} \cdot \mathbf{b})\, dV = 0$$

(integrals are extended to a given volume of magnetofluid, and λ is a Lagrange multiplier). By imposing independent variations we get a force-free solution $\nabla \times \mathbf{b} = \alpha \mathbf{b}$ (α is a constant and $\mathbf{v} = 0$), which means that the kinetic energy decays to zero and the magnetic energy occupies the largest scale. This solution [93] is particularly useful to explain the large-scale behavior of the reversed field pinch.

When we require that the energy assume a minimum value, constrained to the conservation of cross-helicity, we have

$$\delta \int (|\mathbf{v}|^2 + |\mathbf{b}|^2)dV - \lambda \delta \int (\mathbf{v} \cdot \mathbf{b})\, dV = 0,$$

thus obtaining the solution $\mathbf{v} = \sigma \mathbf{b}$ and $\sigma = \pm 1$. Physically this represents a *dynamical alignment* between the velocity and the magnetic field, that is, the system tends dynamically to increase the correlation between the velocity and the magnetic field. The discovery in the solar wind of quite particular Alfvénic fluctuations characterized by a high degree of correlation between velocity and magnetic field was interpreted as a result of this kind of dynamical alignment within a turbulent magnetofluid [24].

It has also been shown that in numerical simulations of 2D and 3D MHD equations there exists systematic behavior of turbulent flows. In particular, relaxation processes which bring the fluid toward those particular states have been found to emerge systematically. In terms of ideal invariants it has been shown that Taylor's solution is obtained, in 2D situations, when the magnetofluid tends to dissipate energy faster than the mean square of the vector potential. The dynamical alignment corresponds to situations in which the energy is dissipated faster than cross-helicity, even if the physical motivations remain controversial (cf. Ref. [97]). When both effects are in competition, the situation is much more complicated. A systematic analysis of a large number of 2D MHD simulations at different resolutions [96] can be summarized by using the time behavior of the system projected on the plane (a, h), where $a = A/E$ and $h = 2H_C/E$. In fact, starting from whatever initial conditions, in that plane the system tends to approach the curve

$$1/a = 2(k_{min}/h)^2[1 \pm (1 - h^2)^{1/2}], \qquad (3.39)$$

where k_{min} represents the minimum wave vector allowed in the simulations. Of course the points $(a, h) = (1/2, \pm 1)$ represent the *attractors* of dynamical alignment solutions, the point $(a, h) = (1, 0)$ represents the attractor of selective decay, and the point $(a, h) = (0, 0)$ represents the attractor when the system behaves like a fluid, with zero magnetic field. These extreme points can be recovered through the minimization of energy subject to the conservation of the other invariants[3]. The entire curve, however, does not represent the locus of the extreme of anything over its entire range of definition.

3.4.8.1 A Minimal Triad-Interaction Model of Relaxation Processes in 2D MHD

As we have seen 2D MHD plays a privileged role, because it represents a good approximation for low-beta plasmas. Here we briefly present how the basic non-chaotic triad-interaction works in 2D MHD, and how this represents a minimal basic model for relaxation processes. Choosing only three wave vectors, namely $\mathbf{k}_1 = (1, 1)$, $\mathbf{k}_2 = (2, -1)$, $\mathbf{k}_3 = (3, 0)$, we obtain a set of 12 ODEs for the complex Fourier modes of both velocity $u_i(t) = u(\mathbf{k}_i, t)/2\sqrt{5}$ and magnetic field $b_i(t) = b(\mathbf{k}_i, t)/2\sqrt{5}$. The system can be further reduced through a projection of equations on a subset of the phase space, that is, by considering only real fields. This can be seen by writing the fields in the form $u_j = |u_j|e^{i\alpha_j}$ and $b_j = |b_j|e^{i\beta}$, and by

---

[3] The point $(a, h) = (0, 0)$ can be recovered by allowing that the kinetic energy decays, subject to the constraint that the kinetic helicity remains constant.

defining real fields through $V_j(t) = |u_j| \cos \alpha_j$ and $B_j(t) = |b_j| \cos \beta_j$ (subject to the conditions $\sin \alpha_j = 0$ and $\sin \beta_j = 0$). In this case we found a set of six ODEs for V_j and B_j, namely

$$(d/dt + 2v)V_1 = 4(V_3 V_2 - B_3 B_2),$$
$$(d/dt + 5v)V_2 = -7(V_3 V_1 - B_3 B_1),$$
$$(d/dt + 9v)V_3 = 3(V_1 V_2 - B_1 B_2),$$
$$(d/dt + 2\mu)B_1 = 2(B_3 V_2 - V_3 B_2),$$
$$(d/dt + 5\mu)B_2 = 5(V_3 B_1 - B_3 V_1),$$
$$(d/dt + 9\mu)B_3 = 9(V_1 B_2 - B_1 V_2).$$

Given its simplicity the model represents the basic system to investigate the structure of nonlinear interactions in 2D MHD and to study the role played by the rugged invariants during the dynamical evolution.

Let us introduce the phase space Ω, of dimension $Dim[\Omega] = 6$, which can be built up by using the Fourier amplitudes as coordinates. A point $\Psi(t) \in \Omega$, defined as $\Psi(t) := \{[u_i(t), b_i(t)] \in \Omega\}$, represents the system at a certain time, and this point moves in Ω according to the flow $T_\tau[\Psi(t)] = \Psi(t + \tau)$ which represents the result of the equation of motion for the system. In the absence of dissipative terms the system satisfies the Liouville theorem $H = \sum_i (\partial/\partial\Psi)(d\Psi/dt) = 0$. The quantity H represents the rate of change of volumes in phase space. If we define the ideal flow T_τ^{id} as the flow T_τ obtained when $v = \mu = 0$, the Liouville theorem means that $T_{\tau \to \infty}^{id}[\Psi(t)] = \Psi(t + \tau)$, that is, volumes in the phase space are conserved. In this case, for a given set of initial values $\Psi(0)$, the point moves on a hyper-surface $S \subset \Omega$ defined by the initial value of the invariants. In the presence of dissipative terms the quadratic invariants decay, and the rate of change of the volume in the phase space is $H = -16(v + \mu) \leq 0$. The condition $H \leq 0$ implies that the dissipative flow pushes the system toward the trivial asymptotic state where all the amplitudes of the fields are zero $T_{\tau \to \infty}^{diss}[\phi(t)] = 0$.

The triad-interaction model is able to capture the dynamics of the quiescent states observed in MHD. We solved our system by starting from random initial conditions uniformly distributed $\Psi(0) \in [-1, 1]$. We used a fourth-order Runge–Kutta scheme, with a time step $\Delta t = 10^{-3}$ and $v = \mu = 0.01$. This value for the dissipative coefficients allows the nonlinear interactions to have sufficient time to set up the dynamical behavior. In Fig. 3.12 we show the curve ϵ along with two ensembles of points. The first ensemble (white circles) represents the set of points (a, h) obtained with some different initial conditions $\Psi(0)$ randomly chosen in the interval $[-1; 1]$. The second ensemble (black circles) represents the set of points (a, h) at time $t = 80$ (in unit of time steps), calculated from the set of fields $\Psi(t)$ which are obtained through the time evolution of the set $\Psi(0)$, that is, $\Psi(t) = T_t^{diss}[\Psi(0)]$. As can be seen all the initial conditions lead to the final state which belongs to the ellipse ϵ.

Since nonlinear interactions in the simple triad-interaction model have the same structure as in the 2D MHD equations, the model is able to capture relaxation

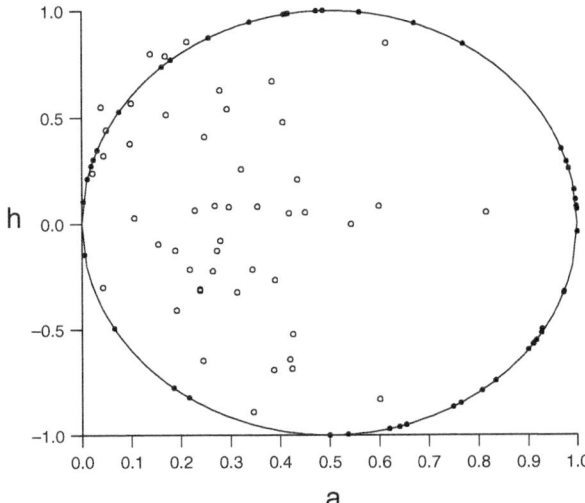

Fig. 3.12 Numerical simulations of the triad-interaction model in the plane (a, h). We show two sets of different solutions at two different times. *White circles* refer to a set of 45 initial values $\Psi(0)$, *black circles* represent the set of point at a given time $t = 80$ (in unit of time steps) obtained from the time evolution of the above initial values $\Psi(t = 80) = T_{t=80}^{diss}[\Psi(0)]$. The solutions represented on that plane are such that, after an initial transient, they fall on the ellipse represented as a full line

properties. These properties are the fixed points and some invariant subspaces. The fixed points of the system can be classified as follows:

(a) two Alfvénic fixed points (say $A^{\pm}$) characterized by $u_i = \pm b_i$;
(b) three fluid fixed points $F(i)$ ($i = 1, 2, 3$) where all variables but the velocity V_i are zero;
(c) three magnetic fixed points $M(i)$ ($i = 1, 2, 3$) where all variables but the magnetic field B_i are zero.

A standard analysis of stability can be performed by linearizing the system around each fixed points. We find that $A^{\pm}$ are always stable, $M(1)$ is the only stable magnetic fixed point, while $F(1)$ and $F(3)$ are stable. The only stable magnetic fixed point $M(1)$ is such that the energy is localized on the minimum wave vector.

Apart from fixed points the system displays some other interesting properties. Looking at the equations it can be easily shown that there exists some 3D subspaces of the 6D phase space which remain invariant under the ideal flow operator. Let us denote by I^{α} the αth invariant subspace, which is then characterized by the fact that if $\Psi(0) \in I^{\alpha}$ then $\Psi(t) = T_t^{id}[\Psi(0)] \in I^{\alpha}$ for each $t > 0$. In other words an ideal invariant subspace is a portion of the phase space where the system lies for all times. The most useful way to classify these structures is through the initial values the rugged invariants assume on them, since under the ideal flow they remain constant:

(a) A fluid subspace F, characterized by $a = A/E = 0$ and $h = 2H_c/E = 0$, which can be recovered by imposing that the magnetic field is always zero, namely $\Psi[F] = (V_1, V_2, V_3, 0, 0, 0)$ is the vector which describes this subspace.
(b) Two Alfvénic subspaces $A^\pm$, characterized by $a = A/E = 0$ and $h = 2H_c/E = 1$, which can be recovered by requiring that $V_i = \pm B_i$ for each $i = 1, 2, 3$. These subspaces are also fixed points of the system.
(c) Three magnetic subspaces H^A, H^B, and H^C characterized by $a \neq 0$ and $h = 0$, i.e., a minimal value for the cross-helicity. These subspaces can be recovered by imposing that the cross-helicity is initially zero over all wave vectors, namely either $V_i = 0$ and $B_i \neq 0$ or vice versa; specifically, $\Psi[H^A] = (0, 0, V_3, B_1, B_2, 0)$, $\Psi[H^B] = (V_1, 0, 0, 0, B_2, B_3)$, and $\Psi[H^C] = (0, V_2, 0, B_1, 0, B_3)$.

The stability properties of each subspace can be investigated numerically [15]. Simulations start by putting the system on a given subspace, and by adding a small perturbation on the complementary manifold. The motion is not limited to the 3D subspace, and the problem of the stability of a particular I^α consists in determining if the perturbed solution remains close to I^α or goes away from it covering all the allowed 6D phase space. For each subspace we can define two pseudo-energies $E_{int}^{(\alpha)}(t)$ (built up with the fields which belong to the αth subspace) and $E_{ext}^{(\alpha)}(t)$ (built up with the fields which do not belong to the αth subspace). The external energy represents the distance $\| \Delta^{(\alpha)}[\Psi(t)] \|$ between the point $\Psi(t)$ and the αth invariant subspace. Since the total energy $E(t) = E_{ext}^{(\alpha)}(t) + E_{int}^{(\alpha)}(t)$ must remain constant under the ideal flow, two situations can arise, namely

(1) During the time evolution both $E_{ext}^{(\alpha)}(t) \simeq E_{ext}^{(\alpha)}(0)$ and $E_{int}^{(\alpha)}(t) \simeq E_{int}^{(\alpha)}(0)$, which means that the solution remains trapped near the invariant subspace. In that case the subspace is ideally stable.
(2) During the time evolution energies become comparable $E_{ext}^{(\alpha)}(t) \simeq E_{int}^{(\alpha)}(t)$, which means that the solution is repelled from the invariant subspace. In that case the subspace is ideally unstable.

Numerical simulations show that the fluid subspace is always stable. As far as the magnetic subspaces are concerned, we find that H^A is always stable, while H^B and H^C are always unstable. More information about the behavior of the system near the invariant subspaces can be obtained by considering the characteristic of solutions when the dissipative coefficients are set different from zero. In that case, the energies decay in time, but the rate of decay is different, thus indicating a kind of selective dissipation. In particular, looking at the time evolution of their ratios $R^\alpha(t) = E_{ext}^{(\alpha)}(t)/E_{int}^{(\alpha)}(t)$, we find that it decays to zero for the ideally stable subspaces, while it settles to a constant value for the unstable subspaces. This means that ideally stable subspaces, namely the Alfvénic, fluid, and magnetic H^A, represent a kind of attractor for the system, while unstable subspaces, say H^B and H^C, repel solutions.

When we start numerical solutions near the stable subspaces, the system is attracted toward the extreme points of the plane (a, h). When we start near the fluid attractor, the system reaches the point $(a, h) = (0, 0)$, and when we start near one of the Alfvénic attractors the system is attracted toward $(a, h) = (1/2, \pm 1)$. Finally when we start near the stable magnetic subspace H^A the system is attracted by the extreme point $(a, h) = (1, 0)$. This last point represents a Taylor regime, say the magnetic field lies on the lowest allowed wave vector. On the contrary, when we start near the unstable magnetic attractors H^B or H^C, the system evolves in an erratic way toward any point (a, h) of the curve (3.39). Each point (a, h) is made by only one mode with wave vector $k_1 = k_{min} = \sqrt{2}$, so that if $\eta = V_1/B_1$ we have $a = 2/(1 + \eta^2)k_{min}^2$ and $h = \eta/(1 + \eta^2)$. These last equations are nothing but the parametric equation of the curve (3.39).

The behavior we have just described can also be recovered when we start the numerical computation with general initial conditions. In that case we can associate with each initial condition an invariant subspace according to the rule that the αth subspace is associated with the initial condition $\Psi(0)$ when the distance $\| \Delta^{(\alpha)}[\Psi(0)] \|$ is the minimum one over α. In other words, we associate an initial condition to the nearest subspace. Numerical simulations show that when the initial conditions are associated with the unstable subspaces H^B and H^C, the final point reached by the system will be any point (a, h) on the curve (3.39). When the initial conditions are associated with one of the stable subspaces, the final state will be one of the extreme points of the curve (3.39).

3.4.9 Large Eddy Simulations

The classical way to model a turbulent flow is to decompose the velocity field into mean $\bar{u}$ and fluctuating u' components and derive the corresponding equations for these two fields. In the latter case, a sub-grid tensor term appears which is

$$\tau_{ij} = \overline{u_i u_j} - \bar{u}_i \bar{u}_j.$$

The simplest way to express τ_{ij} is to make it dissipative, using an eddy viscosity (that can be time, space and/or scale dependent); a common model, now called the Smagorinsky [87, 59] model, writes

$$\tau_{ij} - \frac{\delta_{ij}}{3} \tau_{ll} = -2\nu_{turb} S_{ij},$$

where $S_{ij} = \frac{1}{2}[\partial_i u_j + \partial_j u_i]$ is the symmetrized velocity gradient matrix; $\nu_{turb} = C_S \Delta^2 |\bar{S}|$ with $\bar{S}$ the norm of S_{ij}, Δ the grid filter size; finally, C_S is a constant that can be evaluated dynamically (see, e.g., [43]) in order to avoid having to adjust it for each flow against experimental or numerical data. It can be shown that such a formulation is dimensionally consistent with the Kolmogorov $k^{-5/3}$ spectrum. Improvement to these early formulations of modeling small-scale flows arise when one realizes that near the cut-off scales, the local nonlinear interactions (in Fourier space) are poorly treated, so that the idea of an iteration of the filtering procedure is introduced, with sub-grid and sub-filter scales; such approaches have greatly improved the modeling of turbulent flows widely used in the engineering community.

Many variants of large eddy simulations (LES) have thus been constructed over the years (see, e.g., for recent reviews [60, 83, 62, 52, 75] in the fluid engineering context). Such models can be tested against experiments, or observations in a planetary boundary layer, for which energy transfer between different scales have been evaluated experimentally [91].

In MHD, related models have been devised, starting from two-point stochastic closures (see, e.g., [78, 101]) or one-point variants of the LES closures (see, e.g., [94, 62, 73]), introducing additional sub-grid stress tensors of the form $b_i \bar{b}_j - \bar{b}_i \bar{b}_j$ in the fluid equation and $u_i b_j - b_i u_j - \bar{u}_i \bar{b}_j + \bar{b}_i \bar{u}_j$ in the induction equation.

In an LES the self-similar (velocity) energy spectrum is extended all the way to the numerical cut-off of the computation. One can also think of another way, less extreme, of modeling turbulent flows, a way that can be coined "quasi-DNS"; in such a case, one only aims at gaining a few factors of 2 in numerical resolution (each factor of 2 in resolution is a factor 16 in total costs in 3D, hence non-negligible) by filtering adequately the small scales. One way to do that, in a theoretically consistent manner, has been introduced recently by D. Holm and his collaborators [21], a method also extended to MHD (see, e.g., [42, 61]). Different types of filters can be used, cf. [60, 75, 43].

Finally, stochastic models have been derived recently [56, 22]. For example, in [52] (see also [53]), an extension of rapid distortion theory treats a linear Langevin equation in which the random noise, instead of being Gaussian in space and white in time (and thus not being able to build coherent structures because of the lack of phase relations in the small scales), is computed from what is known of the large scales, allowing for some kind of bootstrapping, with some success [3].

3.5 Theoretical Approaches

3.5.1 Weak Turbulence

3.5.1.1 Introductory Remarks

How are two-point closures in MHD turbulence obtained for both the weak and strong cases? These two regimes arise when the ratio of the characteristic timescales for, on the one hand, Alfvén waves – based on either an external uniform magnetic field or on the large-scale turbulent magnetic field – and, on the other hand, for nonlinear interactions between turbulent eddies and waves is small or not. Such regimes are distinguished by the degree of anisotropy in the fluctuating fields, and by the relative steepness of the distribution of energy among modes. Weak MHD turbulence may be at play in the interstellar medium (ISM) or in the Jovian magnetosphere. In the latter case, an analysis of Galileo data [47] between the orbits of the satellites Io and Callisto may shed some light on the issue of weak MHD turbulence in that it agrees with the predictions of the theory (see below).

3.5.1.2 Conditions for Weak Turbulence

Let us write in symbolic form a simple equation including linear $\mathcal{L}$ and nonlinear (here, assumed quadratic) $\mathcal{N}$ terms, viz.

$$\partial_t u = \mathcal{L}_x u + \epsilon \, \mathcal{N}_x(u, u) \ ,$$

and further assume that, in the strictly linear case ($\epsilon = 0$), the Fourier transform of our velocity field has a purely oscillatory behavior, due to the presence of waves of frequency ω_k at wave number k, i.e., $\hat{u}(\mathbf{k}, t) = \hat{u}_0(\mathbf{k})e^{-i\omega_k t}$. Note that we consider the dissipationless case.

Let us now switch on progressively the nonlinearities; when $\epsilon \ll 1$, there exist two different timescales, and one can write $\hat{u}(\mathbf{k}, t) = a(\mathbf{k}, t)e^{-i\omega_k t}$, with a fast variation in time of an oscillating nature $e^{-i\omega_k t}$, and a slow variation in time of the amplitude a($\mathbf{k}$,t). The essence of the weak turbulence theory is that, because of the existence of resonances, one is led to a clean closure with "kinetic" equations for the various energy spectra. This theory not only takes into account resonant wave interactions but it deals with the temporal behavior not of an individual triad of waves (or of wave packets), but with a wide superposition of waves, leading to the description of the dynamical behavior of cumulants and moments of the stochastic wave field. Numerous texts can be found concerning the weak turbulence theory [5, 102, 63], developed in the context of geophysical flows and of plasma physics in the mid-1960s.

3.5.1.3 The MHD Case

The MHD equations can be linearized in the presence of a strong uniform magnetic field $\mathbf{B}_0$, with the induction equation written as a superposition of a strong uniform magnetic component and a turbulent fluctuation

$$\mathbf{B} = \mathbf{B_0} + \epsilon\mathbf{b} \ .$$

It can be shown that the system supports waves: Alfvén waves propagating at the Alfvén speed $v_A \sim \mathbf{B}_0$ (with uniform density $\rho \equiv 1$ assumed). Hence, two characteristic times can be identified in such a system: the Alfvén time

$$\tau_A(\mathbf{k}) = [\, k_\| \mathbf{B_0} \,]^{-1} = \omega_A^{-1} \tag{3.40}$$

and the time of interaction between waves

$$\tau_{int}(\mathbf{k}) \sim [\mathbf{k}_\perp b_{rms}]^{-1} = \omega_{int}^{-1} \ . \tag{3.41}$$

The small parameter of the problem is thus

$$\frac{\tau_A}{\tau_{int}} \sim \epsilon \ll 1 \; .$$

Using now the interaction representation of a field component f_j

$$f_j(\mathbf{x}, t) = \int A_j(\mathbf{k}, t) e^{i\mathbf{k}\cdot\mathbf{x}} \, d\mathbf{k} = \int a_j(\mathbf{k}) e^{i(\mathbf{k}\cdot\mathbf{x}+s\omega_k t)} \, d\mathbf{k} \; , \qquad (3.42)$$

one can write the integro-differential equation for the temporal evolution of this field amplitude a_j in Fourier space, namely

$$\partial_t a_j(\mathbf{k}) \sim \int a_m(\boldsymbol{\kappa}) a_n(\mathbf{L}) e^{i(-\omega_k - \omega_\kappa + \omega_L)t} \delta(\boldsymbol{\kappa} + \mathbf{L} - \mathbf{k}) \, d\boldsymbol{\kappa} d\mathbf{L} \; . \qquad (3.43)$$

Three-wave interactions (between waves $\mathbf{k}$, $\boldsymbol{\kappa}$, and $\mathbf{L}$) stem from the convolution term, and the advantage of the interaction representation is that, for $\epsilon \equiv 0$, no temporal dynamics would occur, i.e., $\partial_t a_j(\mathbf{k})_{\epsilon=0} \equiv 0$. At this point, no approximation has been performed yet. Starting from the above equation, one can write exactly the equations for the second- and third-order moments and cumulants of the stochastic field components f_j (or equivalently a_j), here the velocity and magnetic field. It is useful to decompose such fields in terms of their poloidal and toroidal components

$$\mathbf{f} = \nabla \times (\Psi_f \mathbf{e}_\parallel) + \nabla \times [\nabla \times (\Phi_f \mathbf{e}_\parallel)]$$

ensuring the divergence-free conditions; thus, auto-correlation and cross-correlation functions of all such fields must be taken into account, rendering the algebra rather complex. The saving grace of such an approach is the following: one can show rigorously that, in Eq. (3.43) and in the related dynamical equations for moments and cumulants, the rapidly oscillating terms $\sim e^{i(-\omega_k - \omega_\kappa + \omega_L)t}$ do not contribute to the dynamical evolution of the cumulants except at resonance. Thus the resonance condition plays an essential role in the weak turbulence closure; it must be stressed that the random phase approximation is not made here.

3.5.1.4 Anisotropic Turbulence of Shear Alfvén Waves

The algebra needed to derive closure equations is strenuous. It can be simplified if one makes some extra assumptions, limiting the range of validity of the approach but, on the other hand, making the structure of the theory clearer.

In weak MHD, this can be accomplished in the limit of strongly anisotropic pulsations. For that case, write the MHD equations for Fourier modes a^s (of Fourier spectra q^s) and suppose that $k_\parallel \ll k_\perp$; hence,

$$a_2^s = -(k_1/k_2)a_1^s - (k_\parallel/k_2)a_\parallel^s \approx -(k_1/k_2)a_1^s$$

and the temporal evolution for a^s becomes

$$\partial_t \, a_1^s(\mathbf{k}) = -i\epsilon \int \frac{k_2}{\kappa_2 L_2 k_\perp^2} \, (\mathbf{k}_\perp \cdot \mathbf{L}_\perp)(\mathbf{k} \times \boldsymbol{\kappa})_\parallel \, a_1^{-s}(\boldsymbol{\kappa}) \, a_1^s(\mathbf{L}) \, e^{-2is b_0 \kappa_\parallel t} \, \delta_{\mathbf{k}, \boldsymbol{\kappa} \mathbf{L}} \, d_{\boldsymbol{\kappa} \mathbf{L}},$$

(3.44)

using the resonance condition $k_\parallel = \kappa_\parallel + L_\parallel$ for Alfvén waves in the exponential.

The applicability conditions for the weak turbulence approximation to occur are that $k_\parallel / k_\perp \gg \epsilon^2$ and that the spectrum must change slowly with $k_\parallel$ in the range

$$-\epsilon^2 k_\perp \ll k_\parallel \ll \epsilon^2 k_\perp$$

(3.45)

at fixed $k_\perp$.

The resulting equation for the total energy spectrum in this approximation is

$$\partial_t \, e^s(\mathbf{k}) = \frac{\pi \epsilon^2}{b_0} \int \frac{(\mathbf{k}_\perp \cdot \mathbf{L}_\perp)^2 \, (\mathbf{k} \times \boldsymbol{\kappa})_\parallel^2}{k_\perp^2 L_\perp^2 \kappa_\perp^2} \, e^{-s}(\boldsymbol{\kappa}) \left[e^s(\mathbf{L}) - e^s(\mathbf{k}) \right] \, \delta(\kappa_\parallel) \, \delta_{\mathbf{k}, \boldsymbol{\kappa} \mathbf{L}} \, d_{\boldsymbol{\kappa} \mathbf{L}}.$$

(3.46)

The above equation has a standard structure found in all such approaches: it contains a geometrical factor linked to the physical nonlinearity considered; the convolution term gives rise to $\sim \delta_{\mathbf{k}, \boldsymbol{\kappa} \mathbf{L}} \, d_{\boldsymbol{\kappa} \mathbf{L}}$. Finally, the two terms proportional to the energy spectrum $e(\boldsymbol{\kappa})$ are called an emission and absorption term, respectively; indeed, the former can be seen as bringing energy into mode $\mathbf{k}$ through mode coupling, whereas the latter gives rise to turbulent viscosity (see, e.g., [76, 77] for further discussions).

Note that two routes can be taken to obtain the above Eq. (3.46): either first taking the weak turbulence limit and computing the kinetic equations for the full case, then taking the anisotropic limit on the resulting kinetic equations or the second path being to make the anisotropic assumption first and doing the weak turbulence closure on these simplified equations. It can be checked that these two limits (anisotropy, weak turbulence) commute. Such kinetic equations have constant flux self-similar solutions of the type $f(k_\parallel) \, k_\perp^{-2}$ (see [32] for details, in particular concerning the temporal precursor to that spectrum).

3.5.1.5 Model of Jovian Turbulence Inspired from the "Maltese Cross" Model for the Solar Wind

In the solar wind, it has been shown that a model, called the Maltese cross, of a superposition of two distinct components for the magnetic field works best in explaining the observed structures of the wind. In the Jovian case, magnetospheric turbulence could also arise from the following superposition:

$$\delta \mathbf{b}(\mathbf{k}) = \delta \mathbf{b}^{slab}(k_\parallel) + \delta \mathbf{b}^{2D}(k_\perp),$$

where the slab part depends only on the $k_\parallel$ wave vector component and hence has only two components because of $\nabla \cdot \delta \mathbf{b} = 0$:

$$\delta b_{\parallel}^{slab}(k_{\parallel}) = 0 ,$$

and where the 2D part of the decomposition has all three components. This simple model allows an investigation of the parallel spectrum of observed fluctuations (P_{zz}) since it stems only from the 2D component; thus, weak turbulence theory applies to P_{zz}, whereas the other two spectra P_{xx} and P_{yy}, when observed, arise from a mixture of a $k_{\parallel}$ dependency (which is not predicted by theory) and $k_{\perp}$ dependency.

With such a model in mind, one can examine data coming from the Galileo spacecraft and look at spectra for the magnetic field [84]. Remarkably, the spectra are compatible with the weak turbulence theory for MHD outlined in the preceeding section, although as one goes further from the planet, spectra are seen to become more shallow; but one also expects that the turbulence strengthens as it has more time to evolve, and therefore this shallowness further from Jupiter is not incompatible with the theories proposed. Such a model also gives a plausible explanation for an acceleration mechanism for the generation of the main auroral oval on Jupiter [85].

3.5.1.6 Where the Weak Turbulence Approximation Breaks Down

For the weak turbulence regime to be valid, the parameter

$$\epsilon = \frac{\tau_A}{\tau_{NL}}$$

must be small. But this parameter can in fact be evaluated "locally" as a function of the scale of structures and eddies:

$$\epsilon^{\ell} = \frac{\tau_A^{\ell}}{\tau_{NL}^{\ell}} = \frac{k_{\perp} v_{k_{\perp}}}{k_{\parallel 0} B_0} \ll 1$$

with $k_{\parallel 0}$ given (no transfer a priori in the parallel direction). But since $E(k_{\perp}) \sim k_{\perp}^{-\alpha}$ or in terms of the velocity itself, $v_{k_{\perp}} \sim k_{\perp}^{\frac{1-\alpha}{2}}$, we see that for $\alpha \leq 3$, the local parameter $\epsilon(k_{\perp})$ grows with $k_{\perp}$. In other words, there is a non-uniformity (in scale) of the weak turbulence approximation.

Define now a critical wave number $k_{\perp}^C$ for which $\epsilon(k_{\perp}) \sim 1$. This wave number $k_{\perp}^C$ is reached in a finite time because the nonlinear coupling of waves is fast. One can then define a criterion of "critical balance" [37] whereby $\epsilon(k_{\perp}) = 1$; this happens for $k_{\perp} v_{k_{\perp}} = k_{\parallel} B_0$, or

$$k_{\parallel} \sim k_{\perp}^{\frac{3-\alpha}{2}} .$$

Giving α the value for K41 spectrum ($\alpha = 5/3$) leads to the relationship $k_{\parallel} \sim k_{\perp}^{2/3}$ as argued in [37]; similarly, for an IK spectrum ($\alpha = 3/2$), one has $k_{\parallel} \sim k_{\perp}^{3/4}$. Numerical simulations, on a regular grid of 256^3 points, seem to agree with the former evaluation of α, i.e., one corresponding to the Kolmogorov case [37]. However, one should keep in mind that error bars are very large for under-resolved numerical simulations at moderate Reynolds numbers and that the spectra will not obey the simple power law solutions given by dimensional analysis because of intermittency.

3.5.1.7 A Closure Problem

What does one do when the theory of weak turbulence breaks down or when there are no waves? The closure problem remains, and approximate solutions have to be devised.

Let us write in symbolic form a series of equations for the successive moments of a vector field **a** of components a_j and obeying a nonlinear quadratic equation. In that case, an equation for the second-order moment involves third-order moments, and the process can be iterated ad infinitum; thus, there will always be one more unknown than the number of equations. Hence the statistical problem for nonlinear physics is not closed, and one must come up with a supplementary relationship or introduce an extra hypothesis. We just saw that in the case of weak turbulence, the resonant interactions dominate and allow naturally for a closure which is thus exact. But what can we do in the strongly nonlinear case or in a case where there are no waves, such as for the incompressible Navier–Stokes equations? Let us write the beginning of the infinite hierarchy of equations for the successive moments:

$$\partial_t \langle a_j a_{j'} \rangle = \langle a_j a_{j'} a_{j''} \rangle, \tag{3.47}$$

$$\partial_t \langle a_j a_{j'} a_{j''} \rangle = \langle a_j a_{j'} a_{j''} a_{j'''} \rangle. \tag{3.48}$$

The fourth-order moment can be written as follows:

$$\langle a_j a_{j'} a_{j''} a_{j'''} \rangle = \sum \langle a_j a_{j'} \rangle \langle a_{j''} a_{j'''} \rangle + \langle a_j a_{j'} a_{j''} a_{j'''} \rangle_C , \tag{3.49}$$

where $\langle a_j a_{j'} a_{j''} a_{j'''} \rangle_C$ is the cumulant, a term that would be equal to zero in the case of Gaussian statistics, and where the sum is taken over all possible configurations of second-order moments, here three.

For small ϵ, i.e., for small nonlinear interactions, we saw that there is no contribution at lowest order of the fourth-order cumulants, a fact which renders the weak turbulence theory equivalent at lowest order to the random phase approximation (without that hypothesis being made though).

There are a variety of closures that have been written over the years for the case of strong ϵ. For example, in the eddy damped quasi-normal Markovian approximation (or EDQNM), one writes somewhat arbitrarily a proportionality relationship between third- and fourth-order cumulants:

$$\langle a_j a_{j'} a_{j''} a_{j'''} \rangle = -\mu_m \langle a_j a_{j'} a_{j''} \rangle ,$$

where μ_m is a characteristic rate for the nonlinear interactions to have a substantial effect. With the choice $\mu_m = \mu_0$ for all scales, one can show that the resulting energy spectrum is proportional to k^{-2}; this model is called the Markovian random coupling model, or MRCM. Such a choice for μ_m, leading to a constant characteristic rate, corresponds to a shock, a singular structure in which all small scales are formed at once, and for which, as in the case of the Burgers equations, a k^{-2} spectrum is expected.

Taking now $\mu_m = \omega_{int}(k)$, the energy spectrum becomes that of Kolmogorov, namely $E(k) \sim k^{-5/3}$. Indeed, if one writes the nonlinear time as $\tau_{NL} = \ell/u_\ell$ (see Eq. (3.41) for the anisotropic version) and plugs in the Kolmogorov spectrum to evaluate the velocity u_ℓ at scale ℓ, one can easily find that $\tau_{NL} \sim \ell^{2/3}$, so that the choice of decorrelation rate is compatible with the Kolmogorov spectrum. When the system supports waves with characteristic frequency $\omega_w(k)$, the above formula for the characteristic damping rate of triple correlations can be written as $\mu_m = \omega_{int}(k) + \omega_w(k)$; in the case of Alfvén waves with $\omega_w(k) = k B_0$, the energy spectrum becomes the IK spectrum $E(k) \sim k^{-3/2}$.

More sophisticated closures arise when an equation for the temporal evolution of triple-correlation damping rates is used, as with the test field model or TFM [49]. Finally, note that intermittency, or the spatio-temporal scarcity of strong localized structures, is a key factor that is not included in closures in general, and yet its effect is an essential part of our understanding of turbulence.

3.6 Conclusions and Perspectives

From a theoretical perspective, real progress has been made following the pioneer work of Kraichnan on the statistics of a passive scalar such as temperature when advected by a velocity field with prescribed power law spectrum in the large scales, Gaussian in space, and white noise in time [51]. It was shown for the first time that structure functions of high order display a departure from linear scaling given by dimensional analysis (see also the review in [28]), thus giving credence to the observations and numerical simulations finding such a departure in real data but with sizable error bars. This same model was in fact proposed independently by Kazantsev [46] in the context of passive vector of the induction field for a kinematic dynamo, leading theoretically to a $k^{3/2}$ spectrum now observed in several numerical simulations, including at (moderately) low magnetic Prandtl numbers.

3.6.1 On the Numerical Front

Numerical simulations (direct or otherwise) have played an essential role, and will continue to do so, in our seeking an understanding of turbulent flows. There are at least two reasons for that.

On the one hand, numerical simulations allow us to reach parameter values that are not available in the laboratory. The best example is probably that of the magnetic Reynolds number. A sphere of mercury of 1 m in diameter weighs roughly 13 tons; it costs a lot of money, it is dangerous to manipulate and you have to impart a velocity of $1\,\mathrm{m\,s}^{-1}$ in order to reach $R_M \sim 1$. Only in plasma experiments [33] or in fusion plasmas in tokamaks or in the liquid metal of breeder reactors can one reach substantially higher R_M. In the context of dynamo problem, there are several experiments aimed at obtaining higher R_M in the laboratory and this is clearly one exciting development in MHD.

On the other hand, numerical data, when sufficiently well resolved, give credible information at every point of the grid on every component of the basic variables, and on their combinations, e.g., vorticity, helicity, or any complicated functionals based on these fields such as fluxes or filtered data. 3D and 4D space–time visualization is also relatively feasible nowadays at least at moderate resolution. However, one of the main challenges remaining is how to handle (store, compress, or transfer) and analyze the huge data files produced by large simulations. For example, the computation of Navier–Stokes turbulence performed recently on the Earth simulator on a grid of 4096^3 points [45] yields at every time slot that is stored a minimum of 2 TB.

Moreover, for a flow to be well resolved, all scales down to the dissipation scale (or Kolmogorov scale $\sim [\epsilon_V/\nu^3]^{1/4}$ in the fluid case) must be resolved; this implies that in 3D, the number of degrees of freedom that one must a priori carry is proportional to $Re^{9/4}$, although the dynamically relevant number of modes may be much less; for example, using both numerical data and experimental data, it is shown in [26] that keeping only 3% of the wavelet coefficients carrying the most energy, one can reproduce the anomalous wings of PDFs of velocity gradients, whereas the rest of the wavelets behave as Gaussian noise. This opens the possibility of modeling turbulent flows with a few dynamical modes and some sort of stochastic forcing: a very promising venue taken by several teams presently (see, e.g., [52, 29, 56]).

Another way to reach higher Reynolds numbers at a given computing cost is to adapt the grid to the structures: grid points are concentrated where shocks or vorticity and current sheets or vortex filaments form, and the grid coarsens where only large scales are present. This technique is well developed for 1D problems (see e.g. [13]) but is more challenging in higher dimensions. Several issues are not settled yet. For example, what is the criterion for refinement (local strong gradients of relevant fields; or numerical errors piling up if the local flow is unresolved, such as can be done with spectral elements [40, 58, 23] or with wavelets [86])? Can one use different time steps in elements of different sizes? Can one maintain load balance between processors as refinement/coarsening occur on the grid? There are several examples of adaptive computations in turbulence, as, for example, in [38, 39] for 2D and 3D MHD turbulence in the non-dissipative case. Obviously, the study of structures such as solar eruptions, coronal mass ejections or magnetic flux tubes, or of the process of star formation and chemistry of molecular clouds in the interstellar medium, as well as the propagation of jets or in cosmology, all are candidates for further progress using adaptive technology.

Finally, numerical modeling (as opposed to physical modeling as discussed previously) can also be implemented. The simplest version of that approach is to integrate numerically the Euler (inviscid) equations with a finite difference method and let the method do its own dissipation; the obvious drawback is that the code itself sets the level of dissipation needed in a given computation, or in other words one does not control the Reynolds number; however, if a finite amount of dissipation occurs in the limit of infinite Reynolds number, through nonlinear coupling, then when the code is sufficiently resolved, results should match with a direct numerical simulation (see, e.g., [79] for the supersonic case or [57] for atmospheric flows).

As a final point, we note that, given the sheer size of a large-scale computation of turbulent flows in astrophysics, a way of the future is to create databases to be shared using the Access Grid technology and information technology tools that allow one to work at a distance, similar to what is being done with remote sensing data, although such an idea and mode of functioning is still in its infancy for turbulence simulations.

3.6.2 Future Space Explorations and Laboratory Experiments

As we said before, measurements in the interplanetary space represent the unique possibility to investigate a wide range of scales of low-frequency turbulence in a magnetized medium. Spacecrafts are probes that investigate in situ the properties of turbulence. The interested readers are strongly encouraged to visit the web pages of each specific space mission to access the complete information about the up-to-date scientific rationale and data sets. As far as laboratory experiments are concerned the interested readers can visit the ocean of web pages on fusion devices.

Among others, the old Helios and Voyager spacecrafts explored, respectively, the inner and outer heliosphere (*http://voyager.jpl.nasa.gov/*), with an almost complete map of the gross features of low-frequency turbulent behavior in plasmas. Currently there are some spacecrafts from which data are currently available. The Ulysses mission (*http://helio.estec.esa.nl/ulysses/*) is investigating heliospheric turbulence out of the ecliptic plane. In particular Ulysses explored the polar zones of the Sun, where high-stream solar winds are born. The results are excellent, so far as turbulence in a nice Alfvénic state is concerned. Because direct injection into a solar polar orbit from the Earth is not feasible, a gravity assist was required to achieve a high-inclination orbit. As a result, Ulysses was launched at high speed toward Jupiter in October 1990, after being carried into low-Earth orbit by the space shuttle Discovery.

Following the fly-by of Jupiter in February 1992, the spacecraft is now traveling in an elliptical, Sun-centered orbit inclined at 80.2° to the solar equator. In the normal operating mode, the scientific data acquired by the Ulysses instruments are stored by a tape recorder on board the spacecraft for approximately 16 h and downlinked to the NASA Deep Space Network once a day together with the real-time data during a nominal 8-h tracking pass. The coverage to date has been excellent, being

$\sim$ 97% on average over the mission. This database represents the most complete set of continuous interplanetary measurements ever recorded. Further details regarding the spacecraft and its scientific investigations can be found in Ref. [103].

A further spacecraft called WIND (*http://www-istp.gsfc.nasa.gov/istp/wind/wind.html*) was launched on November 1, 1994 and is the first of two NASA spacecrafts in the Global Geospace Science initiative and part of the ISTP Project. WIND was positioned in a sunward, multiple double-lunar swingby orbit with a maximum apogee of 250 Earth radii during the first 2 years of operation. This was then followed by a halo orbit at the Earth–Sun L1 point. The high resolution of measurements of turbulence of WIND makes this satellite very useful when small scales must be investigated, where kinetic effects start to play a key role. A different mission, the CLUSTER spacecraft (*http://www.plasma.kth.se/Cluster/*), was launched in 2000. The mission is formed by four identical spacecrafts, and is very interesting as far as spatial properties of the basic fields are concerned. However, the mission flies well within the magnetosphere for almost all times, and only a small amount of time is allowed for solar wind investigations. Moreover the distance between the four spacecrafts has often been small, so that we could investigate turbulence at scales well below that where MHD turbulence is at work.

The above missions have explored much of the available heliosphere. However, two further spacecrafts are planned in the future to explore regions very near the onset of the solar wind. In particular the European Space Agency plans to launch the Solar Orbiter mission (*http://www.orbiter.rl.ac.uk/*) in the next decade (2015). The scientific payload includes remote sensing instruments for EUV/X-ray spectroscopy and imaging of the disk and the extended corona. This will offer us the possibility to investigate, in a natural plasma, features related to spatio-temporal behavior of turbulence. The unique features of Solar Orbiter in the context of the study of turbulence dynamics in natural plasmas are the proximity to the Sun (about 0.21 AU), the high-latitude remote sensing (orbital plane inclined up to 38° with respect to the solar equator), and the co-rotation. The first characteristic will give close-up observations of the solar atmosphere, with the capability of resolving fine structures (below 100 km) in the transition region and corona, representing a major step forward in investigating the occurrence of intermittent structures of turbulence at very small scales. The second will allow one to study the structure and the dynamics of magnetic turbulence near the polar regions, also extending the investigation to higher heliocentric distances through the coordinated use of a UV coronograph. The co-rotation will help to remove the effect of the Sun's rotation, offering the opportunity to study the evolution of spatio-temporal turbulence on longer timescales, and to relate the corresponding heliospheric effects (evolution of turbulence in the solar wind) to their source dynamic events in the lower corona. Another mission, the Solar Probe (*http://solarprobe.gsfc.nasa.gov/*), has been developed by NASA. The mission is planned to fly in the more distant future and to reach a region at about 20 solar radii (0.04 AU), near the sonic point of the solar wind.

A different role concerns experiments in laboratory plasmas. Of course the politics of big experiments is to reach the best performances in terms of temperature and confinement time of the plasma, not to investigate turbulence. However, turbulent fluctuations represent, directly or in an indirect way through anomalous transport [104, 1], the main cause for the disruption of the magnetic confinement [104, 2]. Research groups in laboratory plasmas have started to investigate turbulent properties of the plasma state (e.g., Ref. [2] and references therein). In reversed field pinch configurations we can measure the most intense turbulent fluctuations, but also Tokamak configurations display some low-amplitude fluctuations and some kind of measurements are currently available. Turbulence in plasmas can be perhaps more easily investigated from experiments on small devices. Some universities around the world, and some few research laboratories, have at disposal facilities where data on magnetic fluctuations, or other fluctuating quantities, can be available. However, up to now, a laboratory device explicitly devoted to investigate turbulence properties of plasma does not exist. A similar device should reach a high turbulent state, should be feasible enough to have the possibility to insert probes everywhere, and should be politically accessible to research groups around the world.

Acknowledgments Our special thanks go to Antonio Celani with whom we concocted this review at an early stage. We acknowledge valuable discussions with Vanni Antoni, Roberto Bruno, Sébastien Galtier, Paolo Giuliani, Fabio Lepreti, Pablo Mininni, Giusy Nigro, Jean-François Pinton, Hélène Politano, Yannick Ponty, Luca Sorriso-Valvo, Antonio Vecchio, and Pierluigi Veltri.

Appendix A: Data Analysis of Real Turbulent Field

In this appendix we give some basic concepts of data analysis of some turbulent fields. In general measurements result in some time series of fields. For example, a hot wire placed at a certain height inside the turbulent boundary of the atmosphere gives the vector speed $\mathbf{v}(t)$, $u(t)$ being the streamwise component, and the temperature field $T(t)$. In situ satellite measurements of both velocity and magnetic fields $\mathbf{B}(t)$ can be obtained in interplanetary space, while a turbulent magnetic field can also be measured at the edge of plasma devices in the laboratory, mainly on toroidal reversed field pinch configurations where high-amplitude magnetic fluctuations can be detected. Quantities of interest are the differences at a certain timescale r of the streamwise velocity vector, namely $\delta u_r = u(t + r) - u(t)$, and of the temperature $\delta T_r = T(t + r) - T(t)$. The quantities δu_r are analogous to spatial differences, using the Taylor's hypothesis [92]. As far as the magnetic field is concerned, we use the magnitude of the field, namely $b(t) = \left[\sum_i B_i^2(t) \right]^{1/2}$, thus computing the differences $\delta b_r = b(t + r) - b(t)$. However, we can compute also differences for a single component of the magnetic field, or for the vector field itself [19].

Scaling exponents are obtained from the pth-order structure functions $S_r^{(p)} = \langle \delta u_r^p \rangle$ in the inertial range. According to K41 in fluid flows the inertial range is formally defined where $S_r^{(3)}$ behaves like the separation r. In general the range where this is true is very narrow. In MHD flows the inertial range must be defined as the range of scales where (5.4) or (5.5) are defined [70]. Unfortunately in the solar wind turbulence neither the K41 choice nor the MHD result is satisfied in a significant range of scales r (R. Bruno, private communication). In order to recover scaling laws we are thus led to use the extended self-similarity (ESS), a data analysis technique introduced by Benzi et al. [6]. Using the third-order structure function as a generalized scale, we can plot the pth-order structure functions against the third-order one. In this case the range of scales where a linear relation $S_r^{(p)} \sim [S_r^{(3)}]^{\xi_p/\xi_3}$ exists is extended thus allowing a better determination of the normalized scaling exponents [16]. It is worthwhile noting the fact that scaling exponents obtained in fluid flows using ESS are the same as scaling exponents obtained in the same flow without ESS [6].

From data analysis the structure functions are calculated simply using time averaging, but a word of caution must be given concerning the values of the order p which can be used in the calculation. In fact the more formal definition

$$ S_r^{(p)} = \int_{-\infty}^{\infty} \delta u_r^p P(\delta u_r) d\delta u_r $$

clearly indicates that, in order to get a value $S_r^{(p)}$ which is meaningful, we have to make sure that the function $F_p(\delta u_r) = \delta u_r^p P(\delta u_r)$ be integrable. In general this defines an upper bound on the maximum value of p we can reliably calculate with a certain number of points N in our data set. As a rule of thumb, when we have a data set of N points at our disposal, the maximum allowed p_{max} is of the order of $p_{max} \simeq \log N$ (for stretched-exponential PDFs).

Typical scaling exponents of structure functions for both velocity and magnetic field in the heliosphere are shown in Table 3.2. Data were obtained from the Helios 2 spacecraft, sampling low-speed streams. In the same table, for comparison, we report the scaling exponents for velocity and temperature as obtained in a wind tunnel experiment. We show normalized scaling exponents ξ_p/ξ_3 calculated through the extended self-similarity [6]).

Scaling exponents of magnetic structure functions, obtained from laboratory plasma experiments of a reversed-field pinch at different distances from the external wall [18], are shown in Table 3.3. In laboratory plasmas it is difficult to measure all components of the vector field at the same time. Here we show scaling exponents obtained using differences as calculated from the radial component of the magnetic

Table 3.2 Normalized scaling exponents ξ_p/ξ_3 for velocity and magnetic variables in the solar wind. Errors represent the standard deviations of the linear fitting. As a reference we show the scaling exponents of structure functions for velocity and temperature, as calculated in a wind tunnel

p	$u(t)$ (solar wind)	$B(t)$ (solar wind)	$v(t)$ (fluid)	$T(t)$ (fluid)
1	0.36 ± 0.06	0.56 ± 0.06	0.37 ± 0.01	0.46 ± 0.02
2	0.70 ± 0.05	0.83 ± 0.05	0.70 ± 0.01	0.78 ± 0.01
3	1.00	1.00	1.00	1.00
4	1.28 ± 0.02	1.14 ± 0.02	1.28 ± 0.02	1.18 ± 0.02
5	1.53 ± 0.03	1.25 ± 0.03	1.54 ± 0.03	1.31 ± 0.03
6	1.79 ± 0.05	1.35 ± 0.05	1.78 ± 0.05	1.40 ± 0.03

Table 3.3 Normalized scaling exponents ξ_p/ξ_3 for magnetic fluctuations in a laboratory plasma, as measured at different distances r/R ($R \simeq 0.45$ cm being the minor radius of the torus in the experiment) from the external wall. Errors represent the standard deviations of the linear fitting

p	$r/R = 0.96$	$r/R = 0.93$	$r/R = 0.90$	$r/R = 0.86$
1	0.39 ± 0.01	0.38 ± 0.01	0.37 ± 0.01	0.36 ± 0.01
2	0.74 ± 0.01	0.73 ± 0.02	0.71 ± 0.01	0.70 ± 0.01
3	1.00	1.00	1.00	1.00
4	1.20 ± 0.02	1.24 ± 0.02	1.27 ± 0.01	1.28 ± 0.01
5	1.32 ± 0.03	1.41 ± 0.03	1.51 ± 0.03	1.55 ± 0.03
6	1.38 ± 0.04	1.50 ± 0.04	1.71 ± 0.03	1.78 ± 0.04

Table 3.4 Normalized scaling exponents ξ_p/ξ_3 for Alfvénic, velocity, and magnetic fluctuations obtained from data of high-resolution 2D MHD numerical simulations. Scaling exponents have been calculated from spatial fluctuations; different times, in the statistically stationary state, have been used to improve statistics

p	Z^+	Z^-	v	B
1	0.36 ± 0.06	0.56 ± 0.06	0.37 ± 0.01	0.46 ± 0.02
2	0.70 ± 0.05	0.83 ± 0.05	0.70 ± 0.01	0.78 ± 0.01
3	1.00	1.00	1.00	1.00
4	1.28 ± 0.02	1.14 ± 0.02	1.28 ± 0.02	1.18 ± 0.02
5	1.53 ± 0.03	1.25 ± 0.03	1.54 ± 0.03	1.31 ± 0.03
6	1.79 ± 0.05	1.35 ± 0.05	1.78 ± 0.05	1.40 ± 0.03

field in the toroidal device. As it can be seen, the degree of intermittency increases going toward the external wall. This appears to be similar to what is observed in channel flows where intermittency increases going toward the external wall [74].

Scaling exponents of structure functions for Alfvén variables, velocity, and magnetic variables have been calculated also for high-resolution 2D MHD numerical simulations [70]. These scaling exponents are shown in Table 3.4. Note that, even in numerical simulations, the magnetic field behaves like a passive field; intermittency for magnetic variables is stronger than for the velocity field, even if the difference between scaling exponents for velocity and magnetic fluctuations in numerical simulations is smaller than the differences we found in the solar wind observations.

Appendix B: Transition to Chaos in Low-Order Galerkin Approximations

Low-order approximations of the dynamical equations in Fourier space have been used as a dynamical model to investigate the transition to chaos. This originates with E. N. Lorenz [54] who showed that only three equations, from a severe truncation of equations of motion for convective flows, are enough to get *complex solutions*. A class of dynamical models of the same kind can be obtained from Eqs. (3.38), by retaining only a finite number of modes. Since the basic ingredients of the chaotic machinery are the occurrence of quadratic nonlinearities, as an example we present here a few results from a fluid model where the magnetic variables are set to zero.

Adding magnetic variables does not change the basic quadratic nonlinearities. The dynamical variables of the model are then the Fourier coefficients $u_i(t) = u(\mathbf{k}_i, t)$ of (3.38). By retaining only nine triads of wave vectors, setting $\nu = 1$, using only real dynamical variables, and adding a constant forcing term f to the variable $u_3(t)$ to restore the fields, we obtain the following system of ordinary differential equations:

$$\dot{u}_1 + 2u_1 = 4u_2u_3 + 4u_4u_5,$$
$$\dot{u}_2 + 9u_2 = 3u_1u_3 + 9u_8u_5 + 3u_7u_9,$$
$$\dot{u}_3 + 5u_3 = -7u_2u_1 + (9/\sqrt{5})u_1u_7 - 5u_8u_4 + f,$$
$$\dot{u}_4 + 5u_4 = -u_5u_1 + 5u_8u_3 + 7u_9u_6,$$
$$\dot{u}_5 + u_5 = -3u_1u_4 + \sqrt{5}u_1u_6 - u_8u_2,$$
$$\dot{u}_6 + u_6 = -\sqrt{5}u_1u_5 - 3u_9u_4,$$
$$\dot{u}_7 + 5u_7 = -(9/\sqrt{5})u_1u_3 + u_9u_2,$$
$$\dot{u}_8 + 10u_8 = -8u_2u_5,$$
$$\dot{u}_9 + 8u_9 = -4u_7u_2 - 4u_4u_6. \tag{3.50}$$

This set of equations can be easily solved numerically, thus obtaining the time behavior and trajectories in phase space. As f is increased steadily, the trajectory behavior changes as follows:

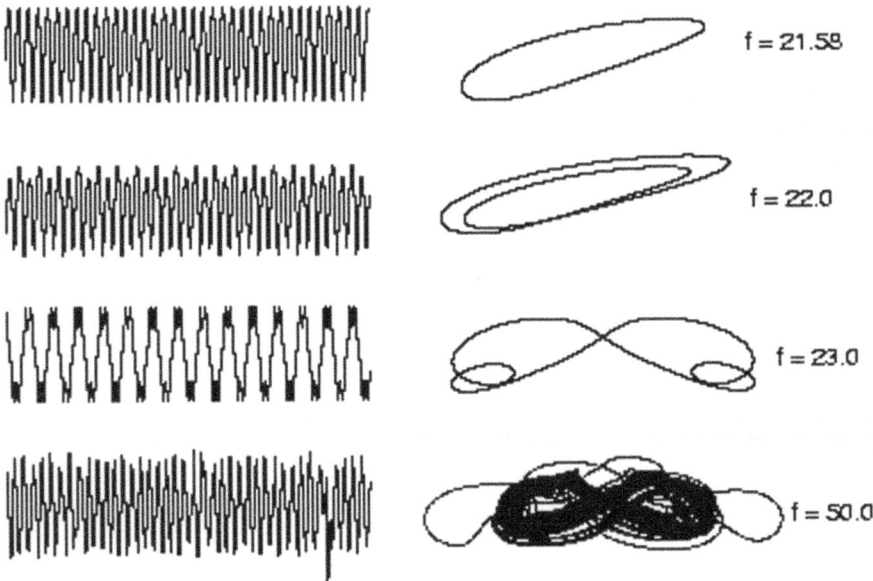

Fig. 3.13 Time evolution of the variable u_1 (*left-hand panels*) and the attractor projected on the plane (u_1, u_3) (*right-hand panels*), for four different values of the external forcing

(1) A single component equilibrium is seen for $f < 5\sqrt{3/2}$, namely the fixed point $u_i = 0$ but $u_3 = f/5$.
(2) A multi-component equilibrium state is observed for $5\sqrt{3/2} < f < f_H = 21.57$. In this state the fixed point emerges with $u_i \neq 0$, depending on the value of f. At $f = f_H$ a Hopf bifurcation gives rise to the birth of a periodic behavior.
(3) For f just above f_H a sequence of periodic/aperiodic states is visible, up to $f \simeq 35$, where a chaotic state settles. The chaotic state is not stable, in the sense that, when f varies, periodic behavior can clearly still be seen even for high f. In Fig. 3.13 we show a sequence of four different states where the sequence periodic $\rightarrow$ aperiodic $\rightarrow$ chaotic behavior is visible. On the left-hand panels we report time series of the variable $u_1(t)$, while on the right-hand panels we report the projection of the system on the plane (u_1, u_3).

Appendix C: About the Word *Intermittency* Encountered in Literature

It is perhaps interesting to write a few words about the different meanings of the word *intermittency* which can be usually encountered in the literature. In fact, the same word usually refers to some different (and often contrasting) kinds of phenomena, and confusion may arise. In a schematic way

- *Intermittency in fully developed turbulence* is what we have just described in the previous sections, that is, the departure from a global self-similarity in fluid or magnetofluid systems (cf. Ref. [27] for references).
- *Intermittent transition to chaos.* The time evolution of some chaotic systems shows alternation between laminar periods and stochastic periods. This happens when the tuning parameter is set close enough to the critical value which define the order-to-chaos transition. As the parameter becomes closer to that critical value, the durations of laminar periods decrease as a power law [72], with universal behavior recognized in some systems.
- *Intermittency in self-organized systems.* Some out-of-equilibrium systems relax through bursts showing typical $1/f^\alpha$ spectra. These systems can be simulated by cellular automata, and from a theoretical point of view they can be seen as self-similar systems at the borderline of chaos in a self-organized critical state. These systems, when continuously perturbed, show a typical sequence of intermittent isolated events. The classical example is the sandpile model [4], where the critical state is represented by a critical slope of the pile. As a perturbation is introduced, in the form of random addition of sand at random sites, and the local slope of the pile exceeds a critical slope, the system relaxes through a burst of activity, that is, an *avalanche*. The ensemble of successive avalanches generates an intermittent sequence of chaotic bursts [4]. Since the critical slope is an attractor, avalanches are Poissonian events, the waiting times between avalanches being distributed according to an exponential function [10].

– *On–off intermittency.* This kind of intermittency is observed in chaotic systems externally driven through, for example, stochastic perturbations. The system, which has been set to lie normally in a laminar state (an attractor), occasionally is pushed away from this state, thus generating intermittent turbulence bursts. The waiting times between successive bursts are distributed as power laws [67], with universal scaling exponents. For a nice and very simple example try the behavior of the logistic map $x_{n+1} = a_n x_n (1 - x_n)$ with variable parameter as, for example, $a_n = 1/2$ with probability p (say $p = 0.34$) and $a_n = 4$ with probability $1 - p$.

References

1. V. Antoni, et al.: Phys. Rev. Lett., **87**, 045001 (2001)
2. V. Antoni, et al.: Contrib. Plasma Phys., **44**, 469 (2004)
3. A. Aringazin, M. Mazhitov: Phys. Rev. E, **69**, 026305 (2004)
4. P. Bak, C. Tang, K. Wiesenfeld: Phys. Rev. Lett., **59**, 381 (1987)
5. D. Benney, A. Newell: Stud. Appl. Math., **48**, 29 (1969)
6. R. Benzi et al.: Phys. Rev. E, **48**, R29 (1993)
7. G. Bertin, D. Farina, R. Pozzoli eds.: *Plasmas in the Laboratory and in the Universe: New insights and new challenges*, AIP Conference Proceedings 703 (2004)
8. L. Biferale: Ann. Rev. Fluid Mech., **35**, 441 (2003)
9. D. Biskamp: *Nonlinear Magnetohydrodynamic.* Cambridge University Press, Cambridge, U.K. (1993)
10. G. Boffetta, V. Carbone, P, Giuliani, P. Veltri, A. Vulpiani: Phys. Rev. Lett., **83**, 4662 (1999)
11. T. Bohr, M.H. Jensen, G. Paladin, A. Vulpiani: *Dynamical system approach to turbulence.* Cambridge University Press, Cambridge, U.K. (1998)
12. A. Brandenburg, G.R. Sarson Phys. Rev. Lett., **88**, 055003 (2002)
13. W. Cao, W. Huang, R. Russell: SIAM J. Sci. Comput., **20**, 1978 (1999)
14. V. Carbone: Phys. Rev. Lett., **71**, 1546 (1993)
15. V. Carbone, P. Veltri: Astron Astrophys., **259**, 359 (1992)
16. V. Carbone, R. Bruno, P. Veltri: Geophys. Res. Lett., **23**, 121 (1996)
17. V. Carbone, P. Veltri, R. Bruno: Phys. Rev. Lett., **75**, 3110 (1995)
18. V. Carbone, et al.: Phys. Rev., E **62**, R49 (2000)
19. V. Carbone, R. Bruno, L. Sorriso–Valvo, F. Lepreti: Planet. Space Sci., **52**, 953 (2004)
20. B. Castaing, Y. Gagne, E.J. Hopfinger: Physica, **46D**, 177 (1990)
21. S.Y. Chen et al: Physica D, **133**, 66 (1999)
22. T. DelSole: Surveys in Geophys., **24**, 107 (2004)
23. M. Deville, P. Fischer, E. Mund: *High-order Methods for Incompressible Fluid Flows.* Cambridge Monographs on Applied and Computational Mathematics, Cambridge University Press (2002)
24. M. Dobrowolny, A. Mangeney, P. Veltri: Phys. Rev. Lett., **45**, 144 (1980)
25. W. Elsässer: Rev. Mod. Phys., **28**, 135 (1956)
26. M. Farge, K. Schneider, N. Kevlahan: Phys. Fluids, **11**, 2187 (1999)
27. U. Frisch: *Turbulence: The legacy of A.N. Kolmogorov.* Cambridge University Press (1995)
28. G. Falkovich, K. Gawedski, M. Vergassola: Rev. Mod. Phys., **73**, 913 (2001)
29. B. Farrell, P. Ioannou: Phys. Fluids, **A 5**, 2600 (1993)
30. P. Frick, D.D. Sokoloff: Phys. Rev. E, **57**, 4155 (1998)
31. S. Galtier, T. Gomez, H. Politano, A. Pouquet: Advances in Turbulence VII, pp. 453–456. In: U. Frisch (ed.) *Fluid Mechanics and its applications*, Kluwer USA (1998)
32. S. Galtier, S. Nazarenko, A. Newell and A. Pouquet: J. Plasma Phys., **63**, 447 (2000)
33. W. Gekelman, R. Stenzel: J. Geophys. Res., **89**, 2715 (1984)

34. P. Giuliani, V. Carbone: Europhys. Lett., **43**, 527 (1998)
35. P. Giuliani: Shell models for magnetohydrodynamic turbulence. In: T. Passot, P.L. Sulem (eds.) *Nonlinear MHD waves and Turbulence* Lect. Notes Phys. Springer, Berlin Heidelberg New York (1999)
36. E.B. Gledzer: Sov. Phys. Dokl., **18**, 154 (1985)
37. P. Goldreich, S. Sridhar: Astrophys. J., **485**, 680 (1997)
38. R. Grauer, C. Marliani, K. Germaschewski: Phys. Rev. Lett., **80**, 4177 (1998)
39. R. Grauer, C. Marliani: Phys. Rev. Lett., **84**, 4850 (2000)
40. R. Henderson: Comput. Methods Appl. Mech. Engrg., **175**, 395 (1999)
41. J. Heyvaerts, E. Priest: Astron. Astrophys., **137**, 63 (1984)
42. D. Holm: Physica D, **170**, 252 (2002)
43. T. Hughes et al: Phys. Fluids, **13**, 505 (2001)
44. P. Iroshnikov: Sov. Astron., **7**, 566 (1963)
45. T. Isihara, Y. Kaneda, M. Yokokawa, K. Itakura, A. Uno: J. Phys. Soc. Japan, **72**, 983 (2003)
46. A. Kazantsev: Sov. Phys. JETP, **26**, 1031 (1968)
47. M. Kivelson, K. Khurana, C. Russell, R. Walker: Geophys. Res. Lett., **24**, 2127 (1997)
48. A.N. Kolmogorov: J. Fluid Mech., **5**, 497 (1941)
49. R. Kraichnan: J. Fluid Mech., **5**, 497 (1959)
50. R. Kraichnan: J. Fluid Mech., **62**, 305 (1974)
51. R. Kraichnan: Phys. Fluids, **11**, 945 (1968)
52. J-P. Laval, B. Dubrulle, S. Nazarenko: Physica D, **142**, 231 (2000)
53. J-P. Laval, B. Dubrulle, J.C. McWilliams: Phys. Fluids, **15**, 1327 (2003)
54. E.N. Lorenz: J. Atm. Phys., **18**, 154 (1965)
55. M.–M. Mac Low, R.S. Klessen: Revs. Mod. Phys., **76**, 125 (2004)
56. A.J. Majda, I. Timofeyev, E. Vanden Eijnden: Physica D, **170**, 206 (2002)
57. L. Margolin, P. Smolarkiewicz, Z. Sorbjan: Physica D, **133**, 390 (1999)
58. C. Mavriplis: Comput. Methods Appl. Mech, Engrg., **116**, 77 (1994)
59. C. Meneveau, T. Lund: Phys. Fluids, **9**, 3932 (1997)
60. C. Meneveau, J. Katz: Annu. Rev. Fluid Mech., **32**, 1 (2000)
61. D. Montgomery, A. Pouquet: Phys.Fluids, **14**, 3365 (2002)
62. W. Müller, D. Carati: Phys. Plasmas, **9**, 824 (2002)
63. A. Newell, S. Nazarenko, L. Biven: Physica D, **152–153**, 520 (2001)
64. K. Ohkitani, M. Yamada: Prog. Theor. Phys., **89**, 329 (1989)
65. F. Okkels: The intermittent dynamics in turbulent shell models, PhD Thesis, unpublished (1997)
66. L. Onsager: Supplemento al vol. VI, Serie IX del Nuovo-Cimento, **2**, 279 (1949)
67. N.A. Platt, E.A. Spiegel, C. Tressar: Phys. Rev. Lett., **70**, 279 (1993)
68. H. Politano, A. Pouquet: Phys. Rev. E Rapid Comm., **57**, R21 (1998)
69. H. Politano, A. Pouquet: J. Geophys. Lett., **25**, 273 (1998)
70. H. Politano, A. Pouquet, V. Carbone: Europhys. Lett., **43**, 516 (1998)
71. H. Politano, T. Gomez, A. Pouquet: Phys. Rev., E **68**, 026315 1 (2003)
72. Y. Pomeau, P. Manneville: Comm. Math. Phys., **74**, 189 (1980)
73. Y. Ponty, J.F. Pinton, H. Politano: Phys. Rev. Lett., **92**, 144503 (2004)
74. S.B. Pope: *Turbulent Flows*. Cambridge University Press, New York (2001)
75. S.B. Pope: New J. of Phys., **6**, 35 (2004)
76. A. Pouquet: Magnetohydrodynamic Turbulence. In: J.P. Zahn and J. Zinn–Justin Les Houches Summer School on Astrophysical Fluid Dynamics, Session **XLVII**. Elsevier, New York 139; (1993)
77. A. Pouquet: Turbulence, Statistics and Structures: an Introduction, V[th] European School in Astrophysics, San Miniato. In: C. Chiuderi & G. Einaudi (eds.) Springer–Verlag, Lect. Notes Phys. "Plasma Astrophysics" **468**, 163 (1996)
78. A. Pouquet, U. Frisch, J. Léorat: J. FLuid Mech., **77**, 321 (1976)
79. D. Porter, A. Pouquet, P. Woodward: Phys. Rev. E, **66**, 026301 (2002)

80. L.F. Richardson: *Wheater Prediction by Numerical Process*. Cambridge University Press, New York (1922)
81. M.N. Rosenblut, D.A. Monticello, H.R. Strauss: Phys. Fluids, **19**, 1987 (1976)
82. G. Ruiz Chavarria, C. Baudet, S. Ciliberto: Europhys. Lett., **32**, 319 (1995)
83. P. Sagaut: *Large Eddy Simulation for Incompressible Flows*. Springer-Verlag, Berlin (2001)
84. J. Saur, H. Politano, A. Pouquet, W. Matthaeus: A & A, **384(2)**, 699 (2002)
85. J. Saur, A. Pouquet, W. Matthaeus: Geophys. Res. Lett., **30(5)**, 1260 (2003)
86. K. Schneider, M. Farge: C.R. Acad. Sci. Paris, **325**, Ser. II 263 (1997)
87. J. Smagorinsky: Month. Weather Rev., **91**, 99 (1963)
88. L. Sorriso–Valvo, V. Carbone, P. Veltri, H. Politano, A. Pouquet: Europhys. Lett., **51**, 520 (2000)
89. L. Sorriso–Valvo, V. Carbone, P. Veltri, G. Consolini, R. Bruno: Geophys. Res. Lett., **26**, 1801 (1999)
90. M. Stix: *The Sun: An Introduction*. Springer–Verlag (1991)
91. P.P. Sullivan, T.W. Horst, D.H. Lenschow, C-H. Moeng, J.C. Weil: J. Fluid Mech., **482**, 101 (2003)
92. G.I. Taylor: Proc. R. Soc. A, **164**, 476 (1938)
93. J.B. Taylor: Phys. Rev. Lett., **33**, 1139 (1974)
94. M. Theoblad, P. Fox, S. Sofia: Phys. Plasmas, **1**, 3016 (1994)
95. J.H. Thomas: Phys. Fluids, **11**, 1245 (1968)
96. A.C. Ting, W.H. Matthaeus, D. Montgomery: Phys. Fluids, **29**, 3261 (1986)
97. C.-Y. Tu, E. Marsch: Space Sci. Rev., **73**, 1 (1995)
98. L. Woltjer: Rev. Mod. Phys., **32**, 914 (1960)
99. W. Woyczynski: Birkhauser, Berlin, Chap. 15, pp. 279–311 (1995)
100. A. Yaglom: Dokl. Akad. Nauk SSSR, **69**, 743 (1949)
101. A. Yoshizawa: Phys. Fluids, **B2**, 1589 (1990)
102. V. Zakharov, V. Lvov, G. Falkovich: *Kolmogorov spectra of Turbulence, I*. Springer, Berlin (1992)
103. K.-P. Wenzel, R.G. Marsden, D.E. Page, E.J. Smith: Astron. Astrophys. Suppl. Ser., **92**, 207 (1992).
104. R.B. White: *The Theory of Toroidally Confined Plasmas*. Imperial College Press, London (2001)

Chapter 4
The Solar Atmosphere

V.H. Hansteen and M. Carlsson

4.1 Introduction

Looking at the solar photosphere, we see the top of the convection zone in the form of granulation: hot gas rising from the solar interior as part of the energy transport process reaches a position where the opacity is no longer sufficient to prevent the escape of radiation. The gas expands, radiates, and cools and in so doing loses its buoyancy and descends. These motions, ultimately driven by the requirement that the energy generated by nuclear fusion in the Sun's core be transported in the most efficient manner, represent a vast reservoir of "mechanical" energy flux.

Looking closer, we see that granulation is not the only phenomenon visible at the solar surface. The quiet and semi-quiet photosphere is also threaded by magnetic fields that appear as bright points, as well as darker micropores and pores. These small-scale magnetic structures are, while able to modify photospheric emission, subject to granular flows and seem to be passively carried by the convective motions.

Convective flows are also known to generate the perturbations that drive solar oscillations. Oscillations, sound waves, with frequencies mainly in the band centered roughly at 3 mHz or 5 min are omnipresent in the solar photosphere and are collectively known as p-modes ("p" for pressure). These p-modes are a subject in their own right and studies of their properties have given solar physicists a unique tool in gathering information on solar structure—the variation of the speed of sound c_s, the rotation rate, and other important quantities—at depths far below those accessible through direct observations. In this chapter we will consider them only insofar as they interact and possibly channel energy into the layers above the photosphere.

The shuffling, buffeting, and braiding of magnetic structures that presumably continues on up into the upper solar layers, the propagation of the higher frequency

V.H. Hansteen (✉)
Institute of Theoretical Astrophysics, University of Oslo, PB 1029 Blindern, 0315 Oslo, Norway,
viggo.hansteen@astro.uio.no

M. Carlsson
Institute of Theoretical Astrophysics, University of Oslo, PB 1029 Blindern, 0315 Oslo, Norway,
mats.carlsson@astro.uio.no

Hansteen, V.H., Carlsson, M.: *The Solar Atmosphere*. Lect. Notes Phys. **778**, 129–155 (2009)
DOI 10.1007/978-3-642-00210-6_4   © Springer-Verlag Berlin Heidelberg 2009

component[1] photospheric oscillations through the chromosphere and into the corona—all may contribute to heating and thus the production of a 1 MK or hotter corona. But in what proportion? And by how much? And what are the observational signatures of the various possible heating sources?

The structure and the topics of this chapter are summarized in Fig. 4.1. The reader is also referred to Chap. 2 (Chiuderi and Velli) for a brief synopsis of the key scientific questions in the solar atmosphere, as well as for a discussion of the equations of magnetohydrodynamics that govern the behavior of the solar plasma. Chap. 2 also addresses the coronal heating problem in an introductory way. Chaps. 6 and 7 should be consulted for a more detailed discussion of how the corona extends into interplanetary space as the solar wind and Chaps. 5 and 8 address activity such as flares in solar and stellar coronae.

We will consider a photosphere in continual motion, threaded by magnetic fields and subject to 5 min. oscillations. Currently, e.g., with the Swedish Solar Telescope on La Palma, and with the JAXA Hinode satellite, we have access to visible wavelength observations of the photosphere and its attendant fields with a spatial resolution of roughly 0.1–0.2 arcsec, equivalent to 75–150 km on the solar surface. Careful analysis of this wealth of observational data on the "lower boundary" of the corona should eventually yield insight into the mechanism(s) heating the corona and chromosphere.

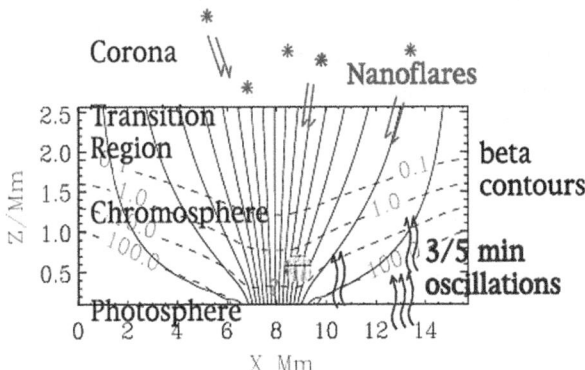

Fig. 4.1 Schematic view of the structure of the solar atmosphere: In the photosphere the gas pressure is larger than the magnetic pressure and photospheric motions are driven by convection—a mainly hydrodynamic phenomenon that brings up new magnetic flux, buffets, and reorganizes existing field. Oscillations of somewhat higher frequency than the photospherically dominant 5 min. oscillations propagate into the chromosphere. Upon reaching the level where the exponentially decreasing gas pressure becomes equal to the magnetic pressure, the surviving sound waves may be converted into other wave modes. It is likely that the processes heating the corona and magnetic chromosphere are episodic; this will likely induce large temperature differences, flows, and other non-steady phenomena that will produce other wave modes, perhaps with observable signatures

---

[1] Waves with frequencies lower than the acoustic cutoff frequency of roughly 5 mHz are evanescent above the photosphere and do not propagate energy into the higher solar atmosphere (unless the magnetic field topology can modify the frequency at which waves become evanescent).

On average the photospheric gas pressure is $p_g = 10^4$ N/m^2 which is much greater than the pressure associated with the average unsigned magnetic field strength of 1–10 Gauss observed. However, in the largely isothermal chromosphere the gas pressure falls exponentially, with a scale height of some 200 km, while the magnetic field strength falls off much less rapidly, even as the field expands and fills all space. Thus, depending on the actual magnetic field topology, the magnetic pressure and energy density will surpass the gas pressure some 1500 km or so above the photosphere in the mid chromosphere. Another 1000 km, or 5 scale heights above the level where $\beta = 1$ (β is defined as the ratio of the gas to magnetic pressures), the plasma's ability to radiate becomes progressively worse, while the dominance of the magnetic field becomes steadily greater. Any given heat input cannot be radiated away and will invariably raise the gas temperature to 1 MK or greater; a corona is formed. A corona that is bound to follow the evolution of magnetic field as the field in turn is bound to photospheric driving.

4.2 The Photosphere

Solar convective motions continuously churn the outer solar atmosphere. Hot, high entropy gas is brought up to the photosphere where the excess energy is radiated away. Cool low entropy material descends into the depths in steadily narrower lanes and plumes as explained in the numerical models of Stein and Nordlund [29]. The result of these motions is, of course, the granular pattern observed on the solar surface and described in countless papers and text-books.

Recounting this wealth of knowledge falls far outside the scope of this chapter. We will rather describe a limited set of recent observations made at the Swedish Solar Telescope with special attention paid to observations of the structure and dynamics of photospheric bright points. These observations are indicative of the quality of data becoming available from the Japanese Hinode satellite from which images and spectra, including vector magnetograms and Dopplergrams from the photosphere with a spatial resolution of roughly 0.2 arcsec are expected.

4.2.1 Recent Observations

In Figs. 4.2 and 4.3 we show typical images of the quiet to semi-quiet photosphere as well as a plage region. These images are made in the so-called G-band centered around 430 nm which is formed some 100 km above the nominal photosphere. The observations were obtained with the Swedish 1-meter Solar Telescope [24] on La Palma. Details of the optical setup and instrumentation are described in [4].

Narrowband interference filters were used for quasi-simultaneous imaging in the line core of the Ca II H-line (396.9 nm, filter passband 0.3 nm), the G-band (430.5 nm), and nearby continuum (436.4 nm, G-cont).

In the work described here seeing effects were reduced using several complementary techniques: adaptive optics [25] utilizing both a tip-tilt mirror and a deformable mirror, real-time frame selection, and postprocessing using the multi object multi

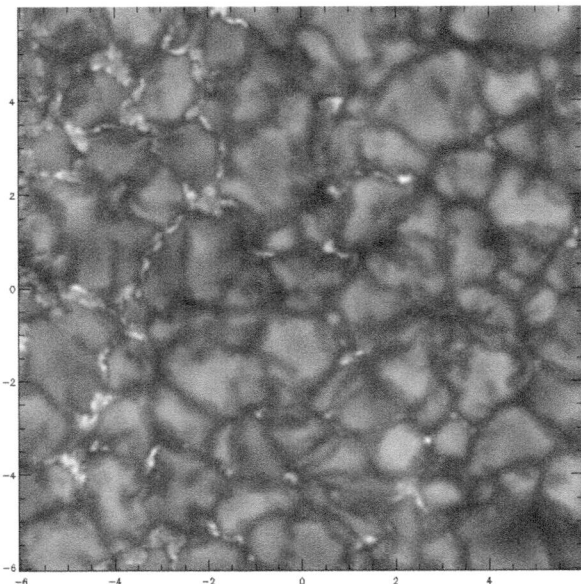

Fig. 4.2 A typical quiet photospheric region observed in the G-band with the Swedish Solar Telescope. This spectral band formed near 430 nm contains several spectral lines (see Fig. 4.4) and notably the lines of the CH molecule is formed near the height where $\tau_{500\ nm} = 1$; granulation and intergranular lanes some 100 km above this height, bright points some 200 km below—as explained in the text in connection with Fig. 4.5. Bright points are regions of enhanced magnetic field concentrated and contained by the granular motions. Notice also that bright points are pulled into ribbons and fill the entire intergranular lane

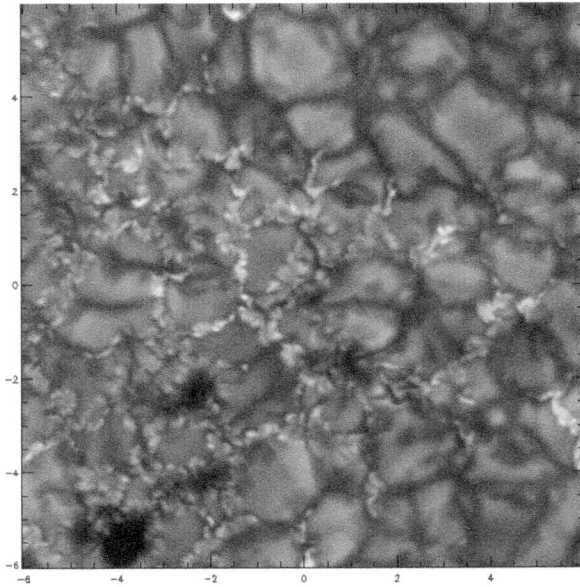

Fig. 4.3 A photospheric plage region observed with the Swedish Solar Telescope in the G-band. Notice the large number of phenomena showing complex structure; ribbons, flowers, micropores, as well as isolated and seemingly simple bright points. The magnetic field is in this image in places strong enough to perturb granulation dynamics and the granules appear "abnormal" while displaying a slower evolution than in the quieter photosphere

frame blind deconvolution (MOMFBD) image reconstruction technique [32]. The latter allows one to build up long-time series of very high-quality data, vital to the imaging of time-dependent phenomena on the solar surface.

In Fig. 4.2 we show examples of simple bright points in a fairly quiet region of the photosphere. Isolated bright points are constrained to intergranular lanes and do not seem to have any internal structure. Isolated bright points appear to be passively advected toward the periphery of supergranular cells [18], where they gather and form the magnetic network.

Several studies of the statistical properties of the dynamical evolution of bright points have appeared in the literature. Berger and Title [2] described a number of effects during the lifetime of bright points. These include shape modifications such as elongation, rotation, folding, splitting, and merging. Significant morphological changes can occur on timescales as short as 100 s and are strongly dependent on the local granular convective flow field. These changes make it difficult to define a typical lifetime of bright points. Berger et al. [3] found an average lifetime of 9.3 min using automated feature tracking techniques on a 70 min sequence of enhanced magnetic network. Some bright points persist up to the entire 70 min, and they experience numerous merging and splitting events but are still regarded as single objects. In a study of isolated bright points in a network region, Nisenson et al. (2003) [19] found a similar mean lifetime of 9.2 min with the longest lived bright point lasting 25 min. These reported values for the mean lifetime are very similar to the evolution timescale of granulation.

Toward the left side of Fig. 4.2 we see examples of bright points forming ribbon-like shapes in the intergranular lanes. In places these ribbons surround entire granules. Circular manifestations of these spread out, ribbon-like structures are dubbed "flowers," of which many examples are found in Fig. 4.3. This figure shows a region of stronger average field strength than does Fig. 4.2. The "flowers" and diverging ribbons have typical scales smaller than 1 arcsec. Flux concentrations with larger spatial extent are embedded in (micropores) with distinctly dark centers. In these more active areas there is a much greater density of bright points. They take on structure and appear to modify the granular flow itself. Granules near network bright points and in plage regions are smaller, have lower contrast and display slower temporal behavior. The granular pattern in these magnetically dominated areas are referred to as "abnormal." Coalescing bright points in plage and network regions form dark centers and thus become micropores.

In order for these magnetic structures to become visible at all they must attain field strengths of sufficient amplitude to perturb the plasma they are embedded in (even while being passively advected). Thus, we expect that there also exists a whole hierarchy of magnetic structures which have not yet attained such a critical field strength. What is the source of this field? One possible scenario is that it is brought up in granules at field strengths too low to leave an observational signature in the photospheric intensity. Granular flow then carries the field to the intergranular lanes and ultimately to the intergranular intersections where it may become compressed and strong enough to make the field visible, i.e., on the order of 1500 Gauss for a photospheric pressure of 10^4 N/m^2. This then would be the reason bright points always seem to appear in intergranular intersections.

4.2.2 Why Bright Points?

Photospheric bright points are correlated with regions of strong magnetic field and
are subject to photospheric motions. Plage regions share similar characteristics but
seem capable of modifying the background flow. Enhanced emission implies high
temperatures, on the other hand, micropores, pores, and sunspots are dark. What is
the relation between magnetic fields and the photospheric and lower chromospheric
temperatures? Why are bright points bright?

One way to answer this question is through atmospheric modeling of the relevant
phenomena using the MHD equations, as described in Chap. 2 and restated here:

$$\frac{\partial \rho}{\partial t} + \nabla \cdot (\rho \mathbf{u}) = 0 \,,$$

$$\frac{\partial e}{\partial t} + \nabla \cdot (e\mathbf{u}) + p\nabla \cdot \mathbf{u} = \nabla \cdot \mathbf{F_r} + \nabla \cdot \mathbf{F_c} + Q_{\text{Joule}},$$

$$\frac{\partial \mathbf{B}}{\partial t} - \nabla \times (\mathbf{u} \times \mathbf{B}) = 0 \,,$$

$$\frac{\partial \mathbf{u}}{\partial t} + (\mathbf{u} \cdot \nabla)\mathbf{u} + \frac{1}{\rho}\nabla p = \frac{1}{\mu_0 \rho}(\nabla \times \mathbf{B}) \times \mathbf{B} - g\hat{\mathbf{z}},$$

$$e = \frac{p}{\gamma - 1} \,,$$

representing conservation of mass, momentum, magnetic flux, and energy in the
usual notation. Here $\mathbf{F_r}$ represents the radiative flux, $\mathbf{F_c}$ represents the conductive
flux, Q_{Joule} is the joule heating and, e is the internal energy per unit volume.

Photospheric convection is driven by the radiative losses from the solar surface.
In order to model granulation correctly it is vital to construct a proper model of
radiative transport. A sophisticated treatment of this difficult problem was devised
by Nordlund [20] in which opacities are binned according to their magnitude; in
effect one is constructing wavelength bins that represent stronger and weaker lines
and the continuum so that radiation in all atmospheric regions is treated to a certain
approximation. In general, the radiative flux divergence from the photosphere and
lower chromosphere is obtained by angle and wavelength integration of the transport
equation. Also, assuming isotropic opacity and emissivity one finds

$$\nabla \cdot \mathbf{F_r} = 4\pi \int_{\lambda} \epsilon_{\lambda} \chi_{\lambda} (B_{\lambda} - J_{\lambda}) d\lambda. \tag{4.1}$$

If one further assumes that opacities are in LTE, the radiation from the photosphere
can be modeled.[2] After binning the opacities at all wavelengths in groups, the group

---

[2] The effects of coherent scattering may be incorporated if one wants to model the lower and mid
chromosphere. The resulting 3D scattering problems are then solved by iteration based on a one-ray
approximation in the angle integral for the mean intensity, a method developed by Skartlien, 2000
[28]

mean opacities are used in calculating a group mean source function from which the emergent intensity, and thus the radiative losses, can be derived by standard methods, such as Feautriers formal solver.

This method of approximating solar radiative transport coupled with a MHD numerical code allows one to model solar convection with a high degree of realism [20]. Carlsson et al. [6] constructed convection simulations such as these in which, in addition, an initial vertical magnetic field of magnitude 250 Gauss was inserted. The equations were discretized on a numerical grid of $253 \times 253 \times 163$ points covering $6 \times 6 \times 2$ Mm^3. With time, the magnetic field was carried along with the convective flow and formed quite complex topologies.

Once the model has evolved to a quasi-steady-state synthetic spectra of the G-band (or any other wavelength band formed in the modeled region) can be computed a posteriori using a radiative transfer code containing more than the essentially four frequency points used in the MHD simulation. In the results presented here, this was done using a spectral model with 2728 frequency points representing emission from 845 lines for the G-band as shown in Fig. 4.4. Note the excellent agreement between the average synthetic spectrum shown in black and an average photospheric spectrum shown in red. The model contains essentially no free parameters (other than the initial magnetic field strength whose value does not change the average computed spectrum to any noticeable degree). This agreement is fairly strong evidence that most of the relevant physics is included in the model.

The advantage of having a model of the phenomena is clear: one can look at any given variable as a function of position and time. Armed with these results one

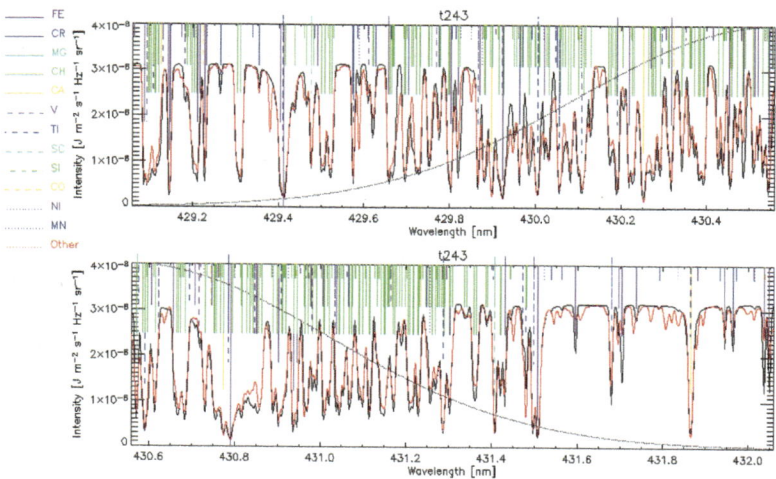

Fig. 4.4 The solar spectrum in the region of the G-band located near 430 nm. The bandpass of the filter used in obtaining Figs. 4.2 and 4.3 is shown by the *dotted line*. A great number of lines constitute this spectral band, lines from the CH molecule are prevalent, but also many other elements are represented. The observed spectrum of the solar photosphere is shown by the *red line*, the modeled spectrum is shown in *black*; for details of the model see the text or Carlsson et al. [6]

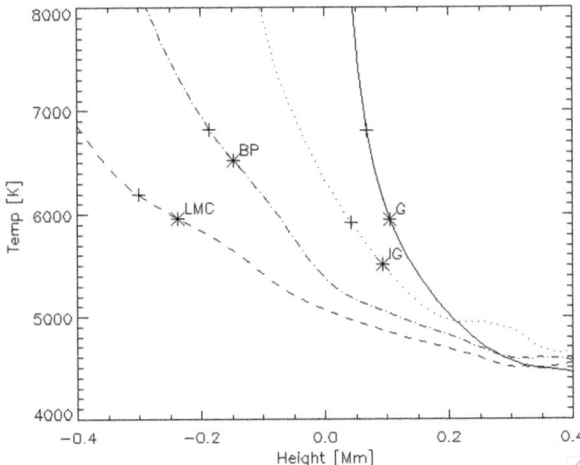

Fig. 4.5 The gas temperature as a function of height for four positions in the model of Carlsson et al. [6]: The curves marked "G" and "IG" represent positions in a typical granule center and intergranular lane. The asterisk show the formation height where the G-band is formed; emergent radiation from the granule center is formed at a higher temperature and thus appears brighter than the intergranular lane. The curves marked "BP" and "LMC" represent a bright point and the region of greatest magnetic field concentration; the bright point appears bright as the opacity is low enough to sample gas as great depths, even though at any geometric point in the simulation the temperature is higher in both granules and intergranular lanes. The same is also true for the region of greatest magnetic field, strong enough to hinder convection, and large enough horizontally to hinder radiative heating from the "walls" of the flux tube

is in a position to describe the emission from various photospheric regions; from a granule, from an intergranular lane, from a bright point, and from the region of strongest magnetic field concentration in the model.

The increased brightness in magnetic elements is due to their lower density compared with the surrounding intergranular medium. One thus sees deeper layers where the temperature is higher. At a given geometric height, the magnetic elements are cooler than the surrounding medium because the magnetic fields prevent convective energy transport from deeper layers. At the edges of the flux concentrations the plasma is radiatively heated by the surrounding hotter, non-magnetic plasma. See Fig. 4.5 for examples of the temperature structure and heights of emergent intensity formation for four different positions in the model.

4.3 The Chromosphere

With the observations and modeling of photospheric granulation and bright points discussed in Sect. 4.2.1 in mind let us turn to the chromosphere. What happens as perturbations implied in photospheric granulation propagate upward? How does the magnetic field expand into the overlying regions? and how does this change the dynamics of the atmosphere the field is expanding into?

Originally the chromosphere, i.e., colored sphere has its name from eclipse observations. One can occasionally observe the strong intensely red light stemming from the H-α line circumscribing the solar limb. The chromosphere is much thicker than the photosphere; it contains roughly 10–12 scale heights of say 200 km each and hence stretches some 2500 km between the photosphere and transition region.

4.3.1 Chromospheric Oscillations

Much work has gone into the elucidation of the thermal structure of the chromosphere and the mechanical heating needed in order to maintain this structure against radiative losses. Seminal in this regard is the work of Gene Avrett and coworkers in the 1980s [33]. However, we would like to approach chromospheric structure from another angle.

Assume that the chromosphere can be considered an initially isothermal slab stratified by a constant gravitational acceleration g. Then the linearized equations of mass, momentum and energy conservation [30] are

$$\frac{\partial}{\partial t}\delta\rho + \rho_0\frac{\partial}{\partial z}\delta u + \delta u\frac{\partial\rho_0}{\partial z} = 0,$$

$$\rho_0\frac{\partial}{\partial t}\delta u = -\frac{\partial}{\partial z}\delta p - g\delta\rho,$$

and

$$\frac{\partial}{\partial z}(\delta p - c_s^2\delta\rho) + \rho_0^\gamma\delta u\frac{\partial}{\partial z}\left(\frac{p_0}{\rho_0^\gamma}\right) = 0,$$

where all quantities have been expanded as $\rho = \rho_0 + \delta\rho$ such that the subscript "0" denotes unperturbed quantities, $\delta\rho$ is the perturbed density, δu is the vertical velocity, while $c_s = \sqrt{\gamma p_0/\rho_0}$ and p_0 are the speed of sound and the gas pressure, respectively.

These equations may be combined into

$$\frac{\partial^2 Q}{\partial t^2} - c_s^2\frac{\partial^2 Q}{\partial z^2} + \omega_a Q = 0, \tag{4.2}$$

where $Q \equiv \rho_0(z)^{1/2}u$ and the acoustic cutoff frequency is $\omega_a \equiv c_s/2H_p = \gamma g/2c_s$. This is a Klein–Gordon equation, the solution of which, after eventual initial transients have died down, is an oscillatory wake with a period close to the acoustic cutoff frequency. If we imagine photospheric dynamics as a driver—a driver with typical frequencies in the 5 min/3 mHz band, along with other excitations due to individual granule dynamics—we therefore expect on the basis of Eq. (4.2) the chromospheric response to be an oscillation with a frequency near 5 mHz corresponding to a period of roughly 3 min. Note that this conclusion is taken without any consideration of effects such as radiative damping or close consideration of the driver spectrum.

Be that as it may, in many cases a chromosphere dominated by 3 min power is indeed what is found. In Fig. 4.6 we see continuum emission in the 104.3 nm band observed with the SUMER instrument aboard the SOHO satellite. The image is made by stacking consecutive exposures of the slit vertically as a function of time. Note the horizontal bands of enhanced emission with horizontal extent on the order of tens of arcseconds that recur with a period of roughly 3 min. These horizontal bands are omnipresent in the image presented here and can be explained as a result of upwardly propagating wave-trains.

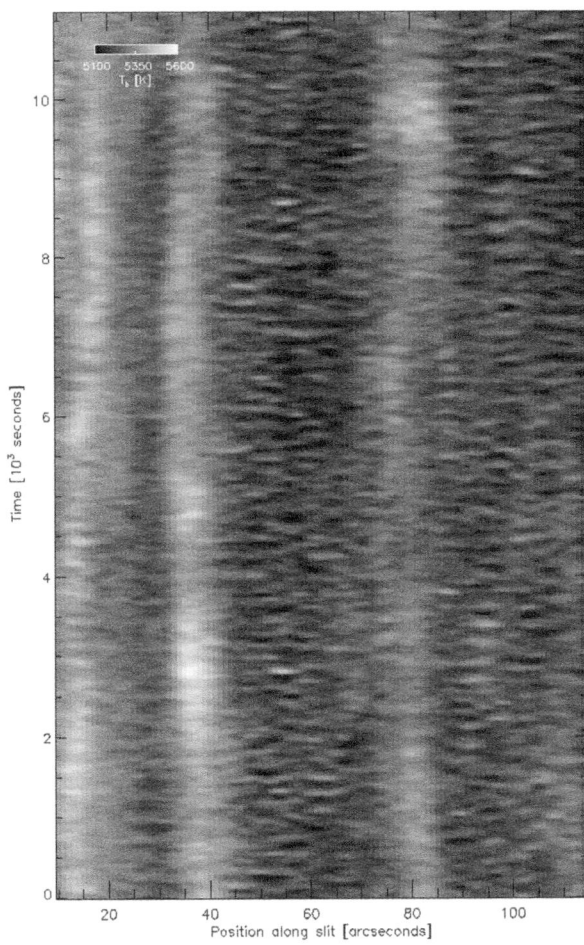

Fig. 4.6 Chromospheric oscillations as seen in the C continuum formed near 104 nm at a height roughly 1100 km above the photosphere. This observation is made by pointing the SUMER slit at a given, quiet sun, location of the sun, and making an exposure every 20 s or so. The image is made by stacking consecutive exposures of the slit vertically to form an image. The brighter vertical bands represent areas of enhanced magnetic field. The horizontal structures most clearly visible between the bright vertical bands can be explained by upwardly propagating 5 mHz oscillations. See Wikstøl et al. [34]

In fact, Carlsson and Stein [7] have shown that many aspects of the internetwork chromosphere can be explained as a result of photospheric excited, upwardly propagating acoustic waves. In their simulations a photospheric driver, a piston taken from the Doppler velocities measured in a Fe I line, formed in the photosphere, is used to excite waves. These waves propagate upward, and as they do so their amplitude grows, steepens, and forms shocks. A self-consistent radiative transfer calculation produces energy transfer in the model as well as diagnostics, especially in the Ca II H-line. The results of the model are shown in Fig. 4.7 which displays the observed and computed Ca II spectra as a function of wavelength and time. The figure shows the general behavior of the line emission in both observation and simulation: enhanced emission arises first in the line wings and thereafter moves in toward the line core. This emission is consistent with an upwardly propagating

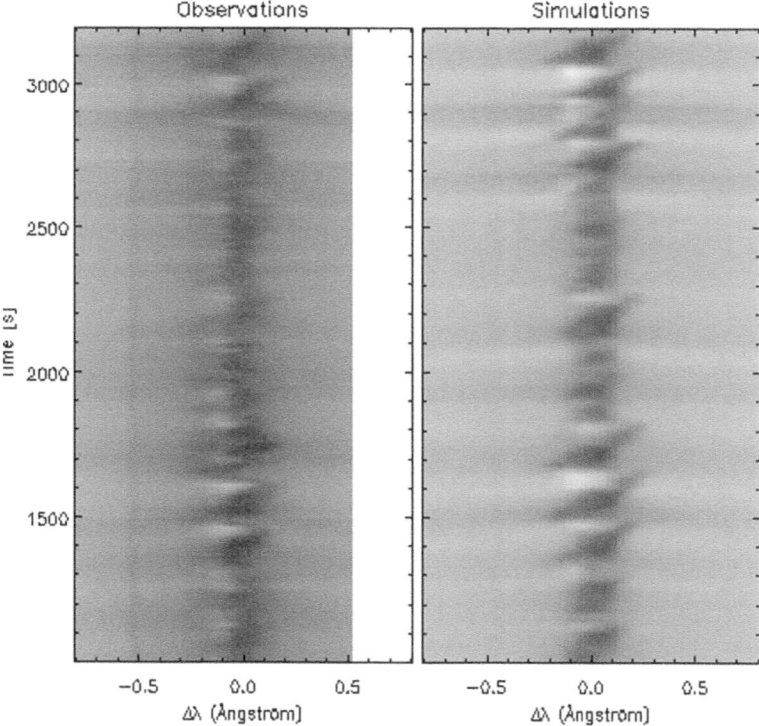

Fig. 4.7 The Ca II H-line profile as a function of time. Multiple repeated exposures are made with a slit fixed at a given position on the sun. The observed line profile in a quiet, internetwork, position is shown in the *left panel*. A 1D radiation-hydrodynamic model due to Carlsson and Stein [7], containing upwardly propagating acoustic waves driven by a piston computed from the Doppler shift measured in an Fe I line in the *blue* wing results in the Ca II H-line profiles shown in the *right panel*. Waves appear in the line wings and propagate toward line center as the acoustic wave moves upward in the atmosphere. A peak brightening on the *violet side* of the line core, formed some 1000 km above the photosphere, indicates that the wave amplitude has grown and that the wave is non-linear

wave; emission formed far out in the line wings is formed at lower heights than that formed near the core. Upon reaching the core the perturbation causes the blue, or violet, peak to become very enhanced, while a corresponding red peak is mostly absent. This anisotropy is a signature of a large velocity gradient in the region of emission formation, i.e., the wave has formed a shock at the height where the K2V peak is formed, roughly 1000 km above the photosphere (where $\tau_{500\ nm} = 1$). It is also worth noticing that even the details of Ca II emission are reproduced in this model. Comparing individual peaks show that they are similar in both timing and intensity in observation and model.

These and similar simulations also show that the results indicated by the linear analysis is essentially correct: waves with frequencies lower than the acoustic cutoff frequency at roughly 5 mHz decay exponentially with height in the chromosphere. As shown above, waves are naturally excited at the cutoff frequency of 5 mHz as wake oscillations. However, the simulations mentioned above show that the most important reason for a powerpeak at 5 mHz in the chromosphere is the exponential damping of the lower frequency, non-propagating waves. A photospheric spectrum dominated by 3 mHz waves with wavepower decreasing with frequency above 3 mHz will change in height to become dominated by the lowest propagating, non-damped frequencies. Waves with frequencies much above 5 mHz could in principle propagate up into the upper chromosphere and play an important role. However, high frequency waves are (a) not strongly excited by photospheric motions and (b) are very strongly radiatively damped as they propagate, the damping increasing with increasing frequency [9]. Thus only waves with frequencies in the range 5 to, say, 10 mHz already present in the photosphere propagate up and dominate internetwork chromospheric dynamics as they steepen and form shocks in the mid chromosphere.

There is therefore very strong evidence that the internetwork chromosphere is very dynamic and that the variations in physical quantities such as the temperature may be as large as the quantities themselves. In addition, these models show the grave danger posed by forming time average models based on diagnostics that have a non-linear response to variations in the atmosphere. There is, for example, no need for a chromospheric temperature rise in order to explain the behavior of the Ca II line emission in the internetwork.

4.3.2 The Chromospheric Network

There are however, several mysteries remaining in explaining chromospheric emission. Let us return to the upper chromosphere as imaged in Fig. 4.6. Not mentioned so far are the bright 10 arsec or so wide vertical bands. These regions coincide with regions of enhanced photospheric magnetic field. In addition, even the background emission in the dark internetwork bands is greater than that which can be explained solely by acoustic waves. Clearly, additional heating is required, a heating which presumably is connected in some way to the magnetic field.

Fig. 4.8 The chromospheric network as observed with the Swedish Solar Telescope in the Ca II H-line band. Most emission in this image is formed some hundreds of kilometers above the photosphere. We see inverse granulation, an image of the "overshoot" of vertical granular flow with horizontal scales identical to that of regular granulation. Magnetic bright points are advected with the granular flow and are transported to the chromospheric network which forms a granular pattern with horizontal size scales of 20 arcsec or so

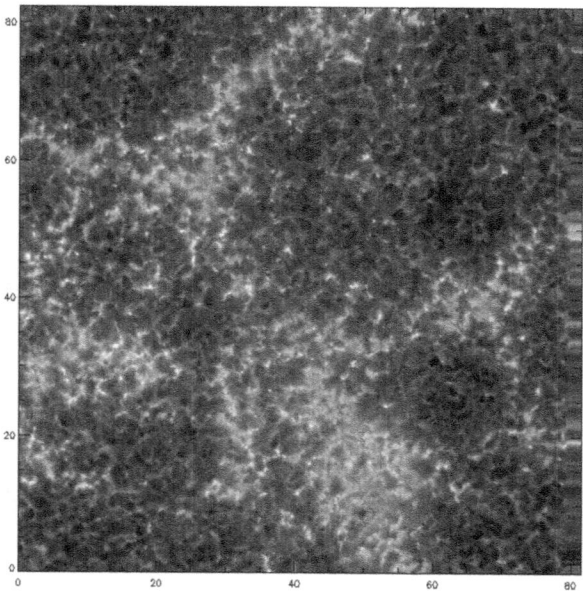

As magnetic bright points are advected with the granular flow they tend to concentrate in a granular pattern with typical horizontal dimensions of 20 arcsec that defines the supergranular flow field. The chromospheric network is the manifestation of this flow field as seen in lines and continua formed in the chromosphere. In Fig. 4.8, we see the chromospheric network clearly outlined at greater scales than that of the granular network seen in the form of inverse granulation in this image. The inverse granulation is a hydrodynamic phenomena caused by the previously hotter granules rising, expanding, and cooling, whereupon the gas is compressed it begins to fall back toward the lower photosphere and converges with the flow from other expended granules. The network emission, on the other hand, seems to be caused by an enhanced heating due to the magnetic field, the nature of which is still unknown.

The magnitude of the magnetic field in the chromosphere (and corona) can be estimated by potential field extrapolations from the measured longitudinal field in the photosphere. This is done by solving the equation

$$\nabla \times \mathbf{B} = 0,$$

with the boundary condition $B_z(x, y, 0) = B_{z0}(x, y)$ using an appropriate method [27]. Alternately one can carry out force-free calculations with a constant α, or nonlinear force-free calculations by various methods as surveyed by Schrijver et al. [26]. Extrapolations show that magnetic field will penetrate to a height that is roughly equal to the separation between sources. Thus, granular scale field will reach heights equivalent to a few arcseconds; i.e., 1000 km or so, while fields on network scales will reach high into the corona, on the order of 20–30 Mm. Typical photospheric field strengths, which could vary from some few Gauss to 1000 Gauss or more, will dominate the gas pressure at heights varying from some few scale heights above the photosphere to a height of 1500 km or more. In other words, we can expect the surface where $\beta = 1$ to be very corrugated.

We have seen that we expect a multitude of waves and wave modes to be excited in the solar convection zone. In most of the convection zone, the excited waves will be predominantly acoustic in nature. When acoustic waves reach the height where the Alfvén speed is comparable to the sound speed, i.e., roughly where $\beta = 1$, they undergo mode conversion, refraction, and reflection. In an inhomogeneous dynamic chromosphere, this region of mode conversion will be very irregular and change in time. We thus expect complex patterns of wave interactions that are highly variable in time and space.

What happens to these acoustic waves as they propagate from the photosphere into the chromosphere? McIntosh and coworkers have found that there is a clear correlation between observations of wavepower in SOHO/SUMER observations and the magnetic field topology as found from potential field extrapolations based on from SOHO/MDI observations of the longitudinal magnetic field [17, 16].

An understanding of the basic phenomena can be built up by studying some simplified cases. Rosenthal and coworkers [23] made 2D simulations of the propagation of waves through a number of simple field geometries in order to obtain a better insight into the effect of differing field structures on the wave speeds, amplitudes, polarization, direction of propagation, etc. In particular, they studied oscillations in the chromospheric network and internetwork. They find that acoustic, fast mode waves in the photosphere become mostly transverse, magnetic fast mode waves when crossing a magnetic canopy where the field is significantly inclined to the vertical, as shown in Fig. 4.9. Refraction by the rapidly increasing phase speed of the fast mode waves results in total internal reflection of the waves.

This work was extended to other field geometries, resembling a sunspot [5]. Four cases are studied; excitation by either a radial or a transverse sinusoidal perturbation and two magnetic field strengths – either an "umbra" at the bottom boundary or a weak-field case. In the strong-field case the plasma β is below unity at the location of the piston and the upward propagating waves do not cross a magnetic canopy. As the field is not exactly vertical at the location of the piston, both longitudinal and transverse waves are excited. The longitudinal waves propagate as slow mode, predominantly acoustic, waves along the magnetic field. The transverse waves propagate as fast mode, predominantly magnetic, waves. These waves are not confined by the magnetic field and they are refracted toward regions of lower Alfvén speed. They are therefore turned around and they impinge on the magnetic canopy in the "penumbral" region. In places where the wave vector forms a small angle to the field lines, the waves are converted to slow waves in the lower region; in places where the attack angle is large there is no mode conversion and the waves continue across the canopy as fast waves. The simulations show that wave mixing and interference are important aspects of oscillatory phenomena.

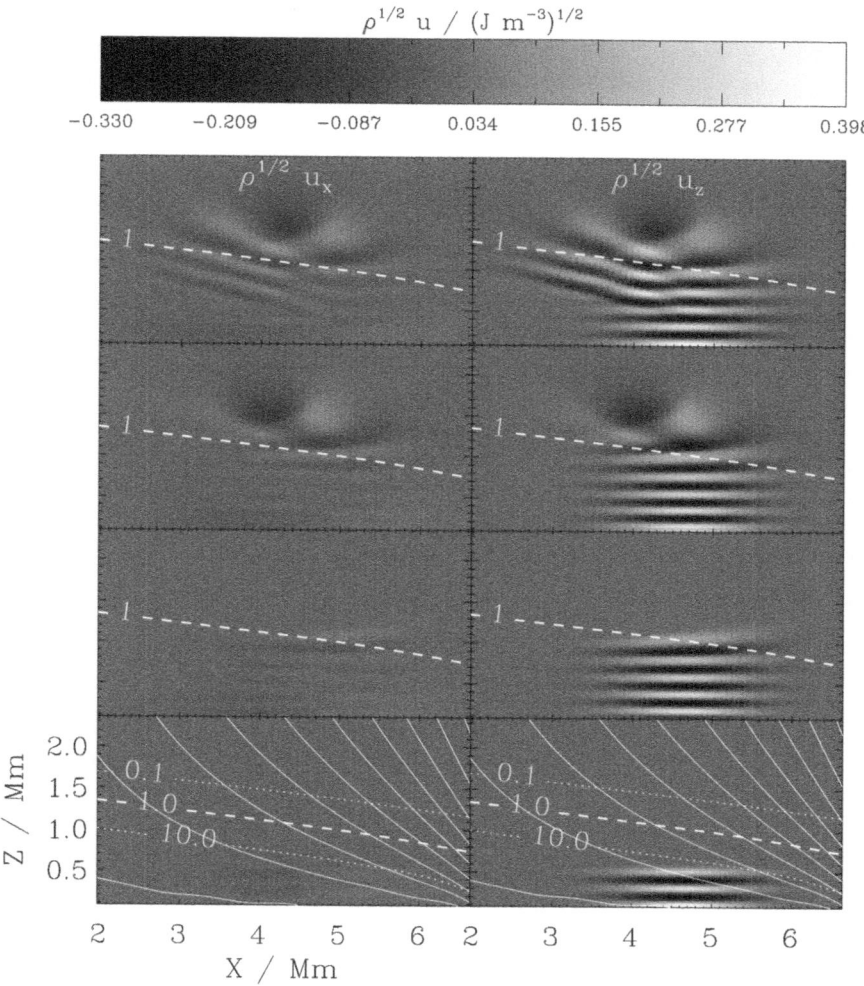

Fig. 4.9 Plots of the horizontal (*left panels*) and vertical velocities scaled with the square root of the density. In this model a schematic isothermal chromosphere with magnetic field structure is shown by the *solid lines* and the position of plasma β by the *dashed lines* in the *lower panels*. A fast mode (acoustic) wave is excited at the lower boundary by a vertically moving piston. The fast mode wave interacts with the magnetic field near $\beta = 1$ such that the wave is refracted and essentially reflected back into the photosphere and toward regions of high plasma β [23, 5]

4.4 Coronal Heating

The problem of coronal heating has existed since the discovery in the 1930s by Edlén and Grotrian that the corona had a temperature of order 1 MK. To this we can also add the problem of heating the network chromosphere, as well as the background emission in the internetwork chromosphere, neither of which can be explained by the action of 3 mHz or higher frequency acoustic waves.

Clearly some "mechanical" form of heat input is necessary to raise the coronal plasma to a temperature much higher than the photospheric radiation temperature. The debate has raged in the decades that have followed: the convection zone produces more than enough mechanical energy flux, but how is this energy flux transported to the corona, and how is it ultimately dissipated? Should one consider the buffeting of magnetic flux tubes on relatively short timescales, causing wave- or so-called AC heating. Or is it more appropriate to see the slower shuffling of flux tubes, causing stresses in the coronal magnetic field to build up, later to be episodically relieved in nanoflares as in the DC-heating scenario.

Even having answered that question, several remain. How is the energy flux ultimately thermalized? Do we need to continually inject new magnetic flux into the photosphere in order to sustain the corona? And if so, how much? Are there robust diagnostics that can separate the various heating mechanisms?

For reasons of personal preference we will unashamedly pursue the nano-flare heating mechanism for the duration of this chapter, but please keep in mind that this is a problem not solved by any means. The focus is also predominately on large-scale numerical modeling. Readers are referred to Chap. 2 (Chiuderi and Velli) for a discussion of other heating scenarios, as well as a discussion of some more historical aspects.

4.4.1 Why Does the Corona Have MK Temperatures?

Before turning to a discussion of coronal modeling and coronal heating it is worth spending a paragraph or so on coronal temperatures. Why is the coronal temperature of order 1 MK? Is achieving such a temperature a robust measure of heating mechanism success? To answer that question it is important to realize that the temperature of a plasma is set not only by the heat dissipated but also by the plasma's ability to lose energy.

The coronal plasma has essentially three possible ways to shed energy:

1. Through optically thin radiation given by

$$n_e n_H f(T_e),$$

   where n_e and n_H are the electron and total hydrogen densities and $f(T_e)$ is a function of temperature dependent mainly on line emission and, at higher temperatures, on thermal bremsstralung.

2. Through thermal conduction along the magnetic field, with a conduction coefficient

$$-\kappa_0 T_e^{5/2} \nabla_\| T_e$$

3. The magnetically open corona can also lose energy through the acceleration of a solar wind. This is a very efficient energy loss mechanism that sets firm limitations on coronal electron and ion temperatures as described by Hansteen, Leer

and coworkers [13, 14], but we will restrict our attention to the magnetically closed corona in the remainder of this discussion. The solar wind is discussed further in Chaps. 6 and 7.

In short, when the plasma is dense, $n_e n_H$ is large and variations in the heat input can be dealt with by small changes in the plasma temperature which will remain on order 10^4 K or less (similar to the photospheric radiation temperature). Conduction, on the other hand, is very inefficient at these temperatures. However, the density drops exponentially with height, with a scale height of only some hundreds of kilometers for a 10^4 K plasma. The efficiency of radiative losses, therefore, drops very rapidly with height and *any* mechanical heat input will raise the temperature of the plasma. The temperature will continue to rise until thermal conduction can balance the energy input. Since thermal conduction varies with a high power of the temperature this does not happen until the plasma has reached 1 MK or so. Thus, we expect any and every heating mechanism to give coronal temperatures of this order, and we must conclude that the amplitude of the observed coronal temperature is not a good guide to the mechanisms heating the corona.

4.4.2 The Transition Region

The argument used above necessarily implies that thermal conduction is the most important energy loss mechanism from the closed corona. This in turn means that the energy flux carried away from the site of coronal heating will mainly be carried by conduction and, therefore, roughly constant. As the temperature falls away from the heating site the plasma's ability to carry a heat flux decreases rapidly (as $T_e^{5/2}$) and the temperature gradient must become large to compensate. This process sets the structure of the transition region; the interface between the hot corona and the much cooler chromosphere is invariably sharp as shown schematically in Fig. 4.10.

With a very small spatial extent, line formation in the transition region becomes particularly simple; the emission is optically thin and confined in space. Observations of transition region lines could, therefore, potentially be both sensitive and understandable in terms of the processes heating both the chromosphere and the corona.

Ions in the transition region will, in general, be in the ground state, excited occasionally by electron collisions followed immediately by a spontaneous de-excitation. Thus the intensity may be written as

$$I_v = \frac{h\nu}{4\pi} \int_0^2 n_u A_{ul} ds = \frac{h\nu}{4\pi} \int_0^s n_l C_{lu} ds,$$

where the integration is carried out along the line of sight, n_u and n_l are the upper and lower level populations of the emitting ion, A_{ul} is the Einstein coefficient, and C_{ul} is the collisional excitation rate. The other symbols retain their usual meanings. The lower level population may be rewritten

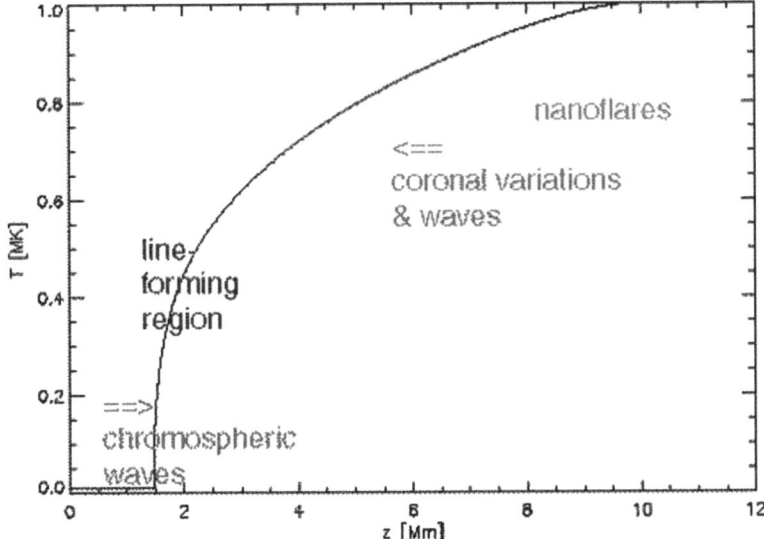

Fig. 4.10 Schematic structure of the transition region between chromosphere and corona: A more or less constant conductive flux from the corona ensures that the temperature gradient must steadily increase with decreasing temperatures until other terms (e.g., radiative losses) in the energy balance become significant. Optically thin spectral lines are formed over a relatively limited range in temperature and can therefore give good diagnostics of how the transition region responds to waves and other dynamic phenomena in the chromosphere below or the corona above

$$n_l = \frac{n_l}{n_i}\frac{n_i}{n_H}n_H = \frac{n_l}{n_i}A_i n_H,$$

where n_i is the total density of atoms of type i, n_l/n_i is the degree of ionization, n_H is the hydrogen number density, and A_i is the element abundance of atom i relative hydrogen. We may rewrite the collisional excitation rate as

$$C_{lu} = n_e C_0 T_e^{1/2}\exp\left(-\frac{h\nu}{kT_e}\right)\Gamma_{lu}(T_e),$$

where C_0 is a constant and $\Gamma(T_e)$ is a slowly varying function of the electron temperature.

Combining the above allows us to rewrite the line intensity as

$$I_\nu = \frac{h\nu}{4\pi}A_i C_0 \int_0^s n_e n_H g(T_e)ds, \tag{4.3}$$

$$\propto E(T_e) \equiv \int_{\Delta T e} n_e n_H (ds/dT_e)dT_e, \tag{4.4}$$

where we have defined the emission measure $E(T_e)$ by noting that the temperature-dependent parts of the intensity may be collected into a rapidly varying function

$g(T_e)$ that is sharply peaked around the region of maximum ion concentration n_l/n_i. The emission measure—as well as the differential emission measure, essentially the integrand of the emission measure, Eq. (4.4)—may be observationally determined from measured line intensities. Likewise given a corona heating model it is relatively straightforward to construct an expected emission measure. Comparisons of observed and predicted emission measures have met with little success; $E(T)$ has proven to be a difficult diagnostic for models to satisfy. In short, most if not all models predict much smaller line intensities for lines formed below 200 kK or so. To quote Grant Athay in a paper [1] written in 1982:

On the other hand, the total failure of all models for $T \lesssim 2.5 \times 10^5 K$ is a clear indication that the models have either a grossly incorrect geometry or they are omitting or misrepresenting a fundamental energy transport process.

Another puzzling observation concerns the line shifts of lines formed in the transition region. Lines formed from the upper chromosphere/lower transition region (e.g., C II 133.4 nm) up to lines formed at temperatures of 500 kK are invariable red-shifted *on average* with the maximum red shift found for lines formed at roughly 100 kK (e.g., C IV 154.8 nm) of 10 km/s or greater. There is no reason to suppose that there is a net flow of material from the corona toward the chromosphere, so some preferential weighting mechanism is implied. This effect is stronger in regions where the magnetic field is assumed strong, such as in the network and in active regions. And it is weaker or may be absent in the internetwork. Upwardly propagating sound waves, where plasma compression and the fluid velocity perturbation are in phase, would result in a net blue shift. Does this imply that the red shift is due to downwardly propagating waves formed, for example, as a result of nanoflare dissipation as suggested by Hansteen [12]? Or is some other mechanism insuring preferential emission of downward moving plasma active as claimed by Peter, Gudiksen and Nordlund [21]?

4.4.3 Forward Modelling

Simple and complex analytical models, semi-empirical modeling, and close analysis of the observations coupled with physical intuition have given researchers important insights into various aspects of the coronal heating question. However, it seems that this is not enough, the convection zone to corona system is of sufficient complexity to confound these methods as to the nature of coronal heating. Perhaps ab initio numerical models can give insight into the problem?

It is only recently that computer power and algorithmic developments have allowed one to even consider taking on this daunting task. And still grave doubts remain on the validity of treating microscopic processes in the corona by the averaging methods inherent in the MHD approximation. The nanoflare scenario is based on photospheric shuffling and braiding resulting in the creation of discontinuities being formed in the coronal magnetic field. This implies that relatively large-scale photospheric dynamics drives the coronal field to steadily smaller scales such that eventually the dissipation scale is reached and energy can be dissipated. Can we trust the results of calculations where this cascade is stopped by dissipation at scales many orders of magnitude larger than those presumably encountered in nature? And if it happens that we create a corona heating mechanism through our numerical modeling: How do we know it is the right one? We will come back to this and other connected issues in the last section of this chapter.

During the last few years the work of Gudiksen and Nordlund [11] has shown that it is possible to overcome the great numerical challenges outlined below to make initial attempts at modeling the photosphere to corona system. In their model a scaled down longitudinal magnetic field taken from an SOHO/MDI magnetogram of an active region is used to produce a potential magnetic field in the computational domain that covers $50 \times 50 \times 30$ Mm^3. This magnetic field is subjected to a parameterization of horizontal photospheric flow based on observations and, at smaller scales, on numerical convection simulations as a driver at the lower boundary. After a period of some 10–15 min solar time (and several months cpu time!) stresses in the simulated corona accumulate and become sufficient to maintain coronal temperatures. Synthetic TRACE images constructed from these models show a spectacular similarity to images such as shown in Fig. 1.1. In addition, other synthetic diagnostics, e.g., of transition region lines show promising characteristics [21].

4.4.3.1 Numerical Challenges

There are several reasons that the attempt to construct forward models of the convection zone or photosphere to corona system has been so long in coming. We will mention only a few:

The size of the simulation. We mentioned in our description of potential field extrapolation that the magnetic field will tend to reach heights of approximately the same as the distance between the sources of the field. Thus if one wishes to model the corona to a height of, say, 10 Mm this requires a horizontal size close to the double, or 20 Mm in order to form closed field regions up to the upper boundary. On the other hand, resolving photospheric scale heights of 100 km or smaller and transition region scales of some few tens of kilometers will require minimum grid sizes of less than 50 km, preferably much smaller. (Numerical "tricks" can perhaps ease some of this difficulty, but will not help by much more than a factor two.) Putting these requirements together means that it is difficult to get away with computational domains of much less than 150^3—a non-trivial exercise even on today's systems.

Thermal conduction. The "Courant condition" for a diffusive operator such as that describing thermal conduction scales with the grid size Δz^2 instead of with Δz for the magnetohydrodynamic operator. This severely limits the time step Δt the code can be stably run at. One solution is to vary the magnitude of the coefficient of thermal conduction when needed. Another is to proceed by operator splitting, such that the operator advancing the variables in time is $L = L_{\text{hydro}} + L_{\text{conduction}}$. Thus the energy equation is solved by discretizing

$$\frac{\partial e}{\partial t} = \nabla \cdot \mathbf{F}_c = -\nabla \kappa_\| \nabla_\| T$$

by the Crank–Nicholson method and then solving the system by, for example, the multigrid method.[3]

Radiative transport. Radiative losses from the photosphere and chromosphere are optically thick and will, in principle, require the solution of the transport equation. This can be done by the methods outlined in Sect. 4.2.2 for the case of the photosphere which is close to LTE. Modeling the chromosphere may require that the scattering of photons is treated with greater care [28], or alternately that one may use methods assuming that chromospheric radiation can be tabulated as a function of local thermodynamic variables a priori.

4.4.4 Convection Zone to Corona

With the proper tools in hand it is very tempting to attempt to model the entire solar atmosphere, from convection zone to corona. In Fig. 4.11 we show the result of such an experiment.

In a box of dimension $16 \times 8 \times 12\,\mathrm{Mm}^3$ with well-established convection, we have inserted a potential magnetic field generated by setting up a source given by a positive and a negative pole with magnitude 1000 Gauss at the lower boundary. The models are convectively unstable due the radiative losses in the photosphere. The average temperature at the bottom boundary is maintained by setting the entropy of the fluid entering through the bottom boundary. The bottom boundary, based on characteristic extrapolation, is otherwise open, allowing fluid to enter and leave the computational domain as required. The magnetic field at the lower boundary is advected with the fluid. As the simulation progresses the field is advected with the fluid flow in the convection zone and photosphere and individual field lines quickly attain quite complex paths through the model as shown in Fig. 4.11.

To prevent immediate coronal cooling the upper temperature boundary was initially set to a given temperature, 800 000 kK, and the models allowed to evolve from their potential state for 20 solar minutes. At this time the upper boundary was set so that the temperature gradient is zero; no conductive heat flux enters or leaves the computational domain. Aside from the temperature, the other hydrodynamic variables and the magnetic field are set using extrapolated characteristics.

---

[3] The general idea behind the multigrid method is to use Jacobi or Gauss – Seidel iterations, which are good at removing errors on small scales and bad at removing errors on large scales. Multigrid methods work by regridding the problem on successively coarser scales, thus converting large scales to small scales. A concise introduction to multigrid methods may be found in Chap. 19 of "Numerical Recipes" [22]. The formulation used in the code described in this chapter is based on the method used by Malagoli, Dubey, Cattaneo as shown at http://astro.uchicago.edu/Computing/On_Line/cfd95/camelse.html.

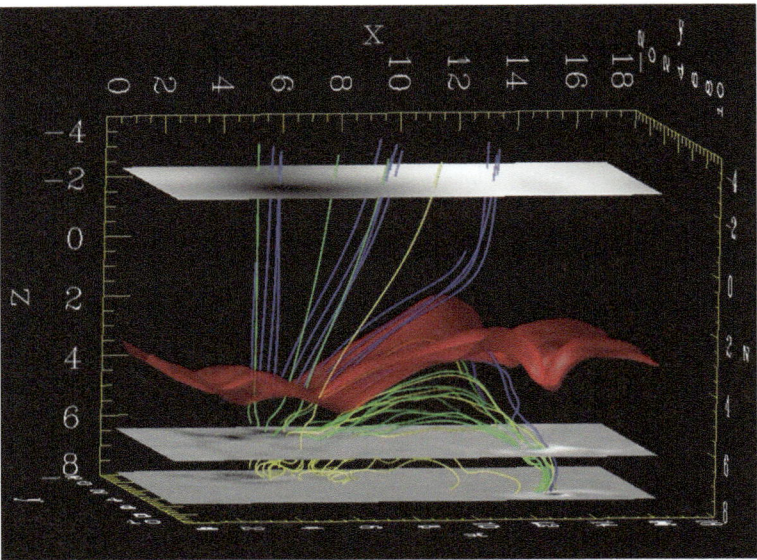

Fig. 4.11 The magnetic field structure as it has developed after some 20 min solar time evolution. Convection zone and photospheric motions have deformed the originally quite simple field. In the photosphere the field is quickly concentrated in intergranular lanes. The *black* and *white slabs* show the vertical magnetic field B_z near the lower boundary and in the photosphere, as well as in the corona. The surface where $T_g = 100$ kK is plotted in *red*. Some field lines are computed, starting from the lower convection zone (*yellow*), from the photosphere/lower chromosphere (*green*), and from the upper chromosphere/lower transition region (*blue*). Note also that the field lines below the photosphere seldom break the surface, rather they become quite tangled as a result of convective buffeting

The temperature structure in the models is shown in Fig. 4.12. We find that the photosphere is found at a model depth of $z = 6$ Mm. The convection zone reaches down to the lower boundary some 2 Mm below, where the temperature is some 16 000 K. Above the photosphere the chromosphere stretches upward to the transition region over a span varying between 1.5 Mm and 4 Mm above the photosphere. The corona fills the remaining 8 Mm or so of the computational domain, but regions of low temperatures are found at all depths upto $z = 0\ Mm$ and we expect that when the simulations have run longer there is no reason to believe that cool regions will not be found all the way up to the upper boundary.

Transition region diagnostics. There are several useful applications a model such as the one described here can be put to. Of these, perhaps the most interesting lies in studying the generation and dissipation of magnetic field stresses in the corona described in Sect. 4.4.5. But there is also potential insight to be gained from studying the chromosphere and transition region is such models; 3D models of the magnetized chromosphere and transition region have been noticeably lacking. As an example, let us consider the emission from the O VI 103.2 nm line formed in regions where the temperature is roughly 300 kK. In Fig. 4.13 we show the average line intensity and average line doppler shift as a function of time taken from a 2D model otherwise equivalent to the 3D model described above. The average magnetic

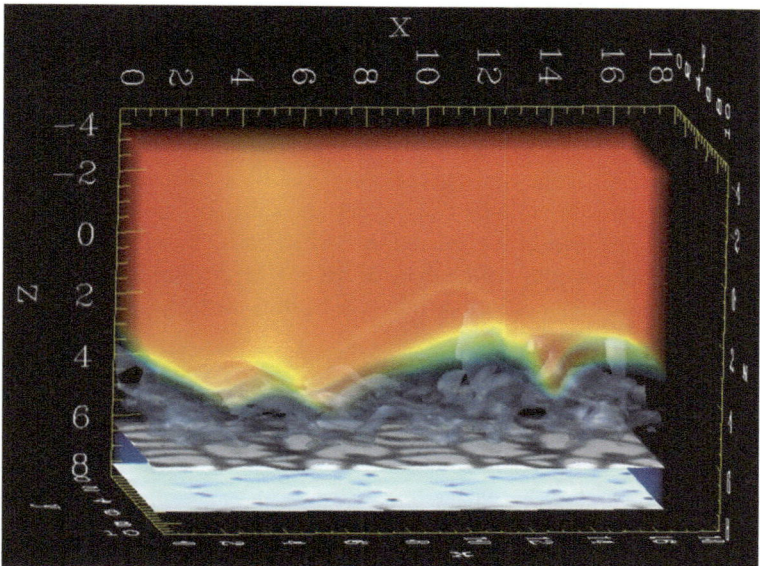

Fig. 4.12 The thermal structure of a 3D model as it has developed after some 20 min solar time evolution. Horizontal slices are shown near the lower boundary (where T_g is on order of 15 000 K and in the photosphere. Isosurfaces of $T_g = 7000$ K are shown in the chromosphere, and coronal temperatures are shown in the colors *blue* (10^4 K)–*red* (6×10^5 K). The corona in this model is still cooling from its initial state where a uniform 8×10^5 K corona was imposed, but heating events are raising the temperature at certain coronal locations as the magnetic field is carried around by convective motions

Fig. 4.13 Simulated observations of the O VI 103.2 nm line formed in regions where the temperature is roughly 300 kK. The *left panel* shows the total intensity in the line, the *right panel* shows the average Doppler shift. These simulated observations are based on a 2D MHD model spanning a region 16 Mm × 10 Mm covering the convection to lower corona. The magnetic topology in this model is similar to that shown in Fig. 4.11: a "loop" with footpoints close to $x = 4$ Mm and $x = 12$ 12 Mm. The periodic oscillations visible are mainly due to upwardly propagating waves generated in the photosphere or below

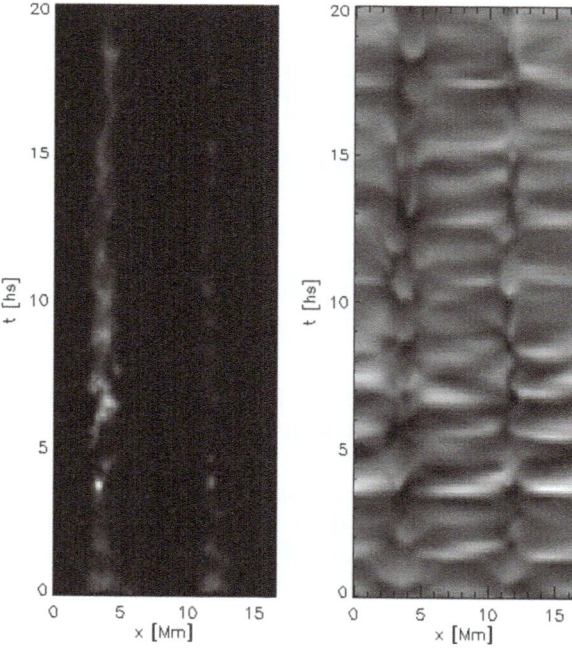

field in this model has a fairly simple structure with a "loop" passing through the transition region with footpoints near $x = 4$ Mm and $x = 12$ Mm. The amplitude of the line emission is strong in regions where the magnetic field is nearly vertical, i.e., in the footpoints, and quite weak where the field is horizontal. The line shifts show that some signal of the chromospheric 3 min oscillations reach the model transition region but at different times in regions of nearly vertical and nearly horizontal field. It is also interesting to note that the net-line shift does not vanish in the footpoints but rather displays an average red shift of 10 km/s. These images are remarkably similar to images constructed from SOHO/SUMER observations of the same line and lead some confidence that this is a fruitful method for interpreting such observations.

4.4.5 Modeling Nanoflares

As the stresses in the coronal field grow so does the energy density of the field. This energy must eventually be dissipated; at a rate commensurate with the rate at which energy flux is pumped in. On the Sun, the magnetic diffusivity η is very small and gradients must become very large before dissipation occurs. In the models presented here we operate with an η many orders of magnitude larger than on the Sun and dissipation starts at much smaller magnetic field gradients. The dissipated energy is

$$Q_{\text{Joule}} = \mathbf{E} \cdot \mathbf{J}, \tag{4.5}$$

where $\mathbf{J} = \nabla \times \mathbf{B}$ is the current density and the resistive part of the electric field is given by

$$E_x^\eta = \left\{ \frac{1}{2}(\eta_y^{(1)} + \eta_z^{(1)}) + \frac{1}{2}(\eta_y^{(2)} + \eta_z^{(2)}) \right\} J_x, \tag{4.6}$$

and similar for E_y and E_z. The diffusivities are given by

$$\eta_j^{(1)} = \frac{\Delta x_j}{\text{Pr}_M}(v_1 c_f + v_2 |u_j|), \tag{4.7}$$

$$\eta_j^{(2)} = \frac{\Delta x_j^2}{\text{Pr}_M} v_3 |\nabla_\perp \cdot \mathbf{u}|_-, \tag{4.8}$$

where Pr_M is the magnetic Prandtl number, c_f is the fast mode speed, v_1, v_2, and v_3 are dimensionless numbers of order unity and the other symbols retain their usual meanings.

The working assumption in these models is then that the artificial magnetic diffusivity used here and the diffusivity found on the Sun differ by many orders of magnitude the total amount of energy actually dissipated in the chromosphere and corona should be similar [10, 15].

The following conclusions are among the results of this modeling effort so far:

- A non-heated non-magnetic corona will cool significantly within 1200–1500 s. Even fairly strong hydrodynamic waves cannot maintain coronal temperatures.
- On the other hand, it seems that coronae threaded by even fairly weak fields can be maintained at temperatures of greater than 700 000 K by the stresses developed through convective and photospheric motions.
- The average coronal temperature (and heating) rises with increasing magnetic field strength. The structure of the field may also have some importance, but perhaps mainly on the location (height) of the heating.
- Magnetic heating rates in the chromosphere are in part high and we expect to see signatures of magnetic chromospheric heating in simulated emission.

4.4.5.1 The Way Forward

It seems, therefore, that the modeling effort so far is very promising; a number of observational characteristics are reproduced in these models. Starting from an observed magnetic field and a parameterization of the solar velocity field in the photosphere as boundaries and drivers, one can reproduce images that look remarkably like those observed in the coronal TRACE bands. In addition, synthetic spectral lines calculated on the basis of these models show characteristics that are very similar to those seen in SUMER and CDS spectra as shown in Fig. 4.13 or by Peter et al. [21]. And this congruence between observations and model is achieved with what is in fact very few free parameters. Can we then conclude that the corona heating problem is solved?

Perhaps a word or two of caution is in order before we celebrate our successes. We do have a promising hypothesis, but the question remains: Are the tests we are subjecting it to, e.g., the comparison of synthetic observations with actual observations actually capable of separating a correct description of the sun from an incorrect one? We have already demonstrated that we expect the solar corona to be heated to roughly 1 MK almost no matter what the mechanism for such heating. Conduction along field lines will naturally make loop-like structures. This implies that reproducing TRACE-like "images" is perhaps not so difficult after all, and possible for a wide spectrum of coronal heating models. The transition region diagnostics are a more discerning test, but clearly it is still too early to say that the only possible coronal model has been identified. It will be very interesting to see how these forward coronal heating models stand up in the face of questions such as: How does the corona react to variations in the total field strength, or the total field topology, and what observable diagnostic signatures do these variations cause?

Another issue is the fact that the treatment of the microphysics of dissipation is demonstrably wrong in these models. As stated in the previous section, the argument can be made that this does not matter, that the only important factor entering the problem is the amount of Poynting flux entering the corona. The way this is dissipated, over what spatial and temporal scale, depends in part on the details of the microphysics, but the total amount of energy flux going into heating the corona remains the same and is independent of the exact physical process that thermalizes the Poynting flux. On the other hand, and persuasive as this argument may seem: It would be a large step forward to find diagnostics that could confirm even more firmly that the model described here is essentially correct.

One could also wonder about the role of emerging flux in coronal heating: How much new magnetic flux must be brought up from below in order to replenish the dissipation of field heating the corona? And what, if any, is the role of an eventual surface dynamo? The thermalization process is itself also of great interest. Is it highly episodic such as found in the work of Einaudi and Velli [8]? Do the electric fields built up by the churning of the magnetic field cause particle acceleration to large energies as claimed by Turkmani et al. [31]? Is there a difference between the coronal heating mechanism and the process heating the network chromosphere? And what happens in open magnetic field regions where stresses built up by photospheric motions are free to propagate out into interplanetary space? There are certainly many open questions to be dealt with in the field of coronal heating, even if it should turn out that the basic scenario is correctly described by the current crop of forward models.

References

1. R.G. Athay: Astrophys. J., **263**, 982 (1982)
2. T.E. Berger, A.M. Title: Astrophys. J., **463**, 365 (1996)
3. T.E. Berger, M.G. Löfdahl, R.S. Shine, A.M. Title: Astrophys. J., **495**, 973 (1998).
4. T.E. Berger et al.: Astron. Astrophys., **428**, 613 (2004)
5. T.J. Bogdan et al.: Astrophys. J., **599**, 626 (2003)
6. M. Carlsson, R.F. Stein, Å Nordlund, G.B. Scharmer: Astrophys. J. Lett., **610**, L137 (2004)
7. M. Carlsson, R.F. Stein: Astrophys. J. Lett. **440**, L29 (1995)
8. G. Einaudi, M. Velli: Phys. Plasmas, **6**, 4146 (1999)
9. A. Fossum, M. Carlsson: Nature, **435**, 919 (2005)
10. K. Galsgaard, Å Nordlund: J. Geophys. Res., **101**, 13445 (1996)
11. B.V. Gudiksen, Å Nordlund: Astrophys. J. Lett., **572**, L113 (2002)
12. V. Hansteen: Astrophys. J., **402**, 741 (1993)
13. V.H. Hansteen, E. Leer: J. Geophys. Res., **100**, 21577 (1995)
14. V.H. Hansteen, E. Leer, T.E. Holzer: Astrophys. J., **482**, 498 (1997)
15. D.L. Hendrix, G. van Hoven, Z. Mikic, D.D. Schnack: Astrophys. J., **470**, 1192 (1996)
16. S.W. McIntosh, P.G. Judge: Astrophys. J., **561**, 420 (2001)
17. S.W. McIntosh et al.: Astrophys. J. Lett., **548**, L237 (2001)
18. R. Muller: Solar Phys., **85**, 113 (1983)
19. P. Nisenson, A.A. van Ballegooijen, A.G. de Wijn, P. Sütterlin: Astrophys. J., **587**, 458 (2003)
20. Å Nordlund: Astron. Astrophys., **107**, 1 (1982)
21. H. Peter, B.V. Gudiksen, Å Nordlund: Astrophys. J. Lett., **617**, L85 (2004)
22. W.H. Press, B.P. Flannery, S.A. Teukolsky, W.T. Vetterling: *Numerical Recipes – Second Edition*. Cambridge University Press, Cambridge (1988).
23. C.S. Rosenthal et al.: Astrophys. J., **564**, 508 (2002)
24. G.B. Scharmer, K. Bjelksjö, T.K. Korhonen, B. Lindberg, B. Petterson: Innovative Telescopes and Instrumentation for Solar Astrophysics. In: Stephen L. Keil, Sergey V. Avakyan (eds.). *Proceedings of the SPIE, Vol. 4853*, p. 341 (2003).

25. G.B. Scharmer, P.M. Dettori, M.G. Löfdahl, M. Shand: Innovative Telescopes and Instrumentation for Solar Astrophysics. In: Stephen L. Keil, Sergey V. Avakyan (eds.). *Proceedings of the SPIE, Vol. 4853*, p. 370 (2003).
26. C.J. Schrijver, M.L. Derosa, T.R. Metcalft, Y. Liu, J. McTiernan, S. Règnier, G. Valori, M.S. Wheatland, T. Wiegelmann: Nonlinear Force-Free Modeling of Coronal Magnetic Fields Part I: A Quantitative Comparison of Methods. Solar Phys., **235**, 161–190 (2006)
27. N. Seehafer: Solar Phys., **58**, 215 (1978)
28. R. Skartlien: Astrophys. J., **536**, 465 (2000)
29. R.F. Stein, Å Nordlund: Astrophys. J., **499**, 914 (1998)
30. A.C. Sterling, J.V. Hollweg: Astrophys. J., **327**, 950 (1988)
31. R. Turkmani, L. Vlahos, K. Galsgaard, P.J. Cargill, H. Isliker: Astrophys. J. Lett., **620**, L59 (2005)
32. M. van Noort, L. Rouppe van der Voort, M.G. Löfdahl: Solar Phys., **228**, 191 (2005)
33. J.E. Vernazza, E.H. Avrett, R. Loeser: Astrophys. J. Supp., **45**, 635 (1981)
34. Ø. Wikstøl, V.H. Hansteen, M. Carlsson, P.G. Judge: Astrophys. J., **531**, 1150 (2000)

Chapter 5
The Solar Flare: A Strongly Turbulent Particle Accelerator

L. Vlahos, S. Krucker, and P. Cargill

5.1 Introduction

The topics of explosive magnetic energy release on a large scale (a solar flare) and particle acceleration during such an event are rarely discussed together in the same article. Many discussions of magnetohydrodynamic (MHD) modeling of solar flares and/or CMEs have appeared (see [142] and references therein) and usually address large-scale destabilization of the coronal magnetic field. Particle acceleration in solar flares has also been discussed extensively [73, 163, 115, 165, 86, 167, 94, 121, 35] with the main emphasis being on the actual mechanisms for acceleration (e.g., shocks, turbulence, DC electric fields) rather than the global magnetic context in which the acceleration takes place.

In MHD studies the topic of particle acceleration is often presented as an additional complication to be addressed by future studies due to (a) its inherent complexity as a scientific problem and (b) the difficulty in reconciling the large MHD and small (kinetic) acceleration spatial and temporal scales. The former point leads to the consideration of acceleration within a framework of simple plasma and magnetic field configurations, with inclusion of the complex magnetic field structures present in the real corona being often deemed intractable. For example, it is often assumed that large monolithic current sheets appear when an eruption drives simultaneously a CME and a flare. The connection of such topologies with the extremely efficient transfer of magnetic energy to high energy particles remains an open question. The

L. Vlahos (✉)
Department of Physics, University of Thessaloniki, 54124 Thessaloniki, Greece
vlahos@astro.auth.gr

S. Krucker
Space Physics Research Group, University of California, Berkeley, USA
krucker@apollo.ssl.berkeley.edu

P. Cargill
Space and Atmospheric Physics, The Blackett Laboratory, Imperial College,
London SW7 2BW, UK
School of Mathematics and Statistics, University of St Andrews, St Andrews, KY16 9SS UK,
p.cargill@imperial.ac.uk

Vlahos, L. et al.: *The Solar Flare: A Strongly Turbulent Particle Accelerator.* Lect. Notes
Phys. **778**, 157–221 (2009)
DOI 10.1007/978-3-642-00210-6_5

latter point is best seen by noting that models of energy release and acceleration require methods that can handle simultaneously the large-scale magnetic field structures ($\sim 10^4$ km) evolving slowly (over the course of hours and days) and the small-scale dissipation regions ($\leq$ km) that evolve extremely rapidly (seconds to minutes).

The issues are well summarized in [142] where it is stated that "*In future, we hope for a closer link between the macroscopic MHD of the flare and the microscopic plasma physics of particle acceleration. The global environment for particle acceleration is created by MHD, but there is a feedback, with the MHD affected by the nature of the turbulent transport coefficients*". We draw attention in particular to the word "feedback": the fundamental question which needs to be fully addressed is the following: *can we disengage the macroscopic MHD physics from the microscopic plasma physics responsible for particle acceleration?* Current observational and theoretical developments suggest that *for the case of explosive energy release in the solar atmosphere, such a separation is not possible.*

The extraordinary efficiency of converting magnetic energy to energetic particles during solar flares (almost 50% of the dissipated magnetic energy will go into energetic particles, see Sect. 5.2) raises questions about the use of macroscopic (ideal or resistive MHD) theories as the description of impulsive energy release. The nonlinear coupling of large and small scales is extremely difficult to handle just by the use of transport coefficients. This is a problem which extends beyond solar physics and is one reason that our progress in understanding solar flares has been relatively slow over the last 100 years [34]. The overall goal of this chapter is to show how alternative approaches to the "flare problem" can begin to show how the integration of large-scale magnetic field dynamics with particle acceleration processes is possible.

In this chapter we present a radically different approach, used less in the current literature, that connects the impulsive energy release in the corona with the *complexity* imposed in active regions by the turbulent photospheric driver [164]. The flare problem is thus posed differently, since it emerges naturally from the evolution of a complex active region. The convection zone actively participates in the formation and evolution of large-scale structures by rearranging the position of the emerged magnetic field lines. At the same time the emergence of new magnetic flux rearranges the existing magnetic topologies in complex ways. 3-D magnetic topologies are thus constantly forced away from a potential state (if they were ever in one at all) due to slow (or abrupt) changes in the convection zone. Within these stressed large-scale magnetic topologies, localized short-lived magnetic discontinuities (current sheets) form spontaneously and dissipate the excess energy in the form of small- or large-scale structures (nanoflares and flares/CMEs). We stress that the concept of the sudden formation of a distribution of unstable discontinuities inside a well-organized large-scale topology is relatively new in the modeling of the solar flare phenomenon (see, for example, [136, 127] for important steps in the development of this approach).

The scenario of spatially distributed self-similar current sheets with localized dissipation evolving intermittently in time is supported by observations which indicate that flares and intense particle acceleration are associated with fragmented energy dissipation regions inside the global magnetic topology [167, 25]. There is strong evidence that narrow-band millisecond spike emission in the radio range is directly associated with the primary energy release. Such emission is fragmented in space and time, as seen in radio-spectrograms and in spatially resolved observations [172]. It can then be suggested that the energy release process is also fragmented in space and time, to at least the same degree as the radio spike emission [24]. Also type III burst radio emission, caused by electron beams escaping from flaring regions, appears in clusters, suggesting that fragmentation is a strong characteristic of the flaring region [23].

One approach which is able to capture the full extent of this interplay of highly localized dissipation in a well-behaved large-scale topology (sporadic flaring) is a special class of models [108, 109, 166, 114, 80, 81] which implement the concept of self-organized criticality (SOC), proposed initially by Bak et al. [19]. The main idea is that active regions evolve smoothly until at some point(s) inside the large-scale structure magnetic discontinuities (of all sizes) are formed and the currents associated with them reach a critical threshold. This causes a fast rearrangement of the local magnetic topology and the release of excess magnetic energy at the unstable point(s). This rearrangement may in turn cause a lack of stability in the immediate neighborhood, and so on, leading to the appearance of flares (avalanches) of all sizes that follow a well-defined statistical law which agrees remarkably well with the observed flare statistics [38].

Based on the current observational and theoretical evidence discussed in this chapter, we suggest that our inability to describe properly the coupling between the MHD evolution and the kinetic plasma aspects of a driven flaring region is the main reason behind our lack of understanding of the mechanism(s) which causes flares and the acceleration of high energy particles. Let us now define the *acceleration problem* during explosive energy release in the Sun: *We need to understand the mechanism(s) which transfer more than 50% of magnetic energy to large numbers $(10^{39}$ particles in total) of energetic electrons and ions, to energies in the highly relativistic regime ($> 100\,MeV$ for electrons and tens of GeVs for ions) on a short timescale (seconds or minutes), with specific energy spectra for the different isotopes and charge states.*

In Sect. 5.2 we briefly describe the key observational constraints. In Sect. 5.3 we present a brief overview of the main theories for impulsive magnetic energy release and in Sect. 5.4 we concentrate on the mechanisms on particle acceleration inside a more realistic and complex magnetic topologies. Finally in Sect. 5.5 we discuss the ability of the proposed accelerators to explain the main observational results and in Sect. 5.6 we report the main points stressed in this review.

5.2 Observational Constraints

5.2.1 X-ray Observations: Diagnostics of Energetic Electrons and Thermal Plasmas

Energetic electrons produce X-ray emission by collisions (the radiation mechanism responsible for the emission is non-thermal bremsstrahlung). The denser the plasma, the more collisions, and the more X-rays are produced (see Fig. 5.1). Therefore,

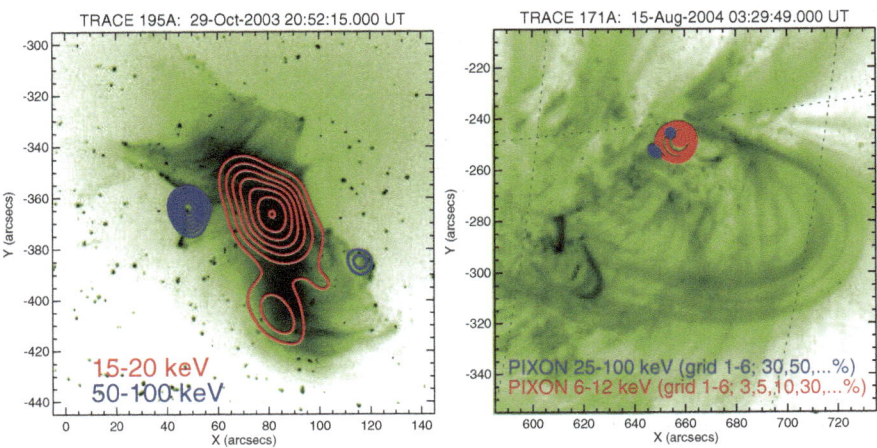

Fig. 5.1 Two examples of X-ray imaging in solar flares: (*left*) a large flare near disk center, (*right*) a small compact flare. Thermal emission in X-rays is shown by *red* contours, while non-thermal emission is shown as *blue*. The *green* images show EUV emission observed by TRACE with *dark colors* corresponding to enhanced intensity

X-rays produced by non-thermal electrons are strongest from the chromospheric footpoints of loops where the density increases rapidly. Indeed as the energetic electrons move into the chromosphere, they eventually lose all their energy through collisions. This scenario is usually called the "*thick-target model*" [30]. X-ray bremsstrahlung emission is in principle also emitted in the corona but the lower density there ($\sim 10^9$ particles cm^{-3}) is not big enough to stop energetic electrons or indeed to make them lose a significant amount of their energy (thin-target model). The estimated mean free path of an electron in the corona is $>10^5$ km. In general present-day instrumentation does not have a high enough signal-to-noise ratio to detect faint thin-target bremsstrahlung emission from the corona next to much brighter footpoints.

Thermal plasmas with temperatures above 1 MK also radiate in X-rays by collisions (thermal bremsstrahlung). Thermal X-ray spectra have a steeply falling continuum component plus some line emissions. In solar flares, thermal emission generally dominates the X-ray spectrum below 10–30 keV. At higher energies, the flare spectra are generally flatter, having power laws with indices between 3 and 5, sometimes with breaks (see Fig. 5.2). This is the non-thermal bremsstrahlung component produced by energetic electrons.

5.2.2 Energy Estimates

Spectral X-ray observations provide quantitative estimates of the energy content. The non-thermal energy (i.e., the energy in the accelerated energetic electrons) can be estimated by inverting the photon spectrum to get the electron spectrum. The total energy is then derived by integrating the electron spectrum above a cutoff energy. The largest uncertainties in this derivation are due to the not-well-known

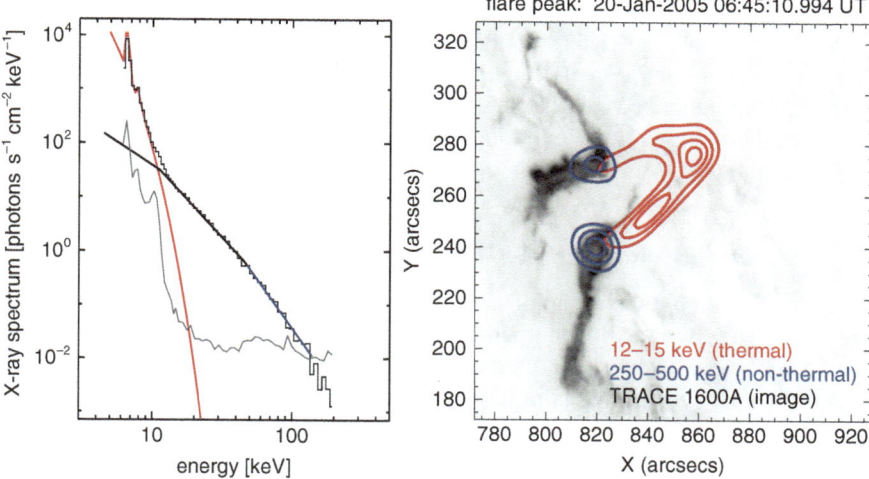

Fig. 5.2 Spectroscopy and imaging in X-rays: (*left*) a spatially integrated X-ray spectrum with a thermal fit in *red* and a broken power law fit (non-thermal emission) in *blue*. The data are shown in *black* and the instrumental background emission is shown in *gray*. (*right*) X-ray imaging with thermal emission in *red* and non-thermal in *blue*

cutoff energy. Often only an upper limit is known, giving lower limits to the non-thermal energy. Current estimates suggest that almost 50% of the total flare energy is deposited in energetic particles [55, 56].

Thermal flare energies are derived by fitting the thermal part of the X-ray spectra with a single temperature model, thus providing estimates of temperature and emission measure $EM \sim n^2 V$. Here V is an estimate of the volume occupied by the thermal plasma, usually obtained from images. From the emission measure and the volume, the number of heated electrons can be determined, each of which contains $1.5kT$ J. Assuming the same number of ions are heated, the total thermal energy becomes $3kT\sqrt{EM/V}$. This energy estimate is equal to the total energy needed to obtain the observed heated flare plasma and does not account for radiative and conductive losses. The derived energies are therefore only lower limits.

In solar flares, the thermal and non-thermal energy estimates are generally correlated and are often the same order of magnitude. This is consistent with the picture that flare energy release first accelerates electrons which later lose their energy by collisions, heating chromospheric plasma (see [148] for recent results and references therein).

5.2.3 Temporal Correlation

5.2.3.1 Neupert Effect

If the flare-accelerated energetic electrons indeed heat the flare plasma, the X-ray time profile of the thermal and non-thermal emission should reflect this: the non-thermal X-ray time profile should be a rough measure of how much energy is released in non-thermal electrons, and this is then the energy available for heating. So the larger the non-thermal X-ray flux, the more the heating expected. The time history of the integrated non-thermal X-ray flux roughly corresponds to the time profile of the thermal X-ray emission (this is called Neupert effect: see [161] for recent results).

5.2.3.2 Spectral Evolution: Soft–Hard–Soft

A very strong temporal correlation is observed between the non-thermal X-ray flux and the power law index of the photon spectrum: for each individual peak in the time series, the spectral shape hardens (flatter spectrum) until the peak and then softens (steeper spectrum) again during the decay (Fig. 5.3). This is referred to as the soft–hard–soft effect [69, 70] and seems to be a specific characteristic of the acceleration process. It is not understood. In some flares, the spectral behavior is different showing a gradual hardening during rise, peak, and decay for each individual burst. These events tend to be large and have a very good correlation with

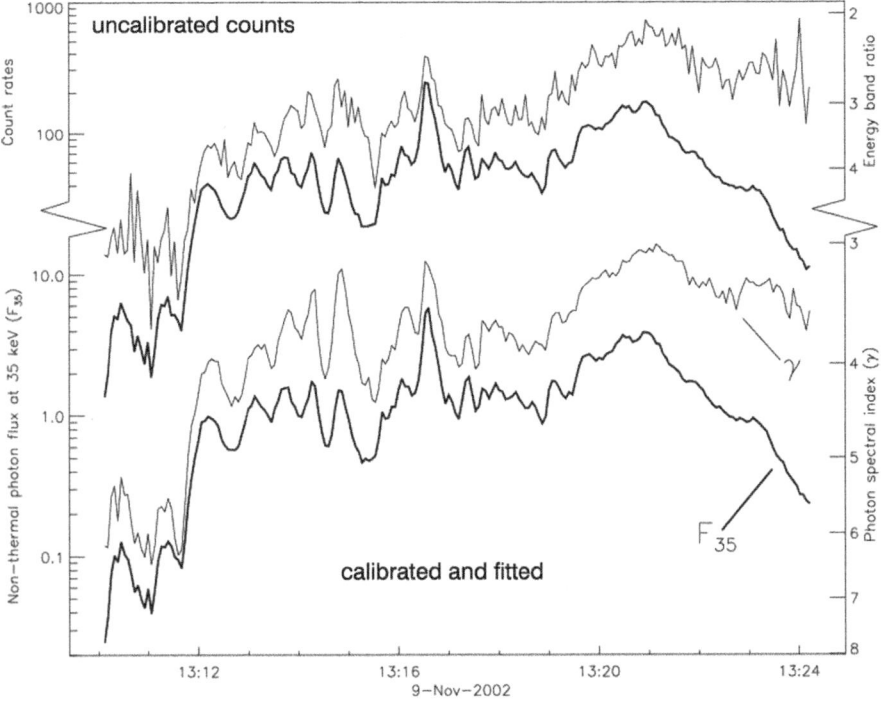

Fig. 5.3 *Top*: The spectral index (*thin line*) and flux (*thick line*) obtained from the uncalibrated total count rates flux in the energy bands 26–35 and 35–44 keV and their ratio. *Bottom*: the spectral index γ (*thin line*) and non-thermal flux F_{35} at 35 keV in photons s^{-1} cm^{-2} keV^{-1} (*thick line*) for the event of November 9, 2002, obtained by spectral fitting [69]

flares related to solar energetic particle (SEP) events [85]. Their behavior is also not understood. Note that spectral hardening can occur if electrons are trapped and low energy particles are lost first.

5.2.4 Location of Energy Release

5.2.4.1 Coronal Hard X-ray (HXR) Sources

Coronal X-ray emission is most often from hot thermal loops as described above. However, for some events an additional X-ray source is observed originating above the thermal X-ray loops [111], called an *"above the loop top source"* (ALT). First observed by Yohkoh, and only seen in a few flares (see Fig. 5.4), these sources are generally fainter with a softer spectrum than X-ray footpoint sources. There is no agreed interpretation of them at this time. If the footpoints and the ALT source are all produced by the same population of energetic electrons, the location of the ALT source indicates that the acceleration does not happen inside the flare loop. Therefore, it is generally speculated that the acceleration occurs above that loop. The relatively small number of flares with ALT sources might be because of the limited dynamic range of the observations alluded to earlier.

Several events do not follow this simple picture and the source is more complicated (see Fig. 5.5). RHESSI observations show several clear examples of ALT sources: the time evolution of these sources shows fast variations with several peaks of tens of second duration. The observed spectra are rather soft with power law indices around 5 and are better represented by non-thermal (power law) spectra than by thermal fits, although multi-thermal fits with temperatures up to 100 MK can represent the data almost as well. The fast time variations are very difficult to explain for a thermal interpretation (i.e., repeated heating to 100 MK and cooling on the same timescale). However, there are also difficulties with the non-thermal

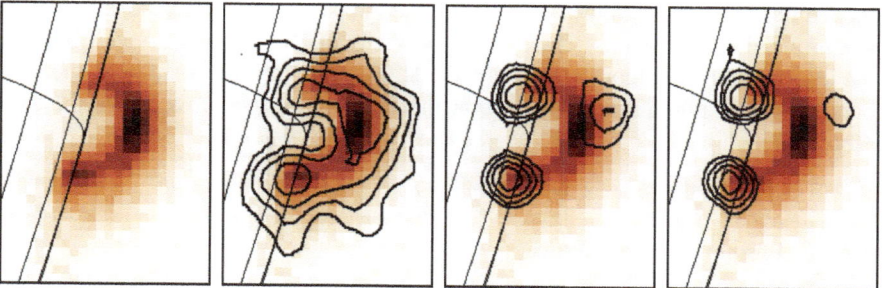

Fig. 5.4 Hard X-ray and soft X-ray images of the January 13, 1992, flare. The leftmost panel shows a soft X-ray image taken with the *Yohkoh*/SXT Be filter at 17:28:07 UT. From *left* to *right*, the remaining three panels show image contours at 14–23, 23–33, and 33–53 keV, respectively, taken from 17:27:35 to 17:28:15 UT by *Yohkoh*/HXT, overlaid on the same soft X-ray image. The contour levels are 6.25, 12.5, 25.0, and 50% of the peak value. The field of view is $59'' \times 79''$ for all panels

Fig. 5.5 X-ray imaging of a complex flare: the image (*red*) shows the thermal emission as seen by RHESSI at 10–15 keV and the non-thermal emission is again given by *blue* contours

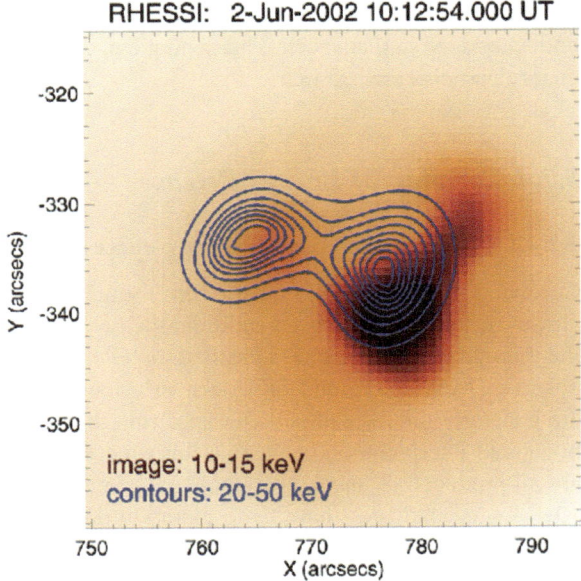

interpretation: the HXR producing electrons should significantly heat the ALT, but the hot thermal loops are observed below it.

5.2.4.2 Time of Flight

Further support for acceleration above the thermal flare loops is provided by timing studies of HXR footpoints at different energies. If energetic electrons at different energies are accelerated almost simultaneously at the same location, time-of-flight effects from the acceleration site to the HXR footpoints should be observed. This is indeed observed and it allows one to estimate the path length from the acceleration site to the HXR footpoints. The derived path lengths are generally longer than half the length of the flare loops connecting the HXR footpoints [16]. Although the error bars are large, this again suggests that the acceleration occurred above the flare loops.

5.2.4.3 Temperature Structure

Recent RHESSI observations also show support for particle acceleration above the main flare loop [156]. Evidence was found for a temperature gradient with decreasing temperatures from the possible coronal acceleration site toward lower and higher altitudes [155]. The hottest flare loops are expected to be the newly reconnected loops at largest altitude. Previously heated flare loops are at lower altitude and have already partly cooled down. For energetic electrons released upward, the opposite is expected with the hottest emission at lowest altitude, as observed. Another explanation could be that there is direct heating at the particle acceleration site (as a by-product of the main acceleration process) that would produce a similar temperature profile.

5.2.5 Footpoint Motions

Standard magnetic reconnection models predict increasing separation of the footpoints during the flare [142] as longer and larger loops are produced. If the reconnection process results in accelerated electrons [130], the HXR footpoints should show this motion. The motion is only apparent; it is due to the HXR emission shifting to footpoints of neighboring newly reconnected field lines. Hence, the speed of footpoint separation reflects the rate of magnetic reconnection and should be roughly proportional to the total HXR emission from the footpoints. Sakao, Kosugi, and Masuda [149] analyzed footpoint motions in 14 flares observed by Yohkoh HXT, but did not find a clear correlation between the footpoint separation speed and the HXR flux. Recently, however, source motion seen in Hα was studied by Qiu et al. [145]. They found some correlation with HXR flux during the main peak, but not before or after.

RHESSI results [58, 90, 91, 70] show systematic but more complex footpoint motions than a simple flare model would predict. Krucker, Hurford, and Lin [92] analyzed HXR footpoint motions in the July 23, 2002, flare (GOES X4.8-class). Above 30 keV, at least three HXR sources are observed during the impulsive phase that can be identified with footpoints of coronal magnetic loops that form an arcade. On the northern ribbon of this arcade, a source is seen that moves systematically along the ribbon for more than 10 min. On the other ribbon, at least two sources are seen that do not seem to move systematically for longer than half a minute, with different sources dominating at different times. The northern source motions are fast during times of strong HXR flux, but almost absent during periods with low HXR emission. This is consistent with magnetic reconnection if a higher rate of reconnection (resulting in a higher footpoint speed) produces more energetic electrons per unit time and therefore more HXR emission. The absence of footpoint motion in one ribbon is inconsistent with simple reconnection models, but can be explained if the magnetic configuration is more complex. Also the motion of the northern footpoint is rather along the ribbon, contrary to the perpendicular motions predicted by simple reconnection models. In some events the motion during the whole flare is clearly along the ribbons [70].

5.2.6 Gamma Rays (Emission Above >300 keV)

5.2.6.1 Electron Bremsstrahlung

The non-thermal electron bremsstrahlung component can extend up to and above 10 MeV. This component is produced in the same way as the emission seen above

20 keV, but from electrons with higher energies. Generally the spectrum shows a hardening (flatter spectrum) above 0.5–1 MeV. Because the spectrum decreases steeply with energy, electron bremsstrahlung in the gamma-ray range is only observed for very large flares. Rarely, however, is it the dominant emission in the gamma-ray range. For most gamma-ray flares, emission produced by energetic ions is present as well.

5.2.6.2 De-Excitation Lines

Flare-accelerated ions (protons, alphas, heavier ions) are responsible for the production of gamma-ray emission when they collide with ambient ions and produce excited nuclei that emit nuclear de-excitation lines (see Fig. 5.6). Again, the emission process depends on the density of the ambient plasma and therefore the emission is expected from dense regions (i.e. the chromosphere). Since the de-excitation is happening almost instantaneously after the collision, these gamma-ray lines are also referred to as prompt lines. Depending on the ratio of the mass of the accelerated ion to the target ion, the line emission can be narrow or broad. Narrow lines are

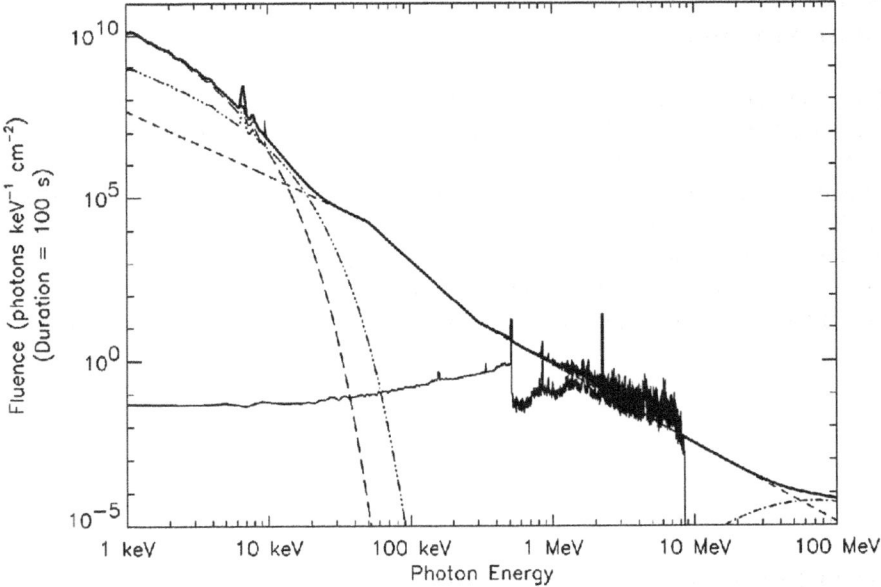

Fig. 5.6 Composite X-ray/gamma-ray spectrum from 1 keV to 100 MeV for a large flare. At energies up to 10–30 keV, emission from hot (10^7) and "superhot" (3×10^7) thermal flare plasmas (the two curves at the *left*) dominates. Bremsstrahlung emission from energetic electrons produces the X-ray/gamma-ray continuum (*straight lines*) up to tens of MeV. Broad and narrow gamma-ray lines from nuclear interactions of energetic ions sometimes dominate the spectrum between 1 and 7 MeV. Above a few tens of MeV the photons produced by the decay of pions (curve at the *right*) dominate. RHESSI observations cover almost 4 orders of magnitude in energy (3 keV to 17 MeV) [103]

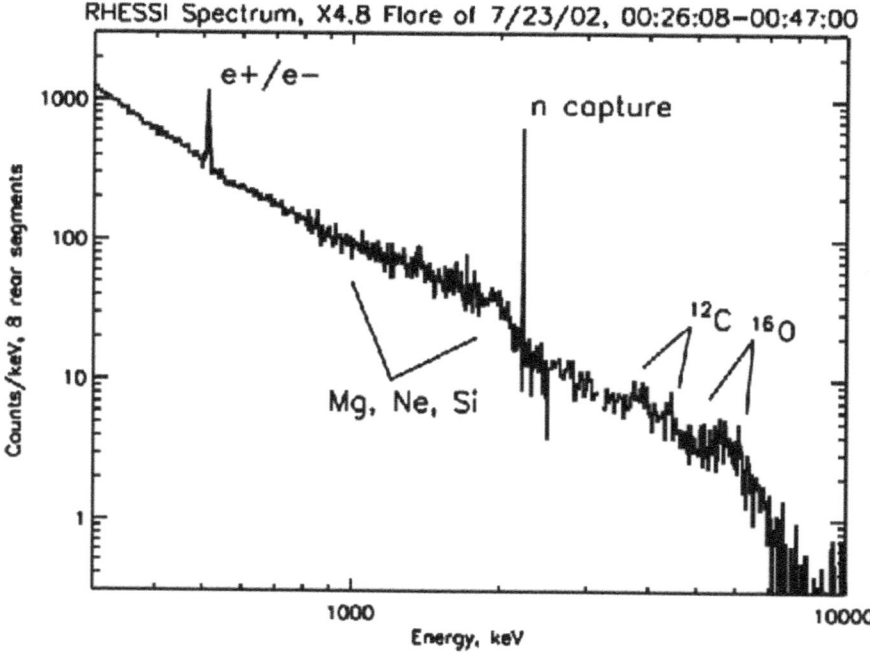

Fig. 5.7 RHESSI gamma-ray spectrum of the July 23, 2002, flare

produced when a flare-accelerated proton or alpha particle hits a heavy ambiention (see Fig. 5.7). The width of the emitted line is then produced by the recoil of the heavy ambient ion, and a narrow line is produced. On the other hand, if a heavy flare-accelerated ion hits an ambient proton or alpha particle, the emitted radiation is Doppler-shifted and therefore broad.

RHESSI provides for the first time spectrally resolved observations of narrow de-excitation lines. Statistics are generally limiting the observations, but the narrow lines can still be fitted and the red shift of the lines can be measured. Heavier nuclei are expected to recoil less and therefore show less redshift [154].

5.2.6.3 Neutron-Capture Line

The most prominent line emission in the gamma ray spectrum is the neutron-capture line at 2.223 MeV. This line is produced by the capture of thermalized neutrons that were produced by nuclear reactions after flare-accelerated ions hit ambient ions (the dominant neutron production at high energies comes from the breakup of He). The thermalized neutrons are captured by ambient protons and a deuterium and a photon at 2.223 MeV are produced. Since the neutrons are thermalized (i.e., have a low velocity) the 2.223 MeV line is very narrow. Since initially the neutrons have to first thermalize before they can be captured, the time profile of the 2.223 MeV line is delayed relative to the prompt lines.

5.2.6.4 Energy Estimates

The different gamma-ray lines can be used to get information about the flare-accelerated ion spectrum. Estimates of the total energy in non-thermal ions can again be derived by integrating over the ion spectrum. The lower energy cutoff, however, is even more uncertain than for the electron spectrum.

5.2.6.5 Comparing Electron and Ions Acceleration

Comparing the fluence of >300 keV emission with the fluence of the 2.2 MeV line, the electron and ion acceleration in flares can be compared. A rough correlation is observed indicating that at least in very large flares (gamma-ray emission from small flares are not detectable with present day instrumentation), both electrons and ions are always accelerated.

5.2.6.6 Gamma-Ray Imaging

RHESSI provides for the first time spatial information of gamma-ray emission, the only direct indication of the spatial properties of accelerated ions near the Sun. The most powerful tool for gamma-ray imaging is the 2.223 MeV neutron-capture line, because of good statistics and a narrow line width which limits the non-solar background to a minimum compared to broad lines. However, the spatial resolution of $35''$ is much poorer than the $2''$ resolution in the hard X-ray range

For the event with best statistics (October 28, 2003 [78]), the 2.223 MeV source structure shows two footpoints similar to the HXR source structure but clearly displaced by $\sim15''$ (see Fig. 5.8). This indicates that electrons and ions are accelerated

Fig. 5.8 Imaging of the 2.223 MeV neutron-capture line and the HXR electron bremsstrahlung of the flare on October 28, 2003. The *red* or *gray circles* show the locations of the event-averaged centroid positions of the 2.223 MeV emission with 1σ uncertainties; the *blue* or *black lines* are the 30, 50, and 90% contours of the 100–200 keV electron bremsstrahlung sources at around 11:06:46 UT. The underlying EUV image is from TRACE at 195Å with offset corrections applied. The gamma-ray and HXR sources are all located on the EUV flare ribbons seen with TRACE

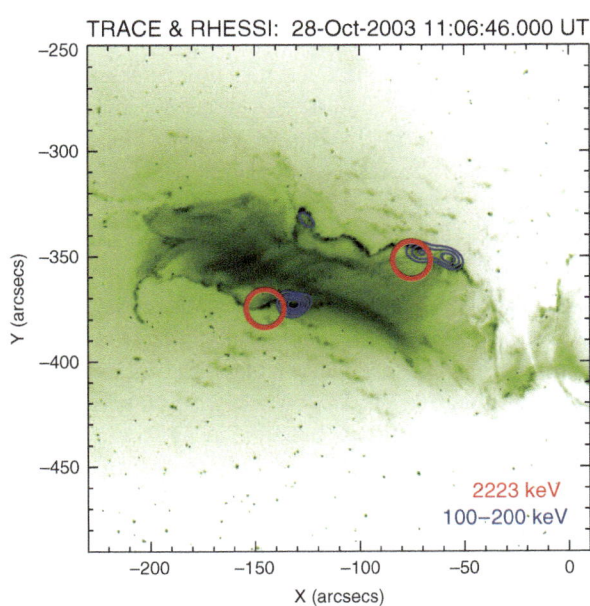

in similar-sized magnetic structures. The displacement could be explained by different accelerator sites for electrons and ions, or by different transport effects from a possibly common acceleration site to the location where the electrons and ions lose their energy by collisions.

5.2.7 Energetic Particles Escaping from the Sun

5.2.7.1 Flare-Accelerated Electrons Escaping the Sun

X-rays are remote sensing diagnostics of energetic electrons that lose their energy by collisions. Upward moving energetic electrons that have access to field lines extending into interplanetary space (often referred to as "open field lines") only suffer a few collisions (the density is decreasing rapidly) and can therefore escape from the Sun and be observed in situ near the Earth with particle detectors. These events show fast rise times with slow decays and are called "impulsive electron events" when observed near the Earth [104]. They are seen with energies from above 1 keV up to the highly relativistic regime. Quite often the first electrons to arrive are observed to travel without suffering any collisions (ballistic transport) and they are therefore referred to as "scatter-free" events. The ballistic transport means that high energy electrons arrive earlier than lower energy ones, indicating that electrons at all energies left the Sun around the same time. The observed dispersion in the onset times of the different energy channels can therefore be used to approximate when the energetic electrons left the Sun. In some case (about one-third of all events) a clear temporal correlation exists with the occurrence of HXR emission during solar flares and the release of energetic electrons into interplanetary space (Fig. 5.9, left).

This indicates that possibly the same acceleration mechanism produces the energetic electrons that create HXR emission in the chromosphere and those energetic electrons that escape into interplanetary space. This picture can be further corroborated by comparing the HXR spectrum with the in situ electron spectrum. If the chromospheric X-ray spectrum is flat (hard), the electron spectrum observed near the Earth is also flat (Fig. 5.9, right).

For particles to escape into interplanetary space, they must have access to open field lines. How that happens is not well understood. In the "classic" flare scenario (e.g., [152]) no open field lines are shown. For flares with a good temporal and spectral correlation with electron events observed in situ, the flare geometry indeed looks different. These events show hot flare loops with HXR footpoints, plus an additional HXR source separated from the loop by 15″ with only little heating. This source structure can be explained by a simple magnetic reconnection model with newly emerging flux tubes that reconnect with previously open field lines, so-called interchange reconnection. The previously open field lines form the flare loops, while the newly opened field lines show less heating since material can be easily lost because the field is open. Upward moving energetic electrons escape along the newly opened field line (see Fig. 5.10).

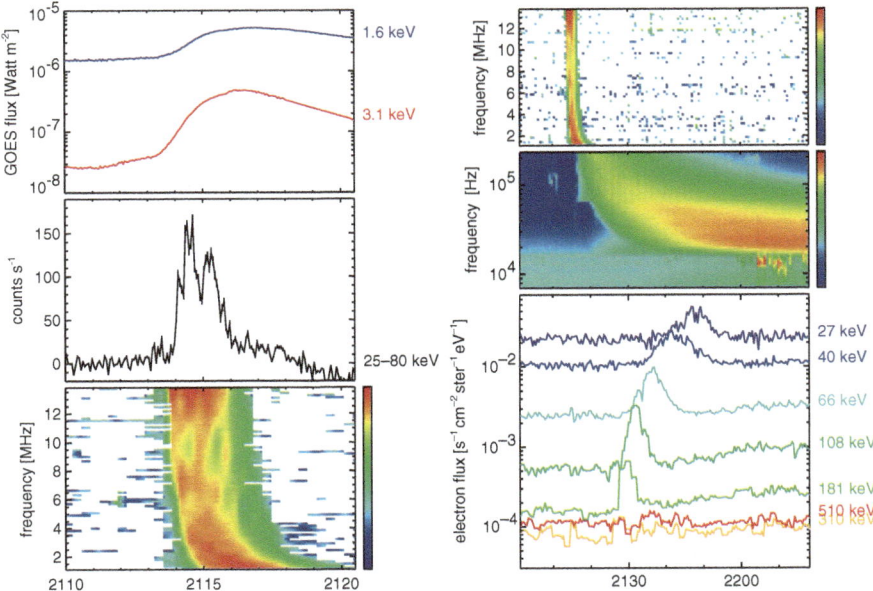

Fig. 5.9 Impulsive electron event observed on October 19, 2002: (*left*) From *top* to *bottom*, GOES soft X-ray light curves, RHESSI 25–80 keV light curve, and WIND/WAVES radio spectrogram in the 1–14 MHz range are shown [29]; (*right*) an expanded view of the WIND/WAVES data including low-frequency observations is presented in the *top two panels*, while the *bottom panel* shows in situ observed energetic electrons from 30 to 500 keV detected by WIND/3DP. This event (like all events selected in this survey) shows a close temporal correlation between non-thermal HXR emission, radio type III emission in interplanetary space, and in situ observed electrons

5.2.7.2 Flare-Accelerated Ions Escaping from the Sun

Temporal and spectral comparisons can also be made for ions escaping from the Sun in a similar way to escaping electrons. However, this is much more difficult to do because of the poorer count statistics in the gamma-ray range.

The timing of escaping ions is sometimes delayed relative to the flare emission, often significantly (1 h) [93]. Generally it is thought that the shocks of coronal mass ejections are mainly responsible for the energetic ions seen near to the Earth. If this is indeed the case, then a spectral comparison between in situ observed ion spectra and gamma-ray line observations should give no correlation. Surprisingly, in the two gamma-ray line flares observed by RHESSI that are magnetically well connected (November 2, 2003 and January 20, 2005), the spectrum of the energetic protons producing the gamma-ray lines was found to be essentially the same as that of the SEP protons observed at 1 AU. These two events had quite different spectral slopes, so this agreement is unlikely to be a coincidence. It suggests that the gamma-ray producing and in situ energetic protons may have the same source (at least in these two events), contrary to the standard two-class paradigm (i.e., flare-accelerated and CME-accelerated ions). These results illustrate the present lack of physical understanding regarding the SEP acceleration process(es).

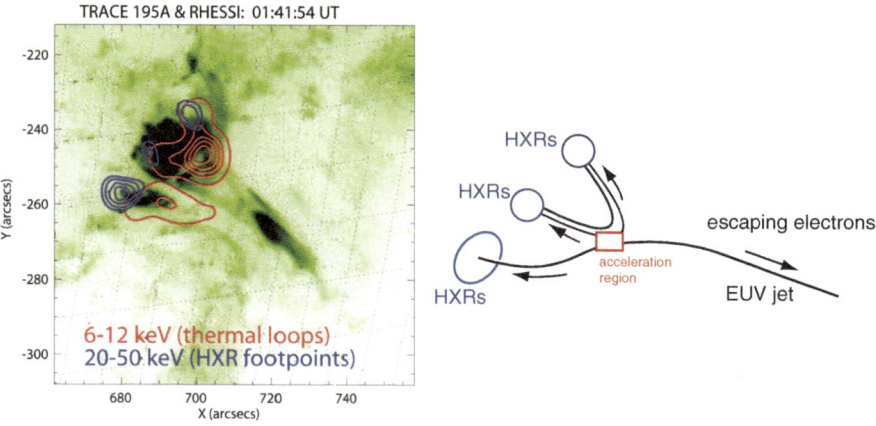

Fig. 5.10 EUV and X-ray sources of a flare that released energetic electrons into interplanetary space that were later observed near the Earth.
Left figure: RHESSI contours at 6–12 keV (*red* or *dark gray*: thermal emission) and 20–50 keV (*blue* or *black*: non-thermal emission) overlaid on a TRACE 195Å EUV image (*dark region* corresponds to enhanced emission). Located at around (700, −245) arcsec, the X-ray emission outlines a loop with two presumably non-thermal footpoints. The strongest footpoint source, however, is slightly to the southeast (683, −257) and shows a surprisingly lower intensity thermal source.
Right figure: Suggested magnetic field configuration showing magnetic reconnection between *open* and *closed* field lines inside the *red* or *dark gray box* marked as the "acceleration region" where downward moving electrons produce the HXR sources and upward moving electrons escape into interplanetary space

5.2.8 Statistical Properties of Flares

Flares are not just simple explosions in the solar atmosphere. Even a single "flare" shows many individual peaks during its evolution [69, 70]. When observing an active region or the whole Sun for a certain period of time, a number of flares with different total energy E (or peak energy E_p) will be recorded. If we define as $F(E)dE$ the fraction of flares which released energy between E and $E + dE$, then a very striking statistical feature of energy release in active regions emerges [38]. The frequency distribution $F(E)$ reconstructed from UV, EUV, and X-ray observations has a simple form (see Fig. 5.11)

$$F(E) = F_0 E^{-a}, \tag{5.1}$$

which holds for 8 orders of magnitude in E. Similar laws are obtained for the peak energy and the flare duration. The value of the exponent is not constant and may range from 1.6 to 2.0, depending on the data set used (see similar results reported in Chap. 8 for stellar flares). Current instruments are not able to observe nanoflares (energies below 10^{24} ergs) and the lower part of the distribution, which plays a crucial role in coronal heating, is uncertain. A key point for our discussion here is that the energy release of the active region is self-similar. This particular feature

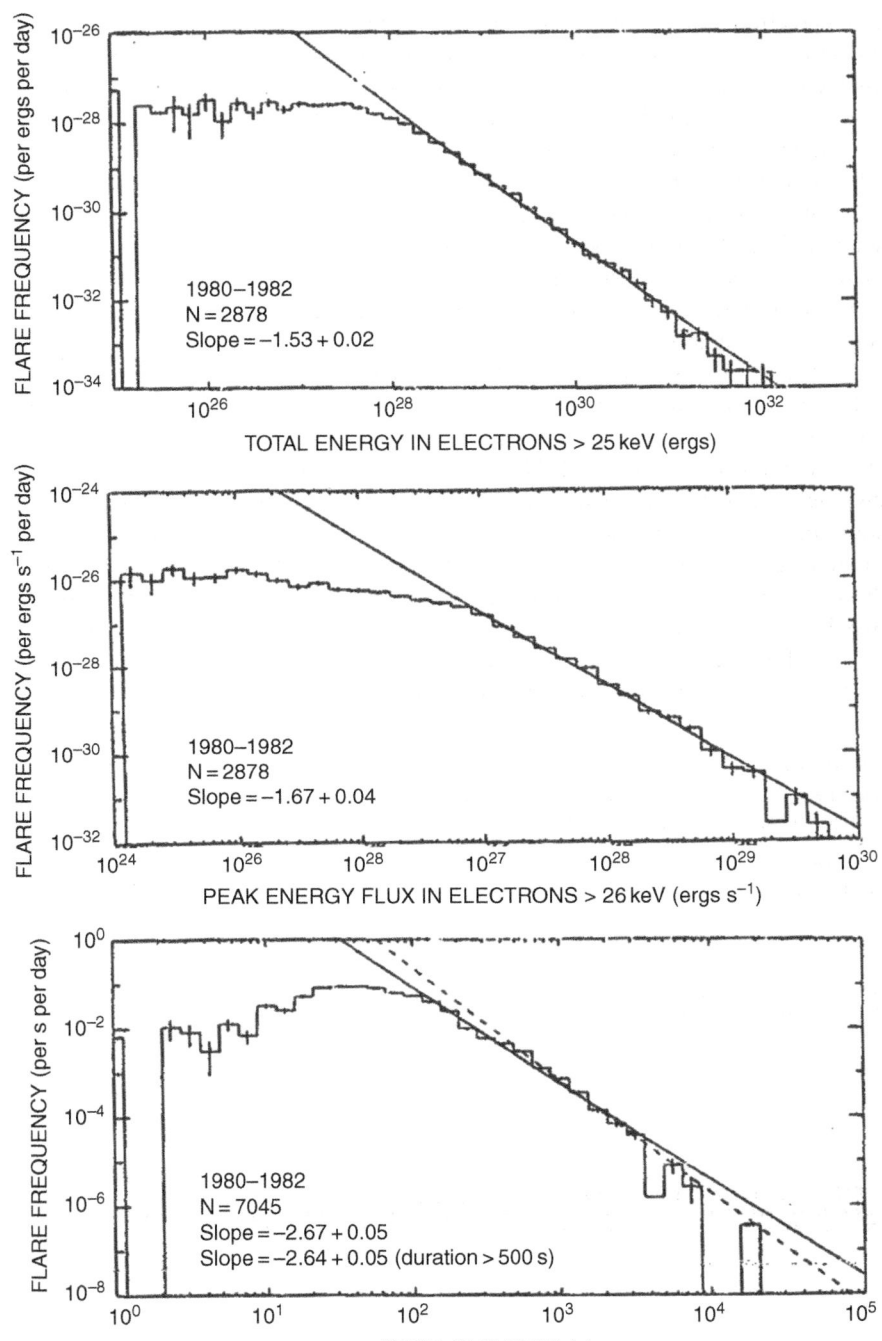

Fig. 5.11 The frequency distribution of total flare energy, peak flux, and duration [38]

of the observed characteristics of flares has created many heated discussions and remain an open and difficult theoretical problem which will be discussed in the next section.

5.2.9 Summary of Observational Constraints and Discussion

We now pull together the above results, and address their implications for understanding flare particle acceleration. Of particular importance are the implications arising from the thick-target model:

1. **The thick-target model for HXR and its theoretical implications:** The theoretical basis of the thick-target model, as originally presented [30] and reiterated recently [31], is based on the assumption that the accelerator is located in the corona and the HXR source in the upper chromosphere. Thus the acceleration region is collision-free and the radiation source is collision-dominated and electrons travel the distance between acceleration region and radiation region ballistically [16]. A large HXR burst flux suggests that the required electron flow rate is $\geq 10^{37}$ electrons s^{-1} with electron energies above 20 keV. This amounts to a total of 10^{39} electrons for a burst lasting several minutes. This result, known as the *number problem*, implies that all the particles inside a very large coronal volume ($\sim 10^{30}$ cm^3, almost the entire corona above an active region) are accelerated within a few minutes and stream toward the chromosphere. Assuming that the acceleration is inside a large-scale current sheet (see Fig. 5.12 and later discussion) with typical dimensions 10^{10} cm $\times$ 10^{10} cm $\times$ 10^5 cm, this monolithic current sheet must accelerate all the particles entering it (the inflow velocity needs to be a fraction of the local Alfvén speed) and remain stable for tens of minutes. We return to these points at the end of the section.

2. **Energetics:** Assuming that $\sim 10^{39}$ electrons are accelerated with a mean energy of 50 keV, the energy they carry is $\sim 10^{31}$ ergs. Since the accelerated particle fill a volume $\sim 10^{30}$ cm^3 and if the mean magnetic field available for dissipation in the corona is 30 G, the available magnetic energy is $\sim 5 \times 10^{31}$ ergs, so a significant fraction of the magnetic energy in this acceleration volume will go to the energetic electrons.

3. **Spectral index and low energy cutoff:** The energetic particles form a thermal distribution up to a critical energy $E_c \sim 1 - 30$ keV and a power law distribution above this energy. The spectral index (δ) varies both in the course of the burst and from event to event but remains within the range 2–6. The presence of multiple breaks at different energies is also observed frequently.

4. **The temporal evolution of the power law index:** The power law index varies during the impulsive phase of the flare, following a specific evolution: soft–hard–soft.

5. **Acceleration time:** The accelerator should start on sub-second timescales and remain on for tenths of minutes for the electrons. Ions are also accelerated in seconds and the accelerator remains active (sometimes) for hours.

6. **Maximum energy:** The maximum energy achieved is close to hundreds of MeV for the electrons and several GeV for the ions.

7. **Flare statistics:** The flares released in a specific active region are not random. They follow a specific statistical law in energy, peak intensity, and duration.

8. **Footpoint motion:** According to the "standard model" (see below) reconnection causes the footpoints to move smoothly away from each other or along the filament. Some observations seem to support this prediction but others not, so the motion of the footpoints is still an open question.

9. **The coronal sources:** Coronal sources at 20–30 keV are hard to confine collisionally, therefore the fact that they persist as isolated blobs in space, their characteristic spectral evolution, and their movement, remain open theoretical challenges.

10. **The close time and spectral evolution of the two footpoints:** When the two footpoints appear (usually in energies above 30 keV), they seem to correlate in temporal and spectral evolution leaving the impression that the accelerated particles moving in them are coming from the same acceleration source.

11. **Interplanetary energetic particles:** There is a close correlation of the HXR index with the properties of energetic particles detected in the interplanetary medium. This appears to need more complex magnetic topologies that currently discussed at the Sun.

12. **High energy ions:** There is an observed shift in the location of ion and electron footpoints. Sometimes, contrary to the electrons, the energetic ions show a single footpoint. The acceleration of ions and electrons in different length loops and the loop anisotropy with the low sensitivity are two explanations offered so far. There is an apparent correlation between electron acceleration above 300 keV and ion acceleration. The correlation of relativistic electron and ions, and the fact that the spectrum of electrons above 300 keV remains a power law with harder spectrum, recalls an older suggestion for two-stage acceleration, where shock acceleration may play an active role in the second stage in some large flares.

From the above summary, several important points arise, many concerning the efficiency requirements of the thick-target model. It is especially interesting to discuss this in the context of what is sometimes referred to the "standard flare model" as shown in Fig. 5.12. This originated in old models for long-decay flares [33, 89], and has been proposed as a generic scenario for coronal flaring. In particular, the model invokes a monolithic current sheet, which, one must assume, is where the particle acceleration takes place. In fact, as we will show in the next section, it is rather difficult to achieve efficient acceleration in simple magnetic topologies.

There are major electrodynamic constraints arising in the thick-target model. The large flux of energetic electrons ($F_{37} \sim 10^{37}$ electrons s^{-1}) flowing through a relatively small area (the observed footpoints are relatively compact with characteristic

Fig. 5.12 This cartoon, suggested several years ago, remains the favorite model and was elevated recently to the "standard flare cartoon". It has been in the literature for many years, it was revised to incorporate more recent observation, and it has been born out in simple 2-D simulations [152]

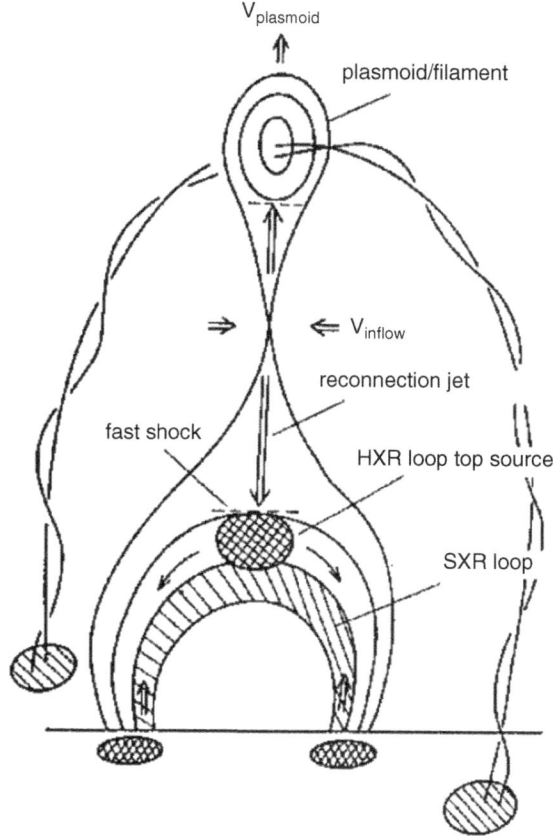

area $A_{17} \sim 10^{17}$ cm^2) suggests that the beam density of the energetic electrons (mean velocity 10^{10} cm s^{-1}) can be as high as $n_b \sim 10^{10}$ electrons cm^{-3}. Assuming that the ambient density at the HXR source is comparable or 1 order of magnitude higher ($n_0 \sim 10^{11}$ particles cm^{-3}), a neutralizing return current is required with a characteristic velocity of $v_r \sim 10^9$ cm s^{-1}. The return current replenishes the already-accelerated particles in the acceleration volume with hot plasma if the acceleration region and radiation region are magnetically connected. The observed hot thermal loops and the Neupent effect can be the observational tests for the reaction of the chromosphere to this intense electron beam injected from the corona. This vital point is not incorporated in current flare models, and the problem of particle replenishment remains an open issue.

We also note that the scenario adopted for the thick-target model for HXR and the "standard flare model" leave number of open questions: (1) How is a correlation between HXR and type III bursts established? (2) The density of the beams driving the normal type III burst ($n_b(III) \sim 10^6$ electrons cm^{-3}) are several orders of magnitude less than the beam density needed to power the HXR through the thick target ($n_b(HXR) \sim 10^{10}$ electrons cm^{-3}). What caused this large imbalance? (3)

The chromospheric evaporation will refill the loops with plasma in seconds, but if the acceleration and the energy release are above the loop(s) and the collapsing process for the formation of loops has been completed, how is the plasma inside the loop is re-accelerated?

We conclude that the standard 2-D flare cartoon shown in Fig. 5.12 and/or 2-D simulations based on the cartoon are not able to handle the relevant physics question. Eruption in 3-D magnetic topologies is still an active research project and the simple magnetic topology presented in Fig. 5.12 can mainly be used to represent an idea of how the overall magnetic structure may respond when energy is released during a CME/flare. We will return on this issue in the next section.

The above constraints for the energy release and the subsequent acceleration of high energy particles during large flares are hard to reconcile with reconnection theory as hosted in a simple magnetic topology and associated with a particular acceleration mechanism (DC electric fields, shocks, or MHD waves). However, this discussion should not be construed as an objection to the role of reconnecting current sheets in flare acceleration and particle trapping per se. It is a characteristic of the magnetohydrodynamic equations that they are "self-similar" over a wide range of scales: in other words the acceleration and heating is not just restricted to the large monolithic sheet, but occurs at current sheets of all sizes. In the following sections we discuss in more detail recent developments in the formation of the magnetic environment for particle acceleration, and try to relate these topologies with mechanisms for particle acceleration.

5.3 Models for Impulsive Energy Release

5.3.1 3-D Extrapolation of Magnetic Field Lines and the Formation of Unstable Current Sheets

The energy needed to power solar flares is provided by photospheric and sub-photospheric motions and is stored in non-potential coronal magnetic fields. Since the magnetic Reynolds number is very large in the solar corona, MHD theory states that magnetic energy can only be released in localized regions where the magnetic field forms small scales and steep gradients, i.e., in thin current sheets (TCSs).

Numerous articles (see recent reviews [45, 106] as well as Chaps. 2 and 4) are devoted to the analysis of magnetic topologies which can host TCSs. The main trend of current research in this area is to find ways to realistically reconstruct the 3-D magnetic field topology in the corona based on the available magnetograms and large-scale plasma motions at the photosphere. One must then search for the location of special magnetic topologies, i.e., separatrix surfaces (places were field lines form null points [98] and bald patches [157]), and more generally quasi-separatrix layers (QSLs) which are regions with drastic changes of the field line linkage [45]. A variety of specific 3-D magnetic configurations (fans, skeletons, etc.) have been analyzed, and their ability to host fast diffusion of the magnetic field lines has also

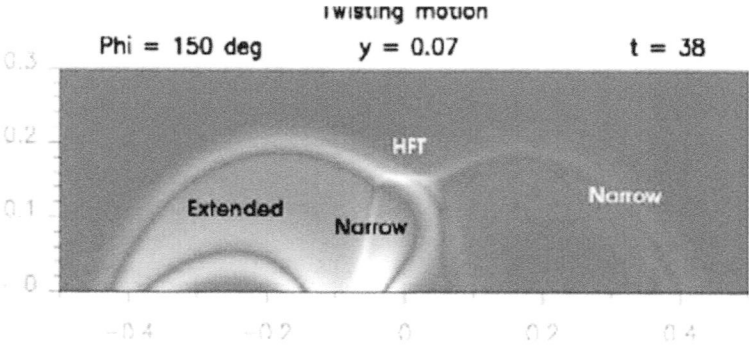

Fig. 5.13 Projected view of the two stressed magnetic field configurations used as initial conditions for the search of QSLs [17]

been investigated [142]. The main analytical and computational approaches through which these structures are analyzed are based on prescribed and simple magnetic structures at the photosphere, e.g., a quadrupole [17] (see Fig. 5.13). A realistic magnetic field generates many "poles and sources" [106] and naturally has a relatively large number of TCSs. We feel that this detailed representation of topological forms of the TCSs is mathematically appealing for relative simple magnetic topologies (dipoles, quadrupoles, symmetric magnetic arcades [17]). When such topological simplicity at the photosphere is broken, for example, due to large-scale sub-Alfvénic photospheric motions or the emergence of new magnetic flux that disturbs the corona, such tools may be less useful. All these constraints restrict our ability to reconstruct fully the dynamically evolving magnetic field of an active region (and it is not clear that such a reconstruction will ever be possible).

Many of the widely used magnetograms measure only the line-of-sight component of the magnetic field. The component of the magnetic field vertical to the surface matches the measured magnetic field only at the center of the disk and becomes increasingly questionable as the limb is approached. Extrapolating the measured magnetic field is relatively simple if we assume that the magnetic field is a force-free equilibrium:

$$\nabla \times \mathbf{B} = \alpha(\mathbf{x})\mathbf{B}, \tag{5.2}$$

where the function $\alpha(\mathbf{x})$ is arbitrary except for the requirement $B \cdot \nabla\alpha(\mathbf{x}) = 0$, in order to preserve $\nabla \cdot \mathbf{B} = 0$. Equation (5.2) is nonlinear since both $\alpha(\mathbf{x})$ and $\mathbf{B}(\mathbf{x})$ are unknown. We can simplify the analysis of Eq. (5.2) when α=constant. The solution is easier still when $\alpha = 0$, which is equivalent to assuming the coronal fields contain no currents (potential field), hence no free energy, and thus uninteresting.

A variety of techniques have been developed for the reconstruction of magnetic field lines above the photosphere and the search for TCSs [106, 112]. It is beyond the scope of this chapter to discuss these techniques in detail. For instructive purposes, we use the simplest method available, a linear force-free extrapolation, and search for "sharp" magnetic discontinuities in the extrapolated magnetic fields. Vlahos and Georgoulis [170] use an observed active region vector magnetogram and then: (i) resolve the intrinsic azimuthal ambiguity of 180° [64] and (ii) find the best-fit value α_{AR} of the force-free parameter for the entire active region, by minimizing the difference between the extrapolated and the ambiguity-resolved observed horizontal field (the "minimum residual" method of [101]). They perform a linear force-free extrapolation [6] to determine the 3-D magnetic field in the active region. Although it is known that magnetic fields at the photosphere are not force-free [66], they argue that a linear force-free approximation is suitable for the statistical purposes of their study.

Two different selection criteria were used in order to identify potentially unstable locations (identified as the aforementioned TCSs) [170]. These are (i) the Parker angle and (ii) the total magnetic field gradient. The angular difference $\Delta\psi$ between two adjacent magnetic field vectors, $\mathbf{B_1}$ and $\mathbf{B_2}$, is given by $\Delta\psi = cos^{-1}[\mathbf{B_1 \cdot B_2}/(B_1 B_2)]$. Assuming a cubic grid, they estimated six different angles at any given location, one for each closest neighbors. The location is considered potentially unstable if at least one $\Delta\psi_i > \Delta\psi_c$, where $i \equiv \{1, 6\}$ and $\Delta\psi_c = 14°$. The critical value $\Delta\psi_c$ is the Parker angle which, if exceeded locally, favors tangential discontinuity formation and the triggering of fast reconnection [135, 136]. In addition, the total magnetic field gradient between two adjacent locations with magnetic field strengths B_1 and B_2 is given by $|B_1 - B_2|/B_1$. Six different gradients were calculated at any given location. If at least one $G_i > G_c$, where $i \equiv \{1, 6\}$ and $G_c = 0.2$ (an arbitrary choice), then the location is considered potentially unstable. When a TCS obeys one of the criteria listed above, it will be transformed to an unstable current sheet (UCS). A steep gradient of the magnetic field strength, or a large shear, favors magnetic energy release in 3-D in the absence of null points [143]. [Note that these thin elongated current sheets have been given different names by different authors, e.g., in [17] they are called hyberbolic flux tubes.]

Potentially unstable volumes are formed by the merging of adjacent selected locations of dissipation. These volumes are given by $V = N\lambda^2 \delta h$, where N is the number of adjacent locations, λ is the pixel size of the magnetogram, and δh is the height step of the force-free extrapolation. The free magnetic energy E in any volume V is given by

$$E = \frac{\lambda^2 \delta h}{2\mu_0} \sum_{l=1}^{N} (\mathbf{B_{ff}}_l - \mathbf{B_{p}}_l)^2, \qquad (5.3)$$

where $\mathbf{B_{ff}}_l$ and $\mathbf{B_{p}}_l$ are the linear force-free and the potential fields at location l, respectively. The assumption used is that any deviation from a potential configuration implies a non-zero free magnetic energy which is likely to be released if certain conditions are met. UCSs are created naturally in active regions even during their formation (Fig. 5.14) and the free energy available in these unstable volumes follows a *power law distribution* with a well-defined exponent (Fig. 5.15). We can then conclude that active regions store energy in many unstable locations, forming UCS

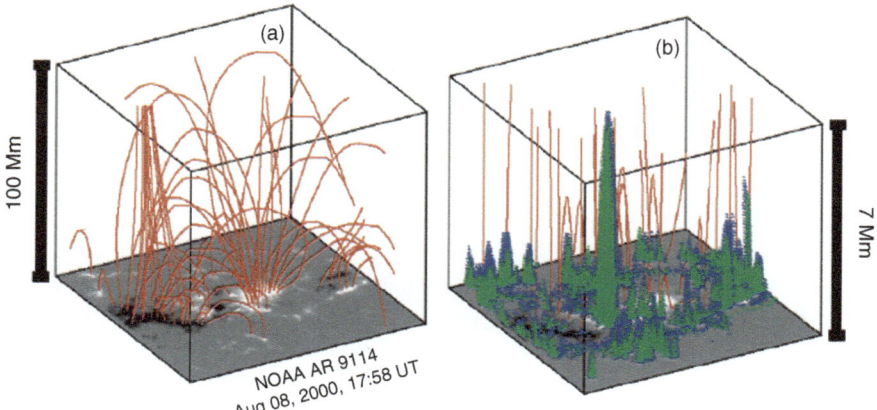

Fig. 5.14 (**a**) Linear force-free field extrapolation in NOAA AR 9114, (**b**) lower part of the AR atmosphere. Shown are the magnetic field lines (*red*) with the identified discontinuities for critical angle 10° [170].

of all sizes (i.e., the UCSs have a self-similar behavior). The UCSs are fragmented and distributed inside the global 3-D structure. Viewing the flare in the context of the UCS scenario presented above, we can expect, depending on the size and the scales of the UCS, to have flares of all sizes. Small flares dominate, and have the potential to heat the corona, and large flares occur when large-scale QSL complexes are formed.

The next step is to analyze the evolution of an isolated UCS. We already stressed above that the method followed by [170] has several weak points, but nevertheless provides a simple tool for the analysis of the statistical behavior of the places hosting UCS and flares (see also [107]). Aulanier et al. [17, 18] started from a carefully prepared magnetic topology in the photosphere (bipolar formed by four flux concentration regions) in which the potential extrapolation contains QSLs, and observed and analyzed the formation and the properties of TCSs. The 3-D magnetic

Fig. 5.15 Typical distribution function of the total free energy in the selected volume, on using a critical angle 14° [170]

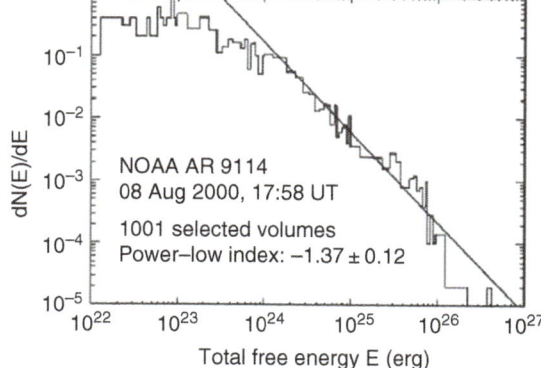

NOAA AR 9114
08 Aug 2000, 17:58 UT

1001 selected volumes
Power–low index: -1.37 ± 0.12

topology was driven by photospheric motions and the end result was the formation of TCSs in the vicinity of the QSLs. Unfortunately no statistical analysis of the characteristics of the TCSs was reported since the MHD codes used do not have the ability to resolve the transition from TCSs to UCSs.

5.3.2 The 3-D Turbulent Current Sheet

Magnetic reconnection is the topological change of a magnetic field by the breaking of the magnetic field lines. It happens in regions where the assumption of flux freezing in ideal magnetohydrodynamics (MHD) no longer holds [141, 26]. Resistivity plays a key role in magnetic reconnection. The classical (Spitzer) resistivity in the solar corona is extremely low ($\sim 10^{-16}$), therefore ideal MHD theory holds in general. Exceptions are the UCS where the resistivity can jump by many orders of magnitude and ideal MHD theory becomes invalid [131, 45]. (Of course the UCS should be analyzed ideally in the framework of 3-D kinetic theory [28, 176].)

Onofri et al. [131] studied the nonlinear evolution of current sheets using the 3-D incompressible and dissipative MHD equations in a slab geometry. The resistive MHD equations in dimensionless form are as follows (see detailed discussion on these equations on Chap. 2):

$$\frac{\partial \mathbf{V}}{\partial t} + (\mathbf{V} \cdot \nabla) \mathbf{V} = -\nabla \left(P + \frac{B^2}{2} \right) + (\mathbf{B} \cdot \nabla) \mathbf{B} + \frac{1}{R_v} \nabla^2 \mathbf{V}, \tag{5.4}$$

$$\frac{\partial \mathbf{B}}{\partial t} = -\nabla \times \mathbf{E}, \tag{5.5}$$

$$\mathbf{j} = \nabla \times \mathbf{B}, \tag{5.6}$$

$$\mathbf{E} + \mathbf{V} \times \mathbf{B} = \frac{1}{R_M} \mathbf{j}, \tag{5.7}$$

$$\nabla \cdot \mathbf{V} = 0, \tag{5.8}$$

$$\nabla \cdot \mathbf{B} = 0, \tag{5.9}$$

where $\mathbf{V}$ and $\mathbf{B}$ are the velocity and the magnetic field, respectively, P is the pressure, and R_v and R_M are the kinetic and magnetic Reynolds numbers with $R_v = 5000$ and $R_M = 5000$, respectively. Here the density has been set to unity (incompressible) and the constant μ_0 absorbed into the magnetic field. The initial conditions were established in such a way as to have a plasma that is at rest in the frame of reference of the computational domain, permeated by an equilibrium magnetic field $\mathbf{B_0}$, sheared along the x-direction, with a current sheet in the middle of the simulation domain:

$$\mathbf{B_0} = B_{yo} \hat{y} + B_{zo}(x) \hat{z},$$

where B_{yo} is constant and B_{zo} is given by

$$B_{zo}(x) = tanh\left(\frac{x}{a}\right) - \frac{x/0.1}{cosh^2(a/0.1)}.$$

In the y- and z-directions, the equilibrium magnetic field is uniform and periodic boundary conditions are imposed, since no boundary effects are expected in the development of the turbulence. In the inhomogeneous x-direction, fixed boundary conditions are imposed. These equilibrium fields were perturbed with 3-D magnetic field fluctuations satisfying the solenoidal condition.

The nonlinear evolution of the system is characterized by the formation of small-scale structures, especially in the lateral regions of the computational domain, and coalescence of magnetic islands in the center. This behavior is reflected in the 3-D structure of the current (see Fig. 5.16), which shows that the initial equilibrium is destroyed by the formation of current filaments, with a prevalence of small-scale features. The final stage of these simulations is a turbulent state, characterized by many spatial scales, with small structures produced by a cascade with wavelengths decreasing with increasing distance from the current sheet. In contrast, inverse energy transfer leads to the coalescence of magnetic islands producing the growth of 2-D modes. The energy spectrum approximates a power law with slope close to 2 at the end of the simulation. Similar results have been reported by many authors using several approximations [52, 118, 97, 153]. It is also interesting to note that similar results are reported from magnetic fluctuations in Earth's magnetotail [177].

It has become apparent over the years that the (theoretical) Ohm's law used in resistive MHD is

$$\mathbf{E} + \mathbf{V} \times \mathbf{B} = \eta\mathbf{j}, \tag{5.10}$$

where $\mathbf{E}$ and $\mathbf{B}$ are the electric and magnetic fields, $\mathbf{V}$ is the fluid velocity, $\mathbf{j}$ is the current, and η is the resistivity, breaks down near reconnection sites. The main

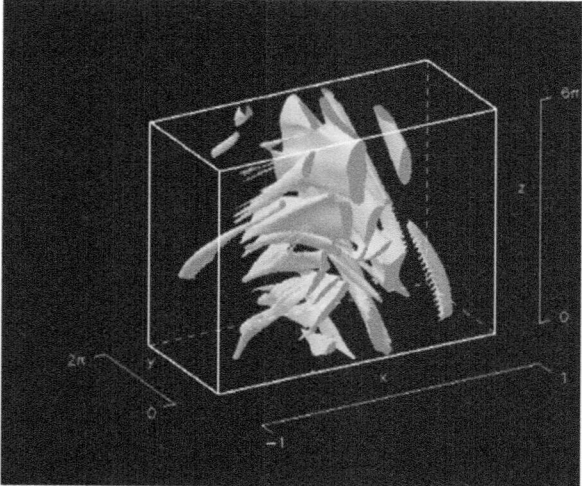

Fig. 5.16 Current isosurfaces showing the formation of current filaments [131]

reason is that the region of electron demagnetization is much smaller than the ion inertial length c/ω_i, where c is the speed of light and ω_i the ion plasma frequency, and so Hall terms in the full version Ohm's law become important:

$$\frac{1}{\omega_e^2}\frac{d\mathbf{j}}{dt} = \mathbf{E} + \mathbf{V} \times \mathbf{B} - \frac{1}{ne}\mathbf{j} \times \mathbf{B} \qquad (5.11)$$

$$+\frac{1}{ne}\nabla \cdot \overset{\Rightarrow}{P_e} - \eta\mathbf{j}, \qquad (5.12)$$

where $\overset{\Rightarrow}{P_e}$ is the electron pressure tensor and n is the plasma density. The proper framework to study magnetic reconnection, including the important contribution of the Hall term in the analysis, is the two-fluid equations. Cessak et al. [36], using a two-fluid code, reported a very interesting scenario for magnetic reconnection. The reconnection proceeds slowly and allows the system to accumulate stresses as it forms TCSs which evolve and remain stable over a long period of time. When the thickness of the TCS reaches a critical value, the system adjusts abruptly to exhibit fast reconnection. The authors called this particular model for reconnection "A catastrophe model for fast reconnection", we have adopted here the term UCS for the fast reconnection and the slow evolving mode is called TCS. Switching from the "stable" TCS to fast reconnection (the UCS) is related to the fact that the anomalous resistivity turns on. The need for a critical threshold is thus crucial for the nonlinear evolution of an active region.

Simulating magnetic reconnection with a 3-D full-particle code is currently an ongoing research project that presents many difficulties (see [144, 28, 176] and references therein). Experimental verification of magnetic reconnection has shown evidence of a positive correlation between the magnitude of magnetic fluctuations up to the lower-hybrid frequency range [84], and in the Hall effect. They also measure short coherent lengths indicating a strongly nonlinear nature of the evolution of the reconnection current sheet. The main difficulty with a realistic analysis of magnetic reconnection using 3-D kinetic models is the wide range of spatial and temporal scales separating the reconnection region from the magnetic fields observed during a flare or a CME.

5.3.3 The Compact Flare

A series of recent studies explored the question "How does a loop respond to a random photospheric driver?" In the past, flares were assumed to be driven by organized and continuous twisting or shearing motions in the photosphere. Galsgaard and Nordlund [61] and Galsgaard [62] explored a different scenario for flare initiation. The 3-D time-dependent MHD equations (Eqs. 5.4, 5.5, 5.6, 5.7, 5.8, 5.9) were solved in a cartesian box with model photospheres at either end. An energy equation with anisotropic heat condition and optical thin radiation is included. Between photosphere and corona there is a stratified atmosphere (the gravitational force is modeled by a sine function, vanishing at the center of the computational box), so that the density profile is a hyperbolic tangent. At the start of each simulation, there is a uniform magnetic field extending between the two photospheric regions. In order to relate the simulations to observed coronal loop structures, the simulation box is 20 times longer than it is wide. The coronal density is 10^3 times smaller than that in the photosphere, implying an Alfvén speed in the corona approximately 30 times larger than in the photosphere.

Solar magnetic flux tubes connect different regions in the photosphere. This initial state is perturbed by imposing simple sinusoidal shear motions on the magnetic field at the two boundaries. Their wavelength is equal to the transverse length, while their phase, orientation, and direction are random. This, in a simple manner, represents the advection of magnetic flux due to convective motions, and injects energy into the corona. The coronal field responds to these boundary motions, with the Lorentz force determining its evolution. After sometime (a few seconds corresponding to the time needed for an Alfvén wave to cross the loop) the stresses are distributed along the entire loop, and coronal current sheet (TCS) formation occurs. As reconnection commences (the sudden formation of UCSs), plasma jets are formed, and eventually their momentum is sufficient to strongly perturb the neighboring plasma, creating secondary current concentrations. A turbulent cascade is thus initiated so that throughout the simulation, energy is injected on large scales, but cascades through a turbulent process to the shortest possible length scale where it is dissipated in numerous small current concentrations randomly distributed throughout the volume (see Fig. 5.17). It was also discovered that the response of the small compact loops (length around $\sim 5 \times 10^9$ cm) is to form fragmented current sheets in the middle part of the loop [62]. For longer loops (length larger than $\sim 10^{10}$ cm) the current sheets form at the footpoints. This particular observation may have important consequences on the interpretation of several observed characteristics of flares.

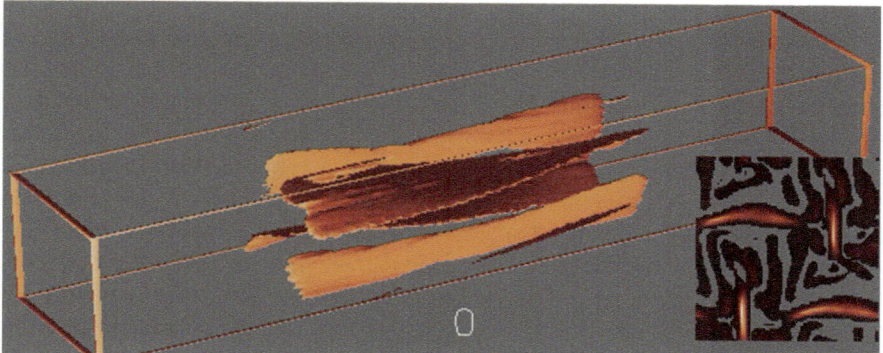

Fig. 5.17 The loop is stressed by random photospheric flows and is led to a state where numerous current sheets are present. A vertical cross section through the middle of the loop shows the formation of current sheets [62]

5.3.4 A Cellular Automata Model for the Energy Release in the Solar Corona

Coronal energy release observed at various wavelengths shows impulsive behavior with events from flares to bright points exhibiting intermittency in time and space. Intense X-ray flare emission typically lasts several minutes to tens of minutes, and only a few flares are recorded in an active region that typically lives several days to several weeks. The flaring volume is small compared to the volume of an active region, regardless of the flare size. Intermittency is the dynamical response of a turbulent system when the triggering of the system is the result of a critical threshold for the instability [36]. In a turbulent system one also expect *self-organization*, i.e., the reduction of the numerous physical parameters (degrees of freedom) present in the system to a small number of significant degrees of freedom that regulates the system's response to external forcing [129]. This is the reason for the success of concepts such as self-organized criticality (SOC) [19, 20] in explaining the statistical behavior of flares discussed in Sect. 5.2.8. Cellular automata (CA) models typically employ one free parameter (the magnetic field, vector potential, etc.) and study its evolution subject to external perturbations. When a critical threshold is exceeded (when the TCS becomes a UCS), parts of the configuration are unstable, and will restructure to re-establish stability. The rearrangement may cause instabilities in adjacent locations, so the relaxation of the system may proceed as an avalanche-type process. In SOC flare models [108, 109, 166] each elementary relaxation is viewed as a single magnetic reconnection event, so magnetic reconnection is explicitly assumed to occur with respect to a critical threshold.

In solar MHD a UCS disrupts either when its width becomes smaller than a critical value [140], or when the magnetic field vector forms tangential discontinuities exceeding a certain angle [134], or when magnetic field gradients are steep enough to trigger restructuring [143]. We notice that a critical threshold is involved in all cases: The first process points to the turbulent evolution in the magnetic field configuration and the onset of anomalous resistivity, while the latter two imply magnetic discontinuities caused either by the orientation of the magnetic field vector or by changes of the magnetic field strength. Magnetic field gradients and discontinuities imply electric currents via Ampére's law, however; so a critical magnetic shear or gradient implies a critical electric current accumulated in the current sheet which in turn leads to the onset of anomalous resistivity [133, 135].

One way of modeling the appearance, disappearance, and spatial organization of UCS inside a large-scale topology is with the use of the extended cellular automaton (X-CA) model [79, 80, 81]. Figure 5.18 illustrates some basic features of the X-CA model. The X-CA model has as its core a cellular automaton model of the sand-pile type and is run in the state of self-organized criticality (SOC). It is extended to be fully consistent with MHD: the primary grid variable is the vector potential, and the magnetic field and the current are calculated by means of interpolation as derivatives of the vector potential in the usual sense of MHD, guaranteeing $\nabla \cdot \mathbf{B} = 0$ and $\mathbf{J} = (1/\mu_0)\nabla \times \mathbf{B}$ everywhere in the simulated 3-D volume. The electric field is defined as $\mathbf{E} = \eta\mathbf{J}$, with η the diffusivity. The latter usually is negligibly small,

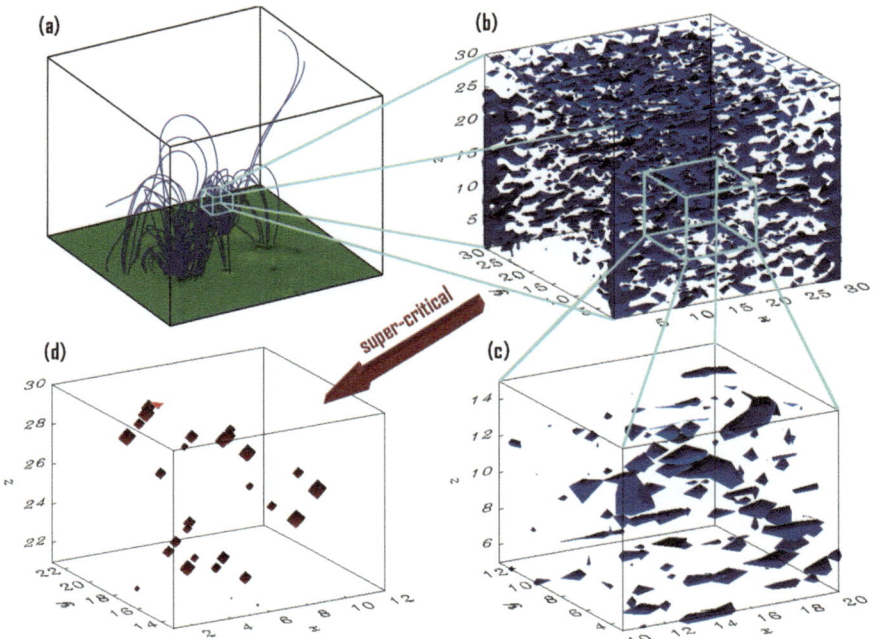

Fig. 5.18 (**a**) Simulated magnetogram of a photospheric active region and force-free magnetic field lines, extrapolated into the corona. (**b**) Sub-critical current iso-surfaces in space, as yielded by the X-CA model, which models a sub-volume of a coronal active region. (**c**) The same as (**b**), but zoomed. (**d**) Temporal snapshot of the X-CA model during a flare, showing the spatial distribution of the UCS (super-critical current isosurface) inside the complex active region [171].

but if a threshold in the current is locally reached ($|\mathbf{J}| > J_{cr}$), then current-driven instabilities are assumed to occur, η becomes anomalous in turn, and the resistive electric field locally increases drastically. These localized regions of intense electric fields are the UCS in the X-CA model.

The X-CA model yields distributions of total energy and peak flux which are compatible with the observations. The UCSs in the X-CA form a set which is highly fragmented in space and time: The individual UCS are small-scale regions, varying in size, and are short-lived. They do not form in their ensemble a simple large-scale structure, but form a fractal set with fractal dimension roughly $D_F = 1.8$ [171]. The individual UCS also do not usually split into smaller UCS, but they trigger new UCSs in their neighborhood, so that different chains of UCS travel through the active region, triggering new side-chains of UCS on their way. It is obvious that the rules of this simulation do not include the fragmentation of the UCS and in many ways the results coincide with the MHD simulations [62].

5.3.5 The Magnetic Coupling of Convection Zone with Corona

Active regions are externally driven (from the turbulent convection zone), dissipative (magnetic energy released in coronal heating, flares, CME), nonlinear dynamical systems [168, 67]. Flux emergence and photospheric boundary flows play the role of the driver. The evolution of an active region is largely dictated by the configuration of the magnetic field vector, which is subject to boundary-induced perturbations. An important question remains open: Is the structure and evolution of magnetograms and the photospheric flows responsible for the activity in active region? In other words, can we predict a flare and/or CME using observations from the photosphere?

A variety of well-established observations have analyzed the characteristics of photospheric magnetograms (see [180, 75] and references therein). The most striking properties are as follows:

1. The active region magnetic fields form self-similar structures, with the area (A). Probability distribution functions (PDF) obeying well-defined power laws $P(A) \sim A^{-1.8}$ and with fractal dimensions ranging approximately between ~ 1.2 and ~ 1.7 are found (see, e.g., [71, 72, 116, 117] and references therein).
2. Numerous studies have revealed the multi-fractal nature of active regions [99, 32, 100, 1, 2] and their structure function [2, 3].

These magnetogram properties are an important diagnostic for the turbulent convection zone dynamics, and as yet are not reproduced in 3-D MHD simulations.

Recently a percolation model was proposed to simulate the formation and evolution of active regions at the photosphere [173, 150]. In this model, the evolution of the magnetograms is followed by reducing all the complicated convection zone dynamics into three dimensionless parameters. The emergence and evolution of magnetic flux on the solar surface using a 2-D cellular automaton (CA) is probabilistic and based on the competition between two "fighting" tendencies: *stimulated* or *spontaneous* emergence of new magnetic flux, and the disappearance of flux due to *diffusion* (i.e., dilution below observable limits), together with random *motion* of the flux tubes on the solar surface. This percolation model explains the observed size distribution of active regions and their fractal characteristics [116, 169]. It was later used for the reconstruction of 3-D active regions using the force-free approximation and many of the observational details reported in [170] were reproduced [60]. The connection of photospheric activity with the statistical properties of flares has also been simulated by several authors and the results are promising [137, 76, 77, 162].

We have a long way to go before we establish a good understanding of the connection of the driver (photosphere) with the coronal part of an active region [67]. One point is worth stressing: The details of the magnetogram and the large-scale sub-Alfvénic photospheric flows hold many of the secrets of the activity of the active region. The formation and the statistical properties of TCSs and UCSs are in many ways connected with the properties of the "driver".

5.3.6 The Eruptive Flare/CME Model

A large number and range of models demonstrate the connection between flares and CMEs [7, 13, 59, 89, 147, 174, 14, 63]. All start from simple (arcade, loop, or emerging flux) magnetic topology (analyzed mostly in 2-D and only recently in 3-D) which is driven to instability by well-described photospheric motions. In most of these models the initial conditions and the photospheric driver are adjusted in such a way so that the magnetic eruption will be unavoidable (see, for example, [59]). However, the inability of the MHD simulations to handle simultaneously the dissipation of magnetic energy (small scales) and mechanisms for heating and acceleration (perhaps on a large scale) led many researchers to sketch the expected radiation signatures using simple cartoons. The cartoon presented by [59], for example, suggest that the high energy particles are confined in a small portion of the total volume related with the erupted structure.

From the theoretical point of view, it is hard to prove that a huge structure with dimensions $\left[10^{10} \text{ cm} \times 10^{10} \text{ cm} \times 10^{5} \text{ cm}\right]$ can remain stable and active for hundreds of seconds. As we have seen in Sect. 5.3.2, the dissolution of the current sheet and the formation of several smaller fragments will be its natural evolutionary path [88, 50, 131].

The 3-D evolution of a simple photospheric magnetic field topologies leads also to the break out model, but the magnetic topology is extremely complex [7]. The formation of a large number of tangential discontinuities (see Fig. 5.19) which will form numerous current filaments may be the answer to the high energy emission observed.

The simple magnetic topology for the current sheet presented earlier and the associated simple accelerators (direct E-field, constant flows, and shocks) are probably replaced in the 3-D magnetic topology with much more complex accelerators as we will see in the next section.

5.3.7 Principal Conclusions Concerning Models for Energy Release in Active Regions

We outline below the main points from this section and how they influence our subsequent discussion of particle acceleration:

1. **The large-scale structure**: The nonlinear extrapolation of observed photospheric magnetic fields gives the basic magnetic field skeleton which hosts the energy release.
2. **Reconstruction of magnetic topologies**: Using quite simple techniques [170], we can demonstrate many interesting properties of 3-D magnetic fields in active regions. The main themes of these approximate extrapolation are "fragmentation and self-organization", both characteristics of driven turbulent systems [170]. It is apparent that the formation of thin current sheets (TCSs) in the vicinity of QSLs is the way flares start in stressed magnetic topologies [17].
3. **The driver**: The detailed structure and sub-Alfvénic flows of the observed photospheric magnetic field, and newly emerging magnetic flux [63, 14], influence the evolution and the activity of the active region. Unfortunately, detailed nonlinear extrapolation of photospheric magnetic fields is impossible at

the present time, presenting a major drawback to our understanding of flares and
CME [45].

4. **Threshold for reconnection and the turbulent current sheet**: Current understanding of magnetic reconnection reveals several important properties. (1) The
   reconnection proceeds in two modes: (a) a slow mode where the TCS continues
   to accumulate stresses and store magnetic energy and (b) a fast mode when the
   TCS reaches a certain threshold when the resistivity suddenly jumps to a high
   value [36]. (2) The current sheet evolves to a "turbulent state" in a relatively
   short time (a few hundred Alfvén times) [131, 118, 97, 50].

5. **Self-organized criticality**: Does the statistical behavior of flares imply that active regions are always in a self-organized critical (SOC) state? Several studies
   suggest that this can occur and is the reason behind the statistical properties of
   flares noted in Sect. 5.2.8.

6. **The appearance of strong turbulence during explosive phenomena**: The fragmentation and self-organization of the turbulent UCS suggest that a flaring active
   region quickly enters into a "turbulent state" during a flare/CME.

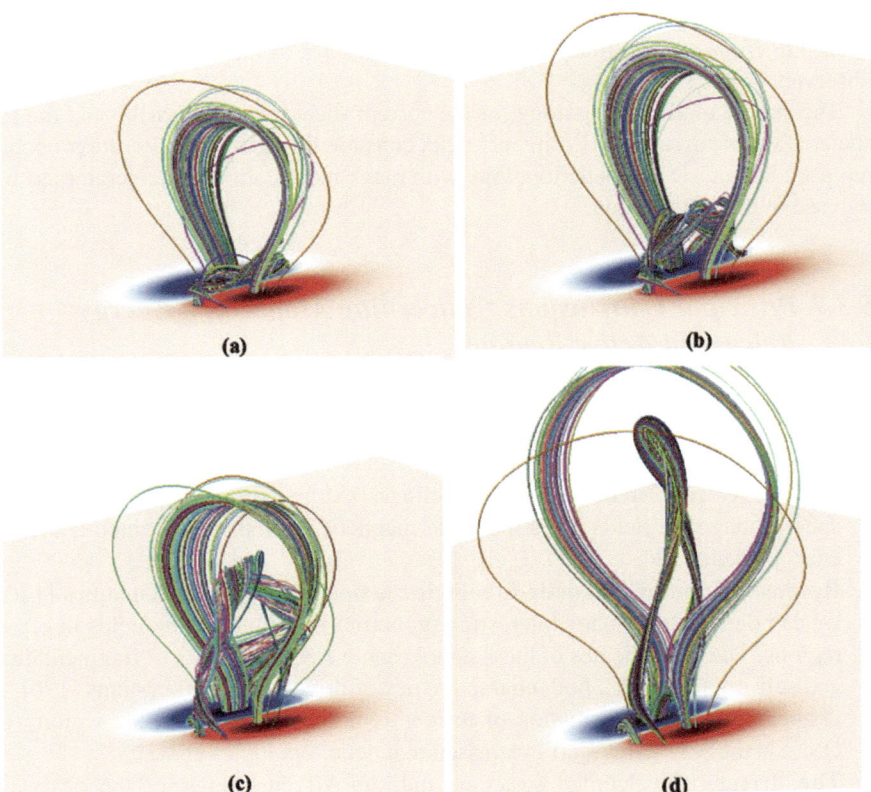

Fig. 5.19 Using the 3-D MHD equations, even by starting from a simple magnetic geometry, the
arcade is stressed and led to the eruption. There are several points in this structure where the
stresses are relatively large leading to reconnection [7]

7. **Two broad classes of flares**: Stressed large-scale magnetic structures (e.g., coronal loops [127, 61, 62]) or eruptive structures forming UCSs [7] everywhere in the stressed structure (see Fig. 5.19). Unfortunately the search for UCS, in analogy with the work reported for the compact loops [62], has not been performed for erupting structures or the interaction of the emerging flux with the pre-existing magnetic fields. The distributed magnetic stresses in large-scale 3-D erupting magnetic topologies remains an unexploited theoretical challenge.

The analysis presented so far in this section suggests that we have a long way to go to understand energy release in flares but many important steps have been made. The key element is the cascade of the UCS of all scales and the inverse cascade of the coalescence of islands form a large turbulent region (with scales of tens of thousands of kilometers) inside the evolving turbulent AR (scales of tenths of millions of kilometers). Particle acceleration mechanisms developed in the next section rely heavily on the concepts presented so far in order to build new "*strong turbulent*" acceleration mechanism for solar flares.

5.4 Particle Acceleration in Turbulent Electromagnetic Fields

5.4.1 Brief Overview of Acceleration Mechanisms

Numerous books and reviews have been devoted to the challenging problem of particle acceleration [73, 163, 115, 165, 86, 167, 94, 121, 35]. The most prominent mechanisms analyzed in-depth so far in the literature are shock waves [74, 27, 54, 43], MHD and higher frequency plasma waves [57, 119], and DC electric fields [22, 123, 124, 110].

Studying a single acceleration mechanism's mechanism (e.g., shock waves, MHD or plasma waves, DC electric fields) in isolation implies that the energy release process favors one specific mechanism over the others. One example of where this holds is in supernova explosions when, at least in the initial stages, diffusive shock acceleration will prevail. Another example is a stable "monolithic" large-scale current sheet, where the direct electric field will dominate. However, as we discuss later, realistic models for the energy release in solar flares may have multiple acceleration mechanisms operating.

There have been a number of investigations of multiple acceleration mechanisms. Decker and Vlahos [42] analyzed shock drift acceleration (SDA) when the shock was surrounded by waves. SDA is fast but not efficient, since the particles drifting along the electric field in the shock surface quickly leave the shock. However, the presence of MHD waves upstream and downstream of the shock sustains the acceleration process by providing a magnetic trap around the shock surface, so forcing a particle to return there many times. Thus, the particle leaves the shock surface, travels a distance s_i inside the turbulent magnetic field, returns back to the shock surface with velocity v_i, drifts a distance l_i along the shock electric field E_{sc}, changing its momentum by $\Delta p_i \sim eE_{sc} \cdot (l_i / v_i)$ (assuming that Δp_i is small). It then escapes again, travels a distance s_{i+1} before returning to the shock and drifting along the electric field: in other words, the acceleration follows a *cyclic process*. The process repeats itself several times before the particle gains enough energy to escape from the turbulent trap.

Let us now note some very important characteristics of this acceleration: (1) The distances s_i traveled by the particle before returning to the shock are only indirectly related to the acceleration, since they basically delay the process and influence the overall timing, i.e., the *acceleration time*, an important parameter of the particle acceleration process. (2) The energy gain depends critically on the lengths l_i that the particle drifts along the shock surface, but in a statistical sense, i.e., on the distribution of the l_i, $i = 1, 2, 3 \ldots$. (3) The times τ_i a particle spends at the shock surface are again crucial for the energy gain, and also, together with the s_i, for the estimation of the acceleration time. (4) In the context of the total acceleration problem, i.e., the energies reached and the times needed to reach them, all three variables, s_i, l_i, τ_i, are of equal importance.

Ambrosiano et al. [8] discussed a similar problem, namely superposing a population of Alfvén waves on a current sheet. Here also the ability of the DC electric field to accelerate particles is enhanced by the presence of the MHD waves. The acceleration process is again cyclic, and is again characterized by the three variables s_i, l_i, τ_i. The turbulent current sheet has several ways to enhance the acceleration efficiency, since the plasma inflow is dynamically driven, and causes a variety of new and still unexplored phenomena. The trapping of the particles inside the turbulent magnetic field gives rise to a new "collision scale", and, in some circumstances, acceleration becomes dependent on an alternative "Dreicer field", in which particle collisions are replaced by collisions with magnetic irregularities. [Indeed diffusive shock acceleration [27, 54] is also of a mixed type, having as elements a shock (moving discontinuity) and "converging" magnetic turbulence. Turbulence plays the role of approaching walls which scatter the particles.] In fact, it seems that most acceleration mechanisms are of a mixed type in some way. We can conclude that the mixture of mechanisms enhances the acceleration efficiency and removes some of the drawbacks attached to different, isolated mechanisms. *Cyclic processes*, e.g., through trapping around the basic accelerator, are important elements – if not the presupposition – of efficient and fast acceleration in space plasmas.

5.4.2 Theoretical Frameworks for the Study of Particle Acceleration

All acceleration mechanisms in space are related to local or global plasma instabilities. The stable plasma, prior to the start of the instability, is usually assumed to be magnetized and in thermal equilibrium. In the stable plasma, the magnetic field $\mathbf{B}_0$ typically is assumed to have a simple topology, the electric field $\mathbf{E}_0$ is zero, the ambient velocity distribution is Maxwellian $f_M(\mathbf{v})$, the ambient particle density is n_0. The unstable plasma is considered as a "perturbation" of the stable state ($\mathbf{B}(\mathbf{r}, t) = \mathbf{B}_0 + \mathbf{B}_1(\mathbf{r}, t)$, $\mathbf{E}(r, t) = \mathbf{E}_1(\mathbf{r}, t)$, $f(\mathbf{r}, \mathbf{v}, t) = f_M(v) + f_1(\mathbf{r}, \mathbf{v}, t)$, $n(\mathbf{r}, t) = n_0 + n_1(\mathbf{r}, t)$). A crucial assumption made in almost all acceleration mechanisms is that $n_1/n_0 << 1$, and the energy carried by the non-thermal particles is small compared to the ambient energy available in the acceleration region. These assumptions are usually correct in most astrophysical systems. Solar flares and gamma-ray bursts (GRB) are two well-documented exceptions where the accelerated particles carry a large fraction of the energy available at the accelerator.

5.4.2.1 Particle Dynamics in Nonlinear Electromagnetic Fields

One important method, used by many researchers to analyze the ability of a nonlinear process to accelerate particles, is the *test particle* approach. While this approach can give many of the important characteristics of the accelerated particles, it is based on the assumptions mentioned above, i.e., that the electromagnetic fields evolve independently of the accelerated particles. The evolution of an ensemble of non-thermal particles is determined from the calculation of the orbits of a large number of particles placed at random places inside the unstable electromagnetic fields. The equations of motion are

$$\frac{d\mathbf{p}_i}{dt} = q_j \mathbf{E}_1 + q_j \left[\mathbf{v}_i \times (\mathbf{B}_0 + \mathbf{B}_1)\right], \tag{5.13}$$

$$\frac{d\mathbf{x}_i}{dt} = \mathbf{v}_i, \tag{5.14}$$

where $i = 1, ..., n_1$ and j denotes the type of particle analyzed, $\mathbf{p}_i = \gamma_i m_j v_i$ is the momentum, v_i is the velocity of the particle, m_j its mass, and $\gamma_i = (1 - v_i^2/c^2)^{-1/2}$ the relativistic factor. The Lorentz force can now be divided into two parts as follows:

$$\frac{d\mathbf{p}_i}{dt} = q_j(\mathbf{v}_i \times \mathbf{B}_0) + q_j[\mathbf{E}_1 + \mathbf{v}_i \times \mathbf{B}_1] = \mathbf{F}_{0i} + \mathbf{F}_{ri}, \tag{5.15}$$

where F_0 is forcing the particle to oscillate around the ambient magnetic field and F_r is a force caused by the nonlinear processes. Its behavior is so complex though that it can be modeled as a random force. Including the collisions of the non-thermal particles with the ambient plasma we get

$$\frac{d\mathbf{p}_i}{dt} = \mathbf{F}_{0i} - \nu \mathbf{p}_i + \mathbf{F}_{ri}, \tag{5.16}$$

where the collision frequency

$$\nu \sim 10^{-11} n_0 (\text{cm}^{-3}) / T^{3/2}(eV) \text{ s}^{-1}.$$

Equation (5.16) is a well-known stochastic differential equation, introduced first by Langevin in 1908 [96]. Most known acceleration mechanisms are stochastic, since even the laminar shock or the monolithic large-scale current sheet effectively introduce a stochastic forcing on the particles in an indirect way. We already mentioned in Sect. 5.4.1 that the acceleration region is finite and the particles spend a random time τ_i there, depending on the position they started, before escaping. Most acceleration processes known today depend critically on the characteristics of the forcing term. Multiplying Eq. (5.16) with the momentum we derive the energy equation

$$\frac{dE}{dt} = -\nu E + \mathbf{F}_{rj} \cdot (\mathbf{p}_i/m_i), \qquad (5.17)$$

with E the kinetic energy. Following hundreds of thousands of test particles with randomly chosen initial conditions inside the acceleration region allows one to recover their statistical characteristics, i.e., injecting initially a Maxwellian distribution in random places inside the unstable plasma we may observe the evolution of the distribution function in time. The wealth of data collected from the evolution of thousands of test particles is much more accurate (but more time-consuming) than the solutions of the Fokker–Planck equation which will be presented next. We will use the test particles approach extensively for the analysis of the acceleration of particles in the environment of fragmented energy release presented in Sect. 5.3.

5.4.2.2 Fermi Acceleration

In 1949 Fermi [57] introduced a prototype stochastic acceleration mechanism to explain the acceleration of cosmic rays. His ideas were the driving force for many well-known acceleration processes today, e.g., for diffusive shock or turbulent wave acceleration. Fermi chose the simplest possible random walk process in velocity space. Assuming that the "scattering centers", moving with constant speed V, are equally spaced (distributed at distances L apart) and that the mean time between collisions is $\tau_{coll} \sim < L/(c \cos a) > \approx 2L/c$, the mean energy gain is

$$\left\langle \frac{dE}{dt} \right\rangle = \frac{1}{\tau_{coll}} \langle \Delta E \rangle = \frac{2c}{3L} \left(\frac{V}{c}\right)^2 E = \frac{E}{\tau_{acc}}, \qquad (5.18)$$

and the mean energy gain after many interactions with the scattering centers is $< \Delta E >= (4/3)(V/c)^2 E$ [57]. Collisions between particles are ignored but particles escape from the spatially restricted acceleration region in a characteristic time τ_{esc}. The simplest way to generalize the ideas of Fermi is the well-known "Fermi map" [102], where

$$v_{n+1} = v_n + 2x_0\omega \sin \omega t_n \qquad (5.19)$$

$$t_{n+1} = t_n + \frac{2L}{v_{n+1}} \qquad (5.20)$$

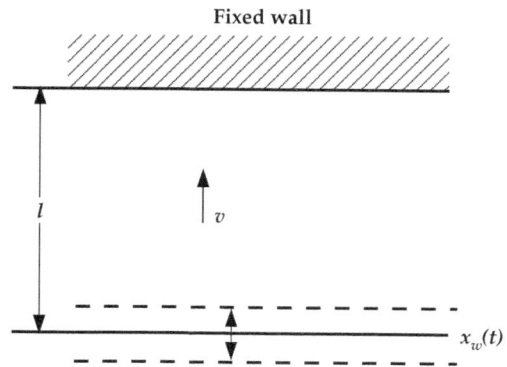

Fig. 5.20 The ball is moving between the walls. The lower wall is oscillating with frequency ω and amplitude x_0. The walls are separated by a distance L [102]

representing a ball moving between two parallel plates (see Fig. 5.20). The kick Δv in the nth step is a periodic function of time, and the time between collisions is inversely proportional to the velocity. The change of the energy is instantaneous.

5.4.2.3 Diffusion Equations

The statistical evolution of a large number of particles inside a collection of equally spaced and slowly moving "scattering centers" can be discussed, under certain constraints, with the use of the Fokker–Planck equation. Particles moving inside stochastic fields follow very complicated orbits. The particles perform, depending on the structure of the fields, strange "walks" inside a turbulent electromagnetic medium. The simplest example is the Brownian particle executing a "random walk", and it represents the motion of a "heavy" particle inside a gas of particles that is in equilibrium. Equation (5.16) can handle these problems when the random force obeys a Gaussian distribution.

Assuming that every step of the "walk" is totally independent (a Markovian process), we can derive formally the Fokker–Planck equation, which has been an important instrument for the study of high energy particles in astrophysics [128, 146]. The derivation of the Fokker–Planck equation is beyond the scope of the present review but it is important to stress that it remains an accurate representation only of phenomena remaining close to equilibrium, and when the forcing term is very weak. It represents a slowly evolving distribution of particles experiencing a weak and rapidly oscillating force. The simplest form of the Fokker–Planck equation is the one describing the energy diffusion (assuming that the particles remain always isotropic, i.e., the scattering process is so frequent that it manages to sustain isotropy all the time):

$$\frac{\partial n_1(E,t)}{\partial t} = \frac{\partial^2}{\partial E^2}(D_{EE}(E)n_1(E,t)) - \frac{\partial}{\partial E}\left[\left(\left\langle\frac{dE}{dt}\right\rangle - \left(\frac{dE}{dt}\right)_l\right)n_1(E,t)\right]$$
$$- \frac{n_1(E,t)}{\tau_{esc}(E)} + Q(E,t), \qquad (5.21)$$

where the term $< dE/dt >$ represents the systematic acceleration,

$$D_{EE}(E) = \frac{< E(t)^2 >}{t}$$

is the diffusion coefficient in energy space [37], $(dE/dt)_l$ represents the energy losses due to Coulomb collisions and radiation, $\tau_{esc}(E)$ is the energy-dependent loss rate of particles out of the finite acceleration system and $Q(E)$ represents the replenishment of particles inside the accelerator (usually taken as a Maxwellian distribution times an injection rate).

Fermi solved Equation (5.21) in its simplest form. Assuming that $< dE/dt >$ is given by Eq. (5.18) and assuming that (1) losses are not important, (2) no source term is included, (3) there is no diffusion in energy (but only systematic acceleration) and searching for a steady-state solution, he found

$$\frac{dn_1(E)}{dE} = -\left(1 + \frac{\tau_{acc}}{\tau_{esc}}\right)\frac{n_1}{E}. \qquad (5.22)$$

The solution is well known,

$$n_1(E) \sim E^{-r}, \qquad (5.23)$$

where $r = 1 + \tau_{acc}/\tau_{esc}$. The solution obtained from the Fokker–Planck equation predicts, in the case of cosmic rays, the correct observed functional form, but the index r is not in agreement with the observations. More sophisticated results can be reached by assuming a spectrum of MHD waves and incorporating the terms dropped by Fermi [160, 95, 4, 138]. Using quasilinear theory, we can also incorporate wave generation, cascade, and dissipation processes [120] and create a self-consistent system of equations. The Fokker–Planck equation is a useful tool for the analysis of particle diffusion in space and energy, but it is very restricted concerning the kind of turbulent environments it can handle. The assumptions behind Eq. (5.21) present a barrier to the analysis of the systems appearing during the *turbulent flare model* presented earlier.

Before closing this section it is worth mentioning a few more points.

- **How can we estimate the transport coefficients (if it is not possible to derive analytical expressions) using Eq. (5.16)?** Following the orbits of many particles and using Eq. (5.16), we can estimate numerically the systematic acceleration $< dE/dt >$ term and the transport coefficient D_{EE}.

- **Is diffusion always normal (Brownian motion is the prototype for normal diffusion)?** The spatial diffusion of a Brownian particle inside a gas in equilibrium follows a simple law $< x^2 >= Dt$, and this type of diffusion is called *normal* [21]. In a turbulent plasma the diffusion processes are much more complicated (they follow strange kinetics in fractal media) and give the relation $< x^2 > \sim t^a$ where the index a can be smaller than 1 (sub-diffusion) or larger than 1 (super-diffusion) [175, 122].

- **If the diffusion is not normal, is the Fokker–Planck equation still valid?** Unfortunately the Fokker–Planck equation is invalid when the diffusion processes are anomalous. More complicated partial differential equations are needed and the derivation and the solution of such equations are not as easy as the standard Fokker–Planck equation. Using more sophisticated numerical methods, e.g., Monte Carlo [68] simulations or the fractional kinetics approach [175, 122], we can obtain more realistic results.

5.4.2.4   Monte Carlo Simulations

A numerical technique, used widely in many astrophysical problems and which will be extremely useful for the results presented in the next section, is the Monte Carlo approach. It is instructive to present this approach in a way that can be used in the context of particle acceleration [68, 125].

Let us assume that a particle starts at a given point $x_{0i}(t = 0)$ in space (this point is randomly selected inside the acceleration region) and with initial velocity $v_{0i}(t = 0)$. The initial velocities are selected from a sample which follows a Maxwellian distribution (so the bulk of the particles are in equilibrium initially). The next step is to move the particle a distance Δ_i till the next "scattering center", i.e., $x_{new} = x_{old} + \Delta_i$, along this path is reached, when the particle loses energy either by collisions or by radiation losses and arrives at the new position with a new velocity, estimated as $v_{new} = v_{old} - v_{loss}\Delta_i$. At the new point the particle enters a "scattering center", gains or loses energy, and departs with a new velocity $v_{new} = v_{old} \pm \Delta_i$. The time has evolved as $t_{new} = t_{old} + \frac{\Delta_i}{v_{new}} + \Delta T_{scat}$. We assume that inside the localized scatterer the particle follows a complicated trajectory (which we do not follow in detail). The particle stops moving when its position is outside the limits of the acceleration region or the energy release time is shorter than t_{new}. Applying all the above to Fermi acceleration is simple since all Δ_i are equal ($\Delta_i = L$), no losses are included, the particles spend no time in the scattering center ($\Delta T = 0$), and Δv_i is given by a simple formula (see the Fermi map).

In more complex environments, the three unknown variables $\Delta_i, \Delta v_i, \Delta T_i$ are considered random and distributed according to probability distributions that should incorporate the statistical properties of the system under consideration. Monte Carlo simulations are a very useful and flexible tool to treat these systems. We will outline a specific example of the Monte Carlo method in solar flares in Sect. 5.4.6.

5.4.3 Turbulent Current Sheets as Particle Accelerators

According to our understanding of magnetic reconnection, several potential mechanisms for particle acceleration co-exist at a UCS. Plasma flows driving turbulence, shock waves, and DC electric fields are expected to appear simultaneously inside and around a driven and evolving UCS. If the UCS is located in the middle of a turbulent magnetic topology, all these phenomena will be enhanced and the sporadic external forcing of the plasma inflow into the UCS will create bursts of sporadic acceleration.

Analyzing the orbits of particles in an isolated current sheet is a very interesting problem and the nonlinear characteristics of the trajectories are impressive (see [53] and references therein). Most studies reported so far use analytical solutions of the static electromagnetic fields for the reconnecting current sheets in 2-D or 3-D [151, 126, 105, 178, 179, 40, 41]. We feel that these studies are interesting but bear little resemblance with the dynamic evolution of the turbulent UCS discussed earlier in Sect. 5.3.2 where the main emphasis was shifted toward the interaction of particles with smaller scale structures *within* the current sheet [87, 51, 132]. The nonlinear evolution of the UCS is characterized by the formation of small-scale structures, especially in the lateral regions of the computational domain, and coalescence of magnetic islands in the center. This behavior is reflected in the 3-D structure of the electric field, which shows that the initial equilibrium is destroyed by the formation of current filaments.

Kliem [87] started off with a 2-D analytical description of the magnetic field topology, which includes two colliding islands. The electric field is derived from the coalescence of the islands moving with characteristic speed $\mathbf{u}$. The acceleration is due to the convective electric field $E_{conv} \sim -\mathbf{u} \times \mathbf{B}$ and happens at the X-line. Electrons reach relativistic energies in a very short time as they move inside these electric fields. Drake et al. [51] also discuss the interaction of particles with contracting magnetic islands. An attempt to draw the analogy with Fermi acceleration was also made and an estimate of the systematic acceleration was

$$\left\langle \frac{dE_\parallel}{dt} \right\rangle = -\frac{E_\parallel}{\tau_{acc}}, \tag{5.24}$$

where

$$\tau_{acc} = 2\frac{u_x}{\delta_x}\frac{B_x^2}{B_0^2},$$

$2\delta_x$ is the length of the island, u_x the velocity of the contracting island (of order the Alfvén speed), B_x, B_0 the reconnecting and the ambient magnetic fields, respectively. Several interesting conclusions were reached, e.g., particles interacting with many islands can easily reach relativistic energies, and the particle distribution (obtained by solving a simplified form of the diffusion equation) was tending toward a power law with index around -1.5 for solar parameters. Onofri [132] used the resistive 3-D MHD equations (see Sect. 5.3.6) to analyze the evolution of a perturbed UCS and the electric field derived from Ohm's law $\mathbf{E} \sim \eta\mathbf{J} - \mathbf{v} \times \mathbf{B}$. Figure 5.21 shows the isosurfaces of the electric field at different times calculated for two different values of the electric field: The red surface represents higher values and the blue surface represents lower values. The structure of the electric field is characterized by small regions of space where the field is stronger, surrounded by a larger volume occupied by lower values. At later times the fragmentation is more evident, and at $t = 400\tau_A$ (where τ_A is the Alfvén time), the initial current sheet has been completely destroyed and the electric field is highly fragmented. The strong electric field regions are acceleration sites for the particles and their distribution

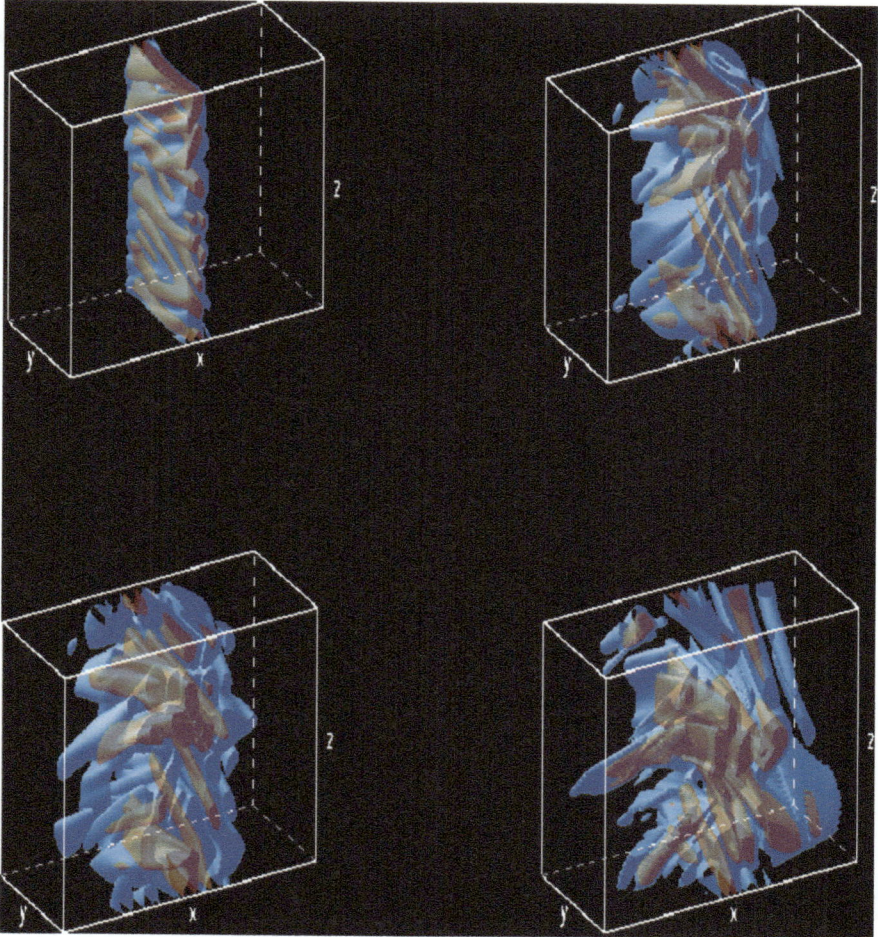

Fig. 5.21 Electric field isosurfaces at $t = 50\tau_A$, $t = 200\tau_A$, and $t = 400\tau_A$ [132]

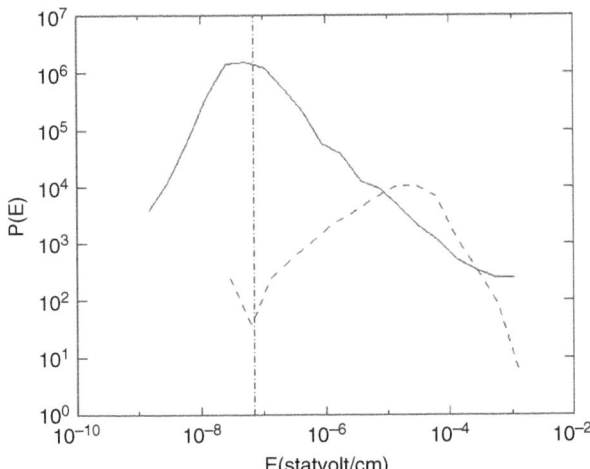

Fig. 5.22 Distribution function of the resistive (*dashed line*) and convective (*solid line*) electric field at $t = 50\tau_A$. The vertical line represents the value of the Dreicer field in the solar corona [132]

in space fills a larger portion of the simulation box at later times, with increasing possibility to accelerate a higher number of particles.

To give a measure of the fragmentation of the electric field, we calculated the fractal dimensions of the fields shown in Fig. 5.21, using the box-counting definition of the fractal dimension [132]. The magnitude E of the electric field at each gridpoint of the simulation domain was estimated, and the distribution function of these quantities was constructed (see Fig. 5.22 for $t = 50\tau_A$.) We separately plot the resistive and the convective components of the electric field. The resistive part is less intense than the convective part, but it is much more important in accelerating particles, as we verified by performing some simulations where only one of the two components was used. Protons and electrons are injected into the simulation box where they move under the action of both components (convective and resistive) electric and magnetic fields, which do not evolve during the particle motion. This is justified by the fact that the evolution of the fields is much slower than the acceleration process, electrons and ions are accelerated on a short timescale to very high energies, and in such short times the fields would not change significantly according to the MHD simulation.

The trajectories of the test particles inside the box are calculated by solving the relativistic equations of motion, using a fourth-order Runge–Kutta adaptive step-size scheme. Since the magnetic and the electric fields are given only at a discrete set of points (the grid-points of the MHD simulation), both fields are interpolated with local 3-D linear interpolation to provide the field values in between grid-points.

For the case of electrons, the particles' energy distribution at different times is shown in Fig. 5.23. Some of the test particles are quickly accelerated to high energies so that the initial Maxwellian distribution changes, developing a tail that grows in time. The kinetic energy of the electrons increases very rapidly, and in a short time

Fig. 5.23 Distribution function of electron kinetic energy at $t = 8 \times 10^{-5} s$ (*solid line*), $t = 3 \times 10^{-5}$ s (*dotted-dashed line*) and the initial distribution (*dashed line*). The electromagnetic field is given at $t = 72$ s [132]

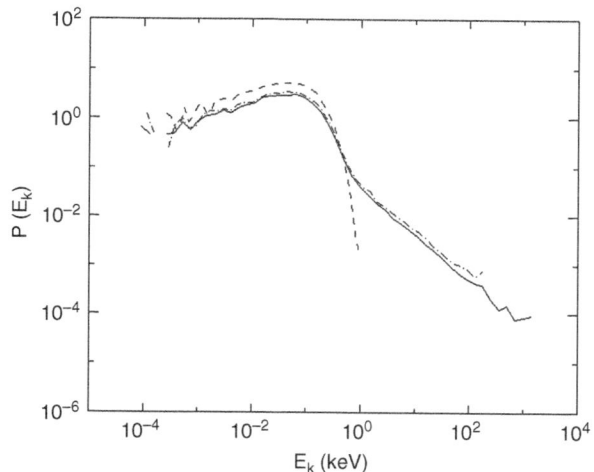

it equals the energy contained in the magnetic field. Since there is no back-reaction of the particles onto the fields, there is no limit to the energy they can gain until they leave the simulation box. For this reason the particle motion was followed only as long as their energy is still less than 50% of the magnetic field energy W_B, which is up to $t_{pe} = 8 \times 10^{-5}$s.

The maximum kinetic energy at the end of the run turns then out to be about 1 MeV. Collisions are not included in the simulations because the collisional time is about $t_c = 5.5 \times 10^{-3}$ s, which is much longer than t_{pe}. In the final distribution, the logarithmic slope of the power law tail is $\simeq 1$. The power law tails of the distributions start at an energy of about 1 keV. The total number of particles contained in the tail of the distributions ($E_K \geq 1$ keV) is $\simeq 6 \times 10^{37}$ for the assumed values of the particle density n_0 and length l_x. Below 1 keV, the electrons have a thermal distribution.

Turning to protons, Onofri et al. find that acceleration is much less efficient than for electrons, only at $t_{pi} = 3 \times 10^{-3}$ s do they reach a maximum kinetic energy of about 1 MeV, with energy distributions that are similar to those of the electrons. Because of the much slower acceleration timescale of the protons, the time limit for our simulation is determined by the electrons, t_{pe} ($t_{pe} << t_{pi}$) — at times as large as t_{pi}, the electrons would have absorbed all the available magnetic energy W_B. At the time limit t_{pe} then, the distribution of the ions has remained close to the initial Maxwellian, with just minor gain in energy.

The results of these simulations show that a decayed and fragmented current sheet can be a very efficient accelerator. The particles absorb a large amount of energy from the magnetic field in a short time, and the magnetic and electric fields lose a large fraction of their energy. However, the back-reaction of the particles is not taken into account in the test particle simulations reported here, which in that sense are not self-consistent. Our results suggest that the lifetime of a current sheet of this size in the solar corona is very short since energetic particles absorb a large fraction of the available magnetic energy. As a consequence of the back-reaction, the magnetic and electric fields would change more quickly than the MHD simulation shows, the acceleration process can be expected to be slower, and the resulting energy distributions will probably be different.

The limitations of this approach, in the case of the electrons, are reflected in the inability of our results to reproduce all the characteristics of the distributions that are observed in solar flares (e.g., variation of power law slopes). In the case of ions, the situation is different. Their maximum energy at the time limit of our simulations is lower than the energy that protons usually reach during solar flares.

We can thus conclude that ions are not accelerated to the high energies observed during solar flares by single, isolated, turbulent current sheets. Stressed and complex large-scale magnetic topologies can though form simultaneously many current sheets [62], it has been shown that the interaction of the ions (and electrons) with many current sheets can be a very efficient accelerator [158, 159].

5.4.4 Acceleration in Stressed Magnetic Fields

The model for the compact flare analyzed in Sect. 5.3.3 is used to examine the acceleration of particles [158, 159]. Many of the techniques used are similar to those in the previous section. Here the MHD simulation was performed on a numerical grid with 200 points between the photospheric boundaries (the x-coordinate), and 60 points in each of the transverse directions (the y- and z-coordinates). Only the coronal portion of the magnetic field is considered, so that in terms of a dimensionless length, the electric and magnetic fields are confined to $\mathbf{L} = (L_x, L_y, L_z) = (1.6, 0.1, 0.1)$. Physical lengths are obtained by multiplying $\mathbf{L}$ by a factor L. Most of the results presented use $L = 10^9$ cm. The value of the background coronal density is taken to be 10^{10} cm^{-3}, the initial background magnetic field is taken to be $B = 100$ G and the coronal plasma beta is equal to 0.04. The electric field (both resistive and inductive) arising in this model, and their potential as particle accelerators, was analyzed. The inductive field appeared to be negligible (especially for high energy particles) because its component parallel to the magnetic field is zero. The resistive electric field is distributed over the domain in the form of a hierarchy of current sheets. Figure 5.24 shows the snapshot of these current sheets from the MHD model that is used in this chapter. It is obvious that a very complex topology is formed. In between the current sheets no electric field exist, while the electric field inside the current sheets takes on values between $\pm 3 \times 10^{-2}$ statV cm^{-1}. The average absolute value of the electric field is 5×10^{-4} statV cm^{-1}.

The output of the MHD model, specifically the 3-D electric and magnetic fields, is used as a basis for studying particle acceleration. Particles were tracked in frozen fields using a similar numerical scheme to that in the previous section, the frozen field being justified by the separation of timescales for acceleration (<1 s) from the characteristic coronal evolution time (>1 s). Acceleration was considered only in

Fig. 5.24 Snapshots of the resistive electric field configurations within the coronal volume, as calculated from the global MHD model. The *blue* and *red* regions represent electric field regions that point toward the left and right footpoints, respectively. The details of the model are described in the text [158]

the coronal part of the model, so current sheets appearing at the footpoints were ignored.

In each example discussed, 30,000 particles are injected with an initial Maxwellian distribution with a temperature of 1.2×10^6 K. The initial positions and pitchangles of the particles are random. The particles are injected in the MHD domain simultaneously and are considered as "lost" when they leave the simulation box and are not replaced. Again, feedback is not included.

In Fig. 5.25a a 1-D sample of the x-component of the electric field along the domain is shown. The distribution function of the values of the magnitude of the electric field is shown in Fig. 5.25b. The distribution has a power law component with an index value of -2.8 which terminates at a cutoff at the highest values.

Figure 5.26 shows the final distribution function at $t = 0.5$ s. The energy used to construct the distribution is either their final energy or that with which they left the domain. This distribution function has three main parts: the thermal part and the two power law components with indexes equal to -0.7 and -2.7, respectively. This distribution is in fact comprised of a number of "classes" of particles which behave differently throughout the simulation. In particular, particles can leave the domain through either the sides, or the ends, or become confined to the corona with or without energization.

The stressed coronal fields are a very effective particle accelerator, with both electrons and protons attaining relativistic energies in a very short time throughout the corona: for example, electrons are accelerated to relativistic energies in milliseconds. The acceleration appears to have four phases, with the maximum energies rising, peaking, and then decaying, as well as there being an extended acceleration phase lasting for almost 1 s. The energy reached scales with the coronal length scale.

Combining the fragmentation of a single current sheet discussed in the previous section and the results presented here, it is easy to conclude that the solar corona forms a multi-scale environment, starting from UCS with characteristic length $> 10^9$ cm, which cascade to very small structures of the order of hundreds of meters. At the same time the unstable current sheets may force other TCSs to

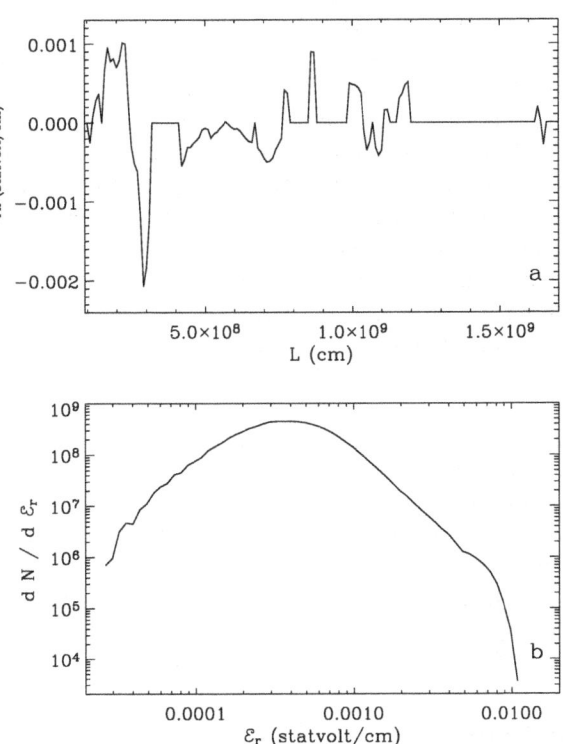

Fig. 5.25 (a) An arbitrary 1-D sample of the resistive electric field along the domain (the x-direction). (b) Distribution function of the resistive electric field [159]

go unstable; therefore, the particle dynamics become extremely complicated. Unfortunately no current code can handle so much complexity and we have to use approximate methods.

5.4.5 Particle Acceleration by MHD Turbulence

5.4.5.1 Low-Amplitude Waves ($\delta B/B \ll 1$)

When accelerating particles by MHD waves one considers the Alfvén branch for ions since it has a resonance below the proton gyrofrequency Ω_H and the fast mode branch (magnetosonic or whistlers) for electrons which has a resonance below the electron gyrofrequency Ω_e. The analysis of the interaction of particles and waves was initially based on the assumption that a large volume was filled with low-amplitude MHD waves, with a power law spectrum $W(k) \sim k^{-q}$. The entire acceleration volume was constantly replenished and the distribution remained isotropic. With these assumptions, the Fokker–Planck equation remains relatively simple (see Eq. 5.21) and in order to make things even simpler, the time evolution of the accelerated particles was ignored. Recently several attempts were made to improve the above scenario and the spectral evolution of the waves was included [120]. This would appear to be an even more complicated project since the wave–wave

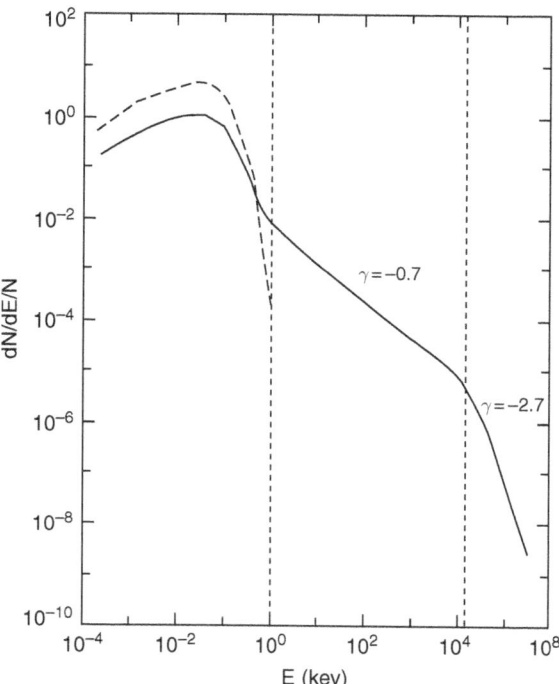

Fig. 5.26 Distribution function for all the particles injected in the domain at $t = 0$ (*dashed curve*) and at the end of the run (*solid curve*). γ is the power law index value [159]

interactions included can hardly capture the fragmentation of the energy release presented in the previous section [139].

The main assumption in these studies is that large-scale current sheets will produce low-amplitude, long-wavelength MHD waves (e.g., Alfvén waves), which will cascade to smaller and smaller scale MHD waves, until they will reach the dissipation scale. It is clear that all these processes will be highly anisotropic in space and time. Damping of the waves is also anisotropic and this will modify the cascade process [139].

We can then conclude that if reconnection is the main mechanism for the energy release in solar flares, attempts to reduce the energy release process to an extremely simplified coupled system of low-amplitude, homogeneous MHD waves which evolve quasi-linearly do not resemble a real flare, especially when 50% of the magnetic energy goes to high energy particles.

5.4.5.2 Large-Amplitude Waves ($\delta B / B \geq 1$)

Dmitruk et al. [46, 47] analyzed the acceleration of particles inside 3-D MHD turbulence. The compressible MHD equations (see Eqs. 5.4, 5.5, 5.6, 5.7, 5.8 and 5.9) were solved numerically. In these simulations the decay of large-amplitude waves was studied. After a very short time (a few Alfvén times) a fully turbulent state with a broad range of scales has been developed (Fig. 5.27).

Fig. 5.27 Visualization of the
turbulent magnetic field | *B* |
(*top*) and electric field | *E* |
(*bottom*) in the simulation
box. High values are in
yellow (*light*) and low values
in *blue* (*dark*) [46]

As in previous sections, the magnetic field is directly obtained from the numerical
solution of the MHD equations with an electric field derived from Ohm's law (see
Eq. 5.10). It is obvious that the electric field is an intermittent quantity with high
values observed in less space filling distribution. Magnetic and electric fields show
a broad range of scales and high degree of complexity. The energy spectrum of the
MHD fields is consistent with a Kolmogorov-5/3 power law. The structure of the
velocity field and the current density along the external magnetic field (J_z) can be
seen in Fig. 5.28. The formation of strong anisotropies in the magnetic field, the
fluid velocity, and the associated electric field is observed. The overall picture is

Fig. 5.28 Cross section of the current density along the external magnetic field in color tones. *Yellow (light)* is positive J_z, *blue (dark)* is negative and the superposed arrows for the velocity field [47]

that current sheet structures along the DC field are formed as a natural evolution of the MHD fields.

Following thousands of particles inside the simulation box we can learn many of the statistical properties of their evolution, e.g., $\sqrt{<\Delta x^2>}$, $\sqrt{<\Delta v^2>}$, the velocity distribution, etc. Electrons and ions are accelerated rapidly, and the non-thermal tails form power law distributions. Most particles seem to escape the volume by crossing only a few of the randomly appearing current sheets. A few particles are trapped in these structures and accelerated to very high energies. The Fokker–Planck equation is not the appropriate way to capture the random appearance of coherent structures inside such a turbulent environment.

Arzner and Vlahos [15] investigated the effect of multiple localized resistive spots on coronal particle acceleration. They considered collisionless test particles in evolved homogeneous MHD turbulence (see Chap. 3 for details) with electromagnetic fields modeled by

$$\mathbf{B} = \nabla \times \mathbf{A}, \tag{5.25}$$

$$\mathbf{E} = -\partial_t \mathbf{A} + \eta(\mathbf{j})\,\mathbf{j}, \tag{5.26}$$

where $\mu_0\mathbf{j} = \nabla \times \mathbf{B}$ and $\eta(\mathbf{j}) = \eta\,\theta(|\mathbf{j}| - j_c)$ is an anomalous resistivity switched on above the critical current $j_c \sim enc_s$ [133] (see also Chap. 1 for a detail analysis of electrostatic turbulence and the role of wave–particle interactions). Here c_s (n) is the sound speed (number density) of the background plasma. The vector potential $\mathbf{A}(\mathbf{x}, t)$ is modeled as a random field, subject to the MHD constraints

$$\mathbf{E} \cdot \mathbf{B} = 0 \quad \text{if} \quad \eta = 0 \quad \text{and} \quad E/B \sim v_A. \tag{5.27}$$

Equation (5.27) can be satisfied in several ways. A spectral representation in axial gauge, $\mathbf{A}(\mathbf{x}, t) = \sum_{\mathbf{k}} \mathbf{a}(\mathbf{k}) \cos(\mathbf{k} \cdot \mathbf{x} - \omega(\mathbf{k})t - \phi_{\mathbf{k}})$ with $\mathbf{a}(\mathbf{k}) \cdot \mathbf{v}_A = 0$ and dispersion relation $\omega(\mathbf{k}) = \mathbf{k} \cdot \mathbf{v}_A$, which is an exact solution of the induction equation with a constant velocity field $\mathbf{v}_A$ was used. For simplicity, $\mathbf{A}(\mathbf{x}, t)$ is taken as Gaussian with random phases $\phi_{\mathbf{k}}$ and (independent) Gaussian amplitudes $\mathbf{a}(\mathbf{k})$ with zero mean and variance

$$\langle |\mathbf{a}(\mathbf{k})|^2 \rangle \propto (1 + \mathbf{k}^T \mathsf{S} \mathbf{k})^{-\nu}. \tag{5.28}$$

A constant magnetic field B_0 along $\mathbf{v}_A$ can be included without violating Eq. (5.27). The total MHD wave velocity is $v_A^2 = \mathcal{B}^2(\mu_0 \rho)^{-1}$ with $\mathcal{B}^2 = B_0^2 + \sigma_B^2$ and $\sigma_B^2 = \frac{1}{2} \sum_{\mathbf{k}} |\mathbf{k} \times \mathbf{a}(\mathbf{k})|^2$ the magnetic fluctuations. The matrix $\mathsf{S} = \mathrm{diag}(l_x^2, l_y^2, l_z^2)$ in Eq. (5.28) contains the outer turbulence scales, and the index ν determines the regularity of the two-point function at short distance. The presented simulations have $\nu = 1.5$, $\mathbf{v}_A = (0, 0, v_A)$, and one turbulence scale is longer by an order of magnitude than the two (equal) others, which describes migrating and reconnecting twisted flux tubes).

The vector potential contains some 100 wave vectors in the inertial shell $\min(l_i^{-1}) < |\mathbf{k}| < 10^{-2} \cdot r_L^{-1}$ with r_L the rms thermal ion Larmor radius. We focus on strong turbulence ($\sigma_B / B_0 > 1$). The rms magnetic field $\mathcal{B}$ is a free parameter, which defines the scales of the particle orbits. The localized enhancement of the resistivity will (1) enhance the local heating inside the unstable current sheet, $Q_j = \eta_j j^2$ forming what we will call here "hot spots" and (2) dramatically enhance the particle (ion and electron) acceleration. The fast heat transport away from the hot plasma will soon transform them into hot loops.

The physical units used in this study [15] are selected to represent the solar atmosphere. In SI units and for typical values $\mathcal{B} \sim 10^{-2}$ T, $n \sim 10^{16}$ m^{-3}, $T \sim 10^6$ K, the reference scales are as follows (electron values in brackets): time $\Omega^{-1} \sim 10^{-6}$ s $(6 \cdot 10^{-10}$ s); length $c\Omega^{-1} \sim 300$ m $(0.17$ m); thermal velocity $\sim 1.2 \cdot 10^5$ m s^{-1} $(5 \cdot 10^6$ m s$^{-1})$; sound speed $c_s \sim 1 \cdot 10^5$ m s^{-1}; Alfvén speed $v_A \sim 2 \cdot 10^6$ m s^{-1}; electron–ion collision time $\tau \sim 0.003$ s; Dreicer field $E_D = ne^3 \ln \Lambda/(4\pi \epsilon_0^2 kT_e) \sim 3 \cdot 10^{-2}$ V m^{-1}. Time is measured in units of $\Omega^{-1} = m/q\mathcal{B}$; velocity is measured in units of the speed of light; distance is measured in units of $c\Omega^{-1}$.

When an initially Maxwellian population is injected into the turbulent electromagnetic field given by Eqs. (5.25) and (5.26), the particles can become stochastically accelerated. Due to their large inertia, protons gain energy in relatively small portions. This is not so for electrons. The momentum evolution of collisionless electrons of the high energy tail of a Maxwellian is shown in [15] (see Fig. 4 in [15]).

Since electrons have much smaller Larmor radius, they follow the field lines adiabatically and gain energy only when dissipation regions are encountered. The resulting orbits then have large energy jumps, so that a Fokker–Planck description is inappropriate [146].

We can then conclude that a large-scale turbulent cascade leads to highly anisotropic structures randomly placed inside the acceleration volume. Electrons and ions evolve inside these structures forming power law energy distributions, and the acceleration time is relatively short. The Fokker–Planck equation is invalid inside such environment and the sudden random formation of DC electric fields along the external magnetic fields is apparent. In these simulations current sheets are part of the turbulent cascade processes.

Collisions were not included in any of the studies reported above [46, 15, 47]; therefore they are applicable on the corona ($n = 10^9 cm^{-3}, T \sim 100 eV$) where the acceleration time is usually milliseconds, much faster than collisional losses.

5.4.6 Particle Acceleration in Complex Magnetic Topologies

Our aim in this chapter is to take advantage of the properties of isolated UCS as accelerators, but at the same time to incorporate the fact that the dissipation happens at *multiple, small-scale* sites. Many attempts have been made in the past to analyze the evolution of a distribution of particles inside a collection of nonlinear dissipation structures [9, 10, 11]. We will attempt here to present all these developments in a more unified way [171].

The 3-D magnetic topology, driven from the convection zone, dissipates energy in localized UCS, which are spread inside an active region, providing a natural fragmentation for the energy release and a multiple, distributed accelerator. In this way, the large-scale magnetic topology acts as the backbone which hosts the UCS, and the spatio-temporal distribution of the latter defines the type of flare, its intensity, the degree of energization and acceleration of the particles, the acceleration timescales, etc. Evolving large-scale magnetic topologies provide a variety of opportunities for acceleration which is not restricted to the impulsive phase, but can also take place before and after it, being just the manifestation of a more relaxed, but still driven topology. Depending on the extent to which the magnetic topology is stressed, particles can be accelerated without a flare, and even long-lived acceleration in non-flaring active regions must be expected. Consequently, the starting point of the model to be introduced below is a driven 3-D magnetic topology, which defines a time-dependent spatial distribution of UCS inside an active region. The details of the mechanisms involved in the acceleration of particles inside the UCS are not essential in a stochastic modeling approach.

Since the global characteristics of the energy release play a crucial role in the acceleration of particles, it is important to make use of the new developments in the theory of SOC models for flares. Also taken into account are ideas from the theory of complex evolving networks [5, 48, 49], adjusted though to the context of plasma physics: the spatially distributed, localized UCS can be viewed as a network, whose "nodes" are the UCS themselves, and whose "edges" are the possible particle trajectories between the nodes (UCS). The particles are moving around in this network, forced to follow the edges, and undergo acceleration when they pass by a node. The network is complex in that it has a non-trivial spatial structure, and it is evolving since the nodes (UCS) are short-lived, as are the connectivity channels, which even change during the evolution of a flare. This instantaneous connectivity of the UCS is an important parameter in our model: it determines to what degree multiple acceleration is imposed onto the system, which in turn influences the instantaneous level of energization and the acceleration timescale of the particles.

The UCS are short-lived and appear randomly inside the large-scale magnetic topology when specific conditions for instability are met. Modeling this dynamic accelerator requires the knowledge of three probability density functions [171]:

- The probability density $P_1(s)$ defines the distribution of the distances a charged particle travels freely in between two subsequent encounters with a UCS. The series of distances $s_1^{(j)}$, $s_2^{(j)}$, ..., $s_n^{(j)}$, ..., generated by the probability density $P_1(s)$ characterizes the trajectory of the particle j in space.

  Every particle follows a different characteristic path, but remains inside the large-scale magnetic topology. The probability density $P_1(s)$ relates the particle acceleration process to the large-scale topology. This part/aspect was never taken into account in previous acceleration models.
- The probability density $P_2(E)$ provides the effective electric field $E(j)_i$ acting on the jth particle for the effective time $\tau(j)_i$ it spends inside the ith UCS. Particles follow very complicated trajectories inside the UCS. They may be accelerated by more than one acceleration mechanisms but what actually is important for our model is the final outcome, i.e., we characterize UCS as a simple input–output system, in which an effective DC electric field is acting. The effective action of a UCS is to increase a particle's momentum by $\Delta \mathbf{p}_i^{(j)} = e\,\mathbf{E}_i^{(j)}\,\tau_i^{(j)}$.
- Finally, the probability density $P_3(\tau)$ gives the effective time $\tau_i^{(j)}$ a UCS interacts with the charged particle.

The above probabilities will define the charged particle dynamics inside the flaring region. The particle j starts with initial momentum $\mathbf{p}_0^{(j)}$ from the initial position $\mathbf{r}_0^{(j)} = 0$ at time $t = 0$. The initial momentum $\mathbf{p}_0^{(j)}$ is such that the corresponding velocity $|\mathbf{v}_0^{(j)}|$ is drawn at random from the tail of a Maxwellian, $|\mathbf{v}_0^{(j)}| \geq v_{th}$, with v_{th} the thermal velocity. The particle is assumed to find itself in the neighborhood of UCS at time $t = 0$, enters it immediately, and undergoes a first acceleration process.

During an interaction with the UCS, the particle's momentum in principle evolves according to

$$\mathbf{p}_{i+1}^{(j)} = \mathbf{p}_i^{(j)} + e\mathbf{E}_i^{(j)} \cdot \tau_i^{(j)}, \tag{5.29}$$

where $\mathbf{E}_i^{(j)}$ and $\tau_i^{(j)}$ have been generated by the corresponding probability densities $P_2(E)$ and $P_3(\tau)$.

After the particle has left the UCS, it performs a free flight until it again meets a UCS and undergoes a new acceleration process (see Fig. 5.29). The probability

Fig. 5.29 Sketch of the basic elements of the model. A particle follows a magnetic field line (*solid line*), although undergoing drifts, and travel in this way freely a distance s_i, until enters (*filled circle*), where it is accelerated by the associated effective DC electric field E_{i+1}. After the acceleration event the particle again moves freely until it meets a new UCS [171]

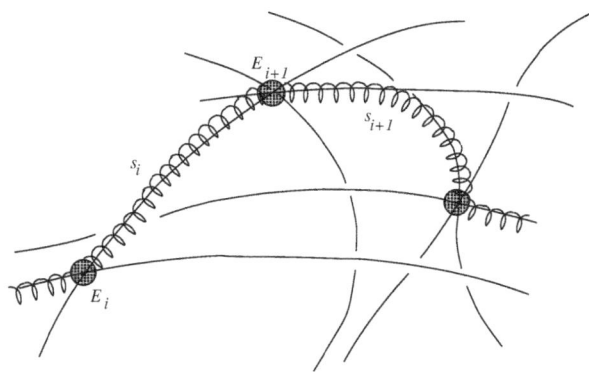

density $P_1(s)$ determines the spatial distance $s_i^{(j)}$ the particle travels before it meets this next UCS, situated at

$$\mathbf{r}_{i+1}^{(j)} = \mathbf{r}_i^{(j)} + s_i^{(j)} \hat{r}_i^{(j)}, \tag{5.30}$$

where $\hat{r}_i^{(j)} \equiv \mathbf{p}_i^{(j)}/|\mathbf{p}_i^{(j)}|$ is a unit vector into the direction of the free flight and $\mathbf{r}_i^{(j)}$ is the location of the previous UCS the particle had met.

The time passed during the acceleration process and the free flight is

$$t_{i+1}^{(j)} = t_i^{(j)} + \tau_i^{(j)} + s_i^{(j)}/v_i^{(j)}, \tag{5.31}$$

where $t_{i+1}^{(j)}$ is the time when the particle enters the $i + 1$th UCS.

The particle starts a new cycle of acceleration and free flight at this point, the process as a whole is a cyclic one with continued probabilistic jumps in position- and momentum space. The system was monitored for times which are relatively short, of the order of 1 s. For such times, the particles can be assumed to be trapped inside the overall acceleration volume $V_{acc} = L_{acc}^3$, an assumption which will be confirmed by the results presented below. A second consequence of the short monitoring times is that the probability density $P_1(s)$, which reflects the magnetic topology and the distribution of the UCS, remains independent of time, since no large-scale changes of the topology are expected for such short times.

Let us now define the probability densities P_1, P_2, P_3 used in this study. The active, flaring region may be assumed to be in a state of MHD turbulence, embedded in a complex, large-scale magnetic topology. The UCS, i.e., the regions of dissipation, are distributed in such a way that they form in their ensemble a fractal set. This claim is based on two facts: (i) flaring active regions have successfully been modeled with self-organized criticality (SOC; [108, 109, 80, 81]). It was demonstrated in [79, 80, 81] that the unstable sites in the SOC models actually represent small-scale current dissipation regions, i.e., they can be considered as UCS. Furthermore, [113, 83] have shown that the regions of dissipation in the SOC models at fixed times form a fractal, with fractal dimension roughly $D_F = 1.8$. (ii) From investigations of hydrodynamic turbulence we know that the eddies in the inertial regime have a scale-free, power law size distribution, making it plausible that at the dissipative scale a fractal set is formed, and indeed different experiments conclude that the dissipative regions form a fractal with dimension around 2.8 (see [12] and references therein).

The particles in this model are thus assumed to move from UCS to UCS, the latter being distributed such that they form a fractal set. Isliker and Vlahos [82] analyzed the kind of random walk where particles move in a volume in which a fractal resides, usually traveling freely but being scattered (accelerated) when they encounter a part of the fractal set. They showed that in this case the distribution of free travel distances r in between two subsequent encounters with the fractal is distributed in good approximation according to

$$p(r) \propto r^{D_F - 3} \qquad (5.32)$$

as long as $D_F < 2$. For $D_F > 2$, $p(r)$ decays exponentially. Given that a dimension D_F below 2 is reported for SOC models, we are led to assume that $P_1(s)$ is of power law form, with index between -1 and -3, preferably near a value of $D_F - 3 = -1.2$ (with $D_F = 1.8$, according to [83, 113]). Not included in the study of [82] are two effects: (a) that the particles do not move on straight-line paths in between two subsequent interactions with UCS, but they follow the bent magnetic field lines and (b) that particles can be mirrored and trapped in some regions, making in this way the free travel distances larger. It is thus reasonable to consider the power law index of $P_1(s)$ as a free parameter.

The freely traveled distances s are distributed according to

$$P_1(s) = A\, s^{-a}, \quad \text{with } l_{min} < s < l_{max}, \qquad (5.33)$$

where $l_{min}(L_{acc})$ and $l_{max}(L_{acc})$ are related to the characteristic length of the coronal active region L_{acc} and A is a normalization constant.

The second probability density determines the effective electric field attached to a specific UCS. Its form should in principle be deduced, either from observations, which is not feasible so far, or from the simulation and modeling of a relevant set-up, which to our knowledge seems not to exist at this time. Two cases of distributions are of particular interest, the "well-behaved" case, where P_2 is Gaussian, and the "ill-behaved" case, where P_2 is of power law form, above all with index between -1 and -3. The Gaussian case is well behaved in the sense that all the moments are finite, and it is a reasonable choice because of the central limit theorem, which suggests Gaussian distributions if the electric field is the result of the superposition of many uncontrollable, small processes. The power law case is ill-behaved in the sense that most moments are infinite. It represents the case of scale-free processes, as they appear, for instance, in SOC models. A characteristic of power law distributions is the importance of the tail, which in fact causes the dominating effects.

Trying also the case of Gaussian distributions and guided by the results, we present in this study only the case where the distribution of the electric field magnitude is of power law form,

$$P_2(E) = B E^{-b}, \quad \text{with } E_{min} < E < E_{max},$$

(5.34)

which shows better compatibility with the observations. We just note that most acceleration mechanisms mentioned earlier have power law probability distributions for the driving quantity. $E_{min}(E_D)$ and $E_{max}(E_D)$ are related to the Dreicer field E_D and B is the normalization constant. The electric field is then determined as $\mathbf{E} = E\,\hat{r}$, where $\hat{r}$ is a 3-D unit vector into a completely random direction.

For the distribution of acceleration times a model would also be needed. Since the acceleration times appear only in combination with the electric fields in the momentum increment, $e\mathbf{E}_i^{(j)} \cdot \tau_i^{(j)}$ (see Eq. (5.29)), we can absorb any non-standard feature, such as scale-freeness or other strong non-Gaussianities, in the distribution of E. This is also reasonable since both τ and E are effective quantities.

It was assumed that the time a particle spends inside a UCS obeys a Gaussian distribution with mean value t_c and standard deviation t_m,

$$P_3(\tau) = C e^{-\frac{(\tau - t_c)^2}{2 t_m^2}}.$$

(5.35)

Defined in this way, the acceleration times are not essential for the acceleration process, they influence through the overall acceleration timescale, i.e., the global timing of acceleration.

The simulations are performed by using 10^6 particles and the system is monitored for 1 s, with the aim of focussing on a short time interval during the impulsive phase (for longer times the loss of particles from the accelerating volume should be included). An extended parametric study was performed in [171]. The particles sustain repeated acceleration events, whose number differs from particle to particle: the minimum number of acceleration events per particle is found to be 1, the maximum is 175, and the mean is 13.4. A substantial fraction of the particles undergo one, initial, acceleration process. To analyze the diffusive behavior of the particles in position space, the mean square displacement $\langle r^2(t) \rangle$ of the particles from the origin as a function of time was determined. For all times the system is monitored, we find strong super-diffusion, $\langle r^2(t) \rangle \propto t^\gamma$, with γ around 3. The behavior is different above and below 4×10^{-3} s, a time which is related to the acceleration time: at $t_c + 3t_m = 4 \times 10^{-3}$ s the vast majority of the particles have finished their first acceleration process (since the acceleration times are Gaussian distributed (see Eq. (5.35)), 99% of the particles have an acceleration time smaller than the time of 3 standard deviations above the mean value). Below 4×10^{-3} s, the particles typically are still in their first acceleration process, whereas above 4×10^{-3} s, some particles are on free flights and others are in new acceleration processes. We cannot claim that the diffusive behavior has settled to a stationary behavior in the 1 s we monitor the system. At 1 s, we find $\langle r^2(t = 10) \rangle \approx 10^{19} = L_{act}^2 / 10$, the particles have diffused a distance less than the active region size.

The diffusive behavior in velocity space means determining the mean square displacement $\langle v^2(t) \rangle$ of the particles from their initial velocity. The particles start with a mean initial velocity, and after roughly 0.001 s, a time slightly earlier than t_m, the mean acceleration time, $\langle v^2(t) \rangle$, starts to increase, i.e., the particles start to feel acceleration. In the range roughly from 0.001 s to 0.02 s, the diffusion in energy is $\langle v^2(t) \rangle \propto t^{0.5}$, and in the range 0.5 s to 1 s, $\langle v^2(t) \rangle \propto t^{0.25}$. The system thus exhibits clear sub-diffusive behavior in velocity space (the latter corresponding to the non-relativistic energy space).

At preordained times, the kinetic energy of the particles was estimated and their histograms constructed, which, normalized to 1, yield the kinetic energy distributions $p(E_{kin}, t)$ shown in Fig. 5.30. The distributions retain a similar shape for the time period monitored, being flatter at low energies, and a power law tail above roughly 5 keV. The low energy part is actually a Maxwellian. The power law index of the high energy tail varies around 4, increasing slightly with time, and the particles also reach higher energies with increasing time. A systematic shift of the Maxwellian toward higher energies was detected, in parallel with the development of a power law tail that extended to higher and higher energies and steepens. At 1 s, the most energetic particles have reached kinetic energies slightly above 1 MeV.

It is of interest to know what will happen to the ions which go through the same kind of processes. Adjusting the particle mass in the model, and keeping all the parameters fixed, the initial distribution of protons is basically unaltered, even for times up to 1000 s. The reason is that the momentum increments are too small for the ions to undergo a visible change in distribution; they need larger momentum increments. The minimum of the electric field distribution was adjusted to $E_{min} = 100\,E_D$, which increases the mean value of the effective electric field and so causes larger momentum increments for the protons. The energy distributions are again Maxwellians with approximate power law tails. The index of the power law tail is around 3.5 at small times. The Maxwellian is shifted to higher energies in the course of time, which corresponds again to heating, as in case of the electrons.

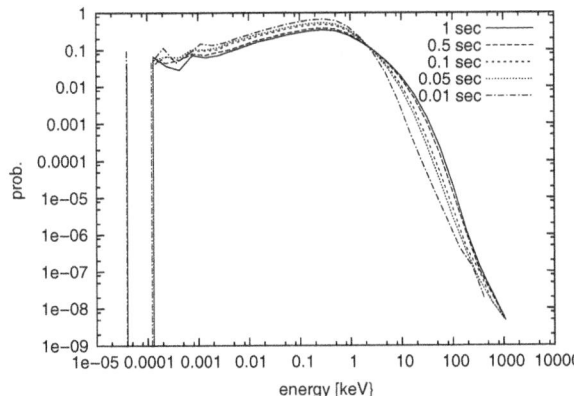

Fig. 5.30 Kinetic energy distributions $p(E_{kin}, t)$ (probability density function, normalized to 1) at times $t = 0.01, 0.05, 0.1, 0.5, 1,$ s. [171]

Fig. 5.31 Kinetic particle distribution obtained with 10,000 UCS, $\zeta = 5/3$, and with a distribution of particles acceleration length give by a power law spectral index $\delta = 3$. The temperature of the injected distribution is 10^6 K [39]

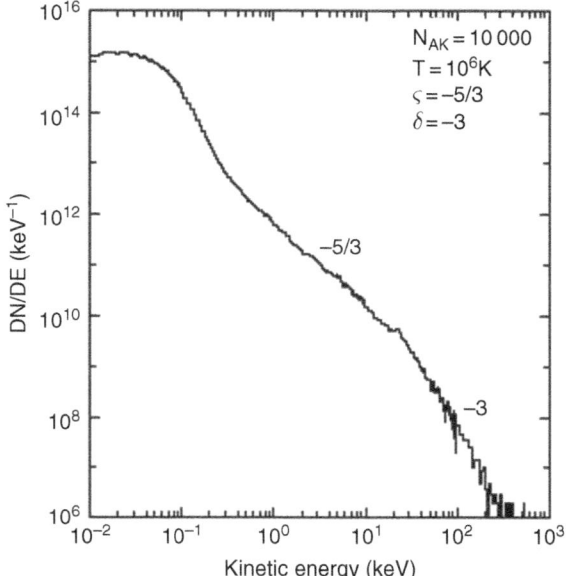

The link of the energy release processes with the acceleration of particles was studied recently in several articles [44, 39]. Dauphin [39] also uses the idea of frag- mented energy release as a starting point of his acceleration model. He assumed that the majority of the particles escape from the UCS after being accelerated only once. A small fraction of the non-thermal particles interact with multiple UCS. The probability distribution function of the acceleration lengths "seen" by a particle is

$$P(\Delta \ell) = k_1 (\Delta \ell)^{-\delta}. \tag{5.36}$$

The idea behind this choice is that in the active region during a flare a large number of UCSs with all scales are present. The electric field distribution is similar to the one reported in Eq. (5.34)

$$P(E_{UCS}) = k_2 (E_{UCS})^{-\zeta}. \tag{5.37}$$

Assuming that the acceleration region is populated by thousands of UCS with different sizes distributed inside the large-scale structure, the kinetic energy distri- bution of the particles was estimated (see Fig. 5.31) and seems to agree remarkably well with the data.

5.4.7 The Strongly Turbulent Accelerator (STA)

In this chapter we have explored the evolution of particles in large-scale stressed magnetic fields, forming UCSs of different characteristic scales and at random points. Starting from the collapse of a turbulent current sheet [51, 132] and progressively moving to larger scales of stressed magnetic topologies [158, 159], a number of UCSs will be present during a flare or/and CME. The random appearance of UCS in the middle of a large-scale MHD structure (compact loops, eruptive structures) which hosts them has been studied in several articles [62, 46, 47, 15]. These UCSs act as local "nodes" of a large evolving network [171]. These accelerators can be examined using a Monte Carlo simulation which link the energy release with the accelerator [171, 44, 39]. We name this new accelerator a *strongly turbulent accelerator (STA)*. This type of accelerator appears more frequently in magnetized strongly turbulent astrophysical systems (AGN, jets, accretion disks, galaxy clusters, black hole accretion, etc.). We have shown that an STA is much more efficient than most known acceleration mechanisms (shocks, MHD waves, etc.) since more than 50% of the available magnetic energy of the system will end up to the high energy particles.

5.5 Discussion of the Global Consideration of Particle Acceleration

In Sect. 5.3 we suggested that there are two broad classes of flares: (1) the compact flare (closed magnetic loops) driven mainly by random photospheric flows and (2) the eruptive structures responsible for the flare/CME driven mainly by flux emergence or unstable magnetic loops. The main theme of this chapter is that flares and CMEs are phenomena which are closely related with the evolution of the active region and not isolated structures. It is worth summarizing the main steps followed by an active region before it reaches the state of "flaring", and the STA turns on.

1. **The formation and the drivers of an active region**: Newly emerged flux and photospheric motions act as the drivers of the active region complex. The non-linear extrapolations of magnetograms remain an open problem, but force-free or simpler forms of extrapolation locate a large number of thin current sheets (TCSs) inside the 3-D magnetic complex [45, 106, 17, 18, 63, 14].
2. **The large-scale 3-D magnetic topology (skeleton) hosting the flares:** The 3-D magnetic structures (compact loops or eruptive structures) form the skeleton in which the explosive phenomena (flares, CMEs) take place.
3. **Storage of magnetic energy inside the large-scale structure:** There is the formation of large QSL or thin current sheets (TCSs) that are stable structures and can store magnetic energy. The energy can be released if the resistivity exceeds a certain value [170, 36].
4. **The trigger of explosions:** The continuous stresses from the driver or the emergence of new magnetic flux forces some TCSs (sometimes even one with very large scale may be enough) to become unstable (a UCS). The appearance of one or more UCS will cause a catastrophic collapse of many current sheets [36, 132].

5. **The formation of a turbulent active region:** The appearance of many UCS in the 3-D large-scale structure forms a strongly turbulent active region with many localized UCS. In some flares the appearance of UCSs may start before any sudden eruption, and there is continued release of energy in repeated explosions for hundreds of seconds after the first major eruption.

6. **The acceleration of particles inside a strongly turbulent active region:** The particle evolution inside a strongly turbulent active region is a very fast and efficient accelerator with many UCS, acting as nodes of acceleration. This accelerator is called an STA [51, 132, 158, 159, 15, 46, 47, 171, 44, 39].

Miller et al. [121] suggested that a model can be a viable particle accelerator for flares, if it has a number of properties based on the observational constraints existing in the mid-1990s. We now assess the ability of the STA to follow these constraints.

- **Can an STA accelerate electrons up to 100 keV (responsible for HXR)?** All UCS, even when isolated, with sizes larger that hundreds of km can accelerate electrons up to 100 keV [151, 126, 105, 178, 179, 40, 41, 51, 132, 88].

- **Can an STA accelerate electrons with energies of tens of MeV?** Electrons interacting with several UCS can reach very high energies. When interacting with a single UCS, it is difficult to reach these energies, but a network of UCS will accelerate electrons up to tens of MeV and ions up to GeV [159, 39, 171].

- **Can the STA accelerate ions?** Yes, very efficiently [159, 39, 171].

- **Is the STA a fast accelerator (seconds)?** Most of the studies done today suggest that it is extremely fast (less than a second) [51, 132, 158, 159, 15, 46, 47, 171].

- **Does the STA reproduce the observed electron distribution?** The models proposed above, either a single collapsing UCS, and especially the network of UCS can reproduce the generic distribution function for electrons [51, 132, 159, 39, 171, 46].

- **Does the STA reproduce the observed proton distribution?** The models proposed above, either the collapsing single UCS or the network of UCS can reproduce the distribution function for protons [159, 171].

- **Does the STA produce a heavy ion enhancement?** It is not clear yet how this can happen.

- **Can the STA accelerate electrons and ions from thermal particles?** Yes the UCSs can accelerate electrons and ions from the thermal background [159, 171, 39, 46, 51, 132].

- **Is filamentation and complexity essential for the STA?** Yes both are essential for this model.

- **Is the mechanism well connected with the global structure?** The STA is based on the energy release (UCS) and the evolution of large-scale structures.

Let us now add to the observational constraints known 10 years ago those reported in Sect. 5.2 based on the recent data collected from RHESSI.

1. **The number problem?** The fact that UCSs are distributed along the large-scale structures will be very helpful. One important factor, still unexplored, is the fact that UCS can be present even in the footpoints of the large-scale structures. The competition between acceleration and collisions will make the number problem less severe. This is an open problem and needs more careful analysis.
2. **Energetics?** The STA is extremely efficient accelerator [51, 132].
3. **Flare statistics?** The turbulent flare, on which the STA is based, is well connected to self-organized criticality theory [171, 39].
4. **Temporal evolution of the Power law index?** Still an open problem.
5. **Coronal X-ray sources?** Particle diffusion inside the STA is not normal, and particles undergo many interactions with the UCSs; therefore, their diffusion along the field lines slows down dramatically. No detailed estimates of the confinement of particles in the turbulent corona are yet available.
6. **The close temporal and spectral evolution of the footpoints?** The fact that UCSs are distributed along evolving large-scale structures suggests that the STA can reproduce this result but more work is needed.
7. **The close correlation of HXR with particles detected in space?** This is closely related to the large-scale topology and the position of the UCS. In principle many simple 3-D topologies can allow this coincidence to exist.

5.6 Summary

In this chapter we have noted the most striking new observational constraints for particle acceleration during solar flares and linked the flare energy release with the acceleration. Before constructing a model for particle acceleration in the Sun we have to understand the formation and the energy release processes in active regions.

We have shown that driven active regions form thin current sheets (TCSs). Large-scale structures are stressed beyond a certain point and one or several TCSs become suddenly unstable current sheets (UCSs). The UCSs collapse and force several other stable TCSs to become UCS (an avalanche). This is the way a turbulent flare will begin inside the large-scale structure of an active region.

Acceleration of particles inside many unstable collapsing UCS has been discussed by many authors, and a new acceleration process has been emerged, the strongly turbulent accelerator (STA). The STA can easily explain most of the constraints reported from the recent data.

There are three major theoretical challenges:

1. The nonlinear extrapolation of observed magnetograms and the detailed formation of the TCS remains an open and challenging problem.
2. Although the evolution of large-scale structures can be analyzed easily with 3-D MHD codes, the evolution of unstable current sheets should be analyzed with 3-D kinetic codes. No current code can handle the simultaneous presence of so many scales.

3. The strange kinetics and the anomalous diffusion of the accelerated particles inside a turbulent flare (an evolving network of UCS) do not follow the Boltzmann equation. Forming fractional kinetic equations for the evolution of particles in strongly turbulent plasmas is not an easy task.

The main observational challenge is the direct observation of the fragmentation of the energy release. Large spatial and temporal resolution is needed and the emphasis should be shifted to smaller flares and their statistical properties. The Ellerman bombs is an example of "micro-flaring" [65] with many interesting points along the lines discussed here.

Acknowledgments We would like to thank our colleagues Drs. A. Anastasiadis, K. Arzner, H. Isliker, K. Galsgaard, M. Georgoulis, F. Lebreti, M. Onofri, R. Turkmani, and Mr. T. Fragos for many useful conversations.

References

1. V.I. Abramenko, V. Yurchyshyn, H. Wang, P.R. Goode: Solar Phys., **201**, 225 (2001)
2. V.I. Abramenko, V. Yurchyshyn, H. Wang, T.J. Spirock, P.R. Goode: Astrophys. J., **577**, 487 (2002)
3. V.I. Abramenko, V. Yurchyshyn, H. Wang, T.J. Spirock, P.R. Goode: Astrophys. J., **597**, 1135 (2003)
4. A. Achterberg: Astron. Astrophys., **97**, 259 (1981)
5. R. Albert, A.L. Barabasi: Rev. Modern Phys., **47**, 47 (2002)
6. C.E. Alissandrakis: Astron. Astrophys., **100**, 197 (1981)
7. T. Amari, J.F. Luciani, J.J. Aly, Z. Mikic, J. Linker: Astrophys. J., **595**, 1231 (2003)
8. J. Ambrosiano, W.H. Matthaeus, M.L. Goldstein, D. Plante: J. Geophys. Res., **93**, 14383 (1988)
9. A. Anastasiadis, L. Vlahos: Astron. Astrophys., **245**, 271 (1991)
10. A. Anastasiadis, L. Vlahos: Astrophys. J., **428**, 819 (1994)
11. A. Anastasiadis, L. Vlahos, M. Georgoulis: Astrophys. J., **489**, 367 (1997)
12. F. Anselmet, R.A. Antonia, L. Danaila: Planetary Space Sci. **49**, 1177 (2001)
13. S. Antiochos, C.R. DeVore, J.A. Klimchuk: Astrophys. J., **510**, 485 (1999)
14. V. Archontis, A.W. Hood, C. Brady: Astron. Astrophys., **466**, 367 (2007)
15. K. Arzner, L. Vlahos: Astrophys. J. Lett., **605**, L69 (2004)
16. M.J. Aschwanden, T. Kosugi, H.S. Hudson, M.J. Wills, R.A. Schwartz: Astrophys. J., **470**, 1198 (1996)
17. G. Aulanier, E. Pariat, P. Demoulin: Astron. Astrophys., **444**, 961 (2005)
18. G. Aulanier, E. Pariat, P. Demoulin: Solar Phys., **238**, 347 (2006)
19. P. Bak, C. Tang, K. Wiesenfeld: 1987, Phys. Rev. Lett., **59**, 381 (1987)
20. P. Bak: *How Nature Works*. Springer Verlag, New York: (1996)
21. R. Balescu: *Statistical Dynamics: Matter out of Equilibrium*. Imperial College Press, London (1997)
22. S.G. Benka, G.D. Holman: Astrophys. J., **391**, 854 (1992)
23. A.O. Benz: Space Sci. Rev., **68**, 135 (1994)
24. A.O. Benz: Energy Conversion and Particle Acceleration in the Solar Corona. In: K.L. Klein (ed.) Springer, Berlin, p. 80 (2003)
25. D. Biskamp: *Magnetohydrodynamic turbulence*. Cambridge University Press, Cambridge (2003)

26. D. Biskamp: *Magnetic Reconnection in plasmas*. Cambridge University Press, Cambridge (2000)
27. R. Blandford, D. Eichler: Phys. Rep., **154**, 1, (1987)
28. D. Borgogno, et al.: Phys. Plasmas, **12**, 32309 (2005)
29. J.L. Bougeret, et al.: Space Sci. Rev., **71**, 231 (1995)
30. J.C. Brown: Solar Phys., **18**, 489, (1971)
31. J.C. Brown, A.G. Emslie, E.P. Kontor: Astrophys. J. Lett., **595**, L115 (2003)
32. A.C. Cadavid, J.K. Lawrence, A.A. Ruzmaikin, A. Kayleng-Knight: Astrophys. J., **429**, 391 (1994)
33. P.J. Cargill, E.R. Priest: Astrophys. J., **266**, 383 (1983)
34. P.J. Cargill: SOLMAG: Magnetic coupling of the solar atmosphere. In: G. Tsiropoula, U. Schuhle (eds.) *ESA SP-505*, p. 245 (2002)
35. P.J. Cargill, L. Vlahos, R. Turkmani, K. Galsgaard, H. Isliker: Space Sci. Rev., **124**, 246 (2006)
36. P.A. Cassak, M.A. Shay, J.F. Drake: Phys. Rev. Lett., **95**, 235002 (2005)
37. S. Chandrasekhar: Rev. Mod. Phys., **15**, 1 (1943)
38. N.B. Crosby, M.J. Aschwanden, B.R. Dennis: Solar Phys., **143**, 275 (1993)
39. C. Dauphin: Astron. Astrophys., **471**, 993 (2007)
40. S. Dalla, P.K. Browning: Astron. Astrophys., **436**, 1103 (2005)
41. S. Dalla, P.K. Browning: Astrophys. J. Lett., **640**, L99 (2006)
42. R.B. Decker, L. Vlahos: Astrophys. J., **306**, 710 (1986)
43. R.B. Decker: Space Sci. Rev., **48**, 195 (1988)
44. N. Décamp, F. Malara: Astrophys. J. Lett., **637**, L61 (2006)
45. P. Démoulin: Adv. Space Res., **39**, 1367 (2007)
46. P. Dmitruk, W.H. Matthaeus, N. Seenu, M.R. Brown: Astrophys. J. Lett., **597**, L81 (2003)
47. P. Dmitruk, W.H. Matthaeus, N. Seenu: Astrophys. J., **617**, 667 (2004)
48. S.N. Dorogovsky, J.F. Mendes, A.N. Samuhhin: Phys. Rev. Lett., **85**, 4633 (2000)
49. S.N. Dorogovtsev, J.F. Mendes: Adv. Phys. **51**, 1079 (2002)
50. J.F. Drake, M.A. Shay, W. Thongthai, M. Swisdak: Phys. Rev. Lett., **94**, 95001 (2005)
51. J.F. Drake, M. Swisdak, H. Che, M.A. Shay: Nature, **443**, 553 (2006)
52. J.F. Drake, M. Swisdak, K.M. Schoeffler, B.N. Rogers, S. Kobayashi: J. Geophys. Res., **33**, L13105 (2006)
53. C. Efthimiopoulos, C. Gontikakis, A. Anastasiadis: Astron. Astrophys., **443**, 643 (2005)
54. D.C. Ellison, R. Ramaty: Astrophys. J., **298**, 400 (1985)
55. A.G. Emslie, et al.: J. Geophys. Res., **109**, A10104 (2004)
56. A.G. Emslie, B.R. Dennis, G.D. Holman, H.S. Hudson: J. Geophys. Res., **110**, A11103 (2005)
57. E. Fermi: Phys. Rev., **75**, 1169 (1949)
58. L. Fletcher, H.S. Hudson: Solar Phys., **210**, 307 (2002)
59. T.G. Forbes, E.R. Priest: Astrophys. J., **446**, 377 (1995)
60. T. Fragos, M. Rantziou, L. Vlahos: Astron. Astrophys., **420**, 719 (2004)
61. K. Galsgaard, A. Nordlund: J. Geophys. Res., **101**, 13445 (1996)
62. K. Galsgaard: SOLMAG: Magnetic Coupling of the Solar Atmosphere. In: G. Tsiropoula, U. Schuhle (eds.) *ESA SP-505* (2002)
63. K. Galsgaard, V. Archontis, F. Moreno-Insertis, A.W. Hood: Astrophys. J., **666**, 516 (2007)
64. M.K. Georgoulis, B.J. LaBonte, T.R. Metcalf: Astrophys. J., **602**, 447 (2004)
65. M.K. Georgoulis, D.M. Rust, P.N. Bernasconi, B. Schmieder: Astrophys. J., **575**, 506 (2002)
66. M.K. Georgoulis, B.J. LaBonte: Astrophys. J., **615**, 1029 (2004)
67. M.K. Georgoulis: Solar Phys., **228**, 5 (2005)
68. H. Gould, J. Tobochnik: *An Introduction to Computer Methods*, 2nd Edition, Adison-Wesley, The Netherlands (1996)
69. P.C. Grigis, A.O. Benz: Astron. Astrophys., **426**, 1093 (2004)
70. P.C. Grigis, A.O. Benz: Astron. Astrophys., **434**, 1173 (2005)

71. K.L. Harvey: PhD thesis, University of Utrecht, The Netherlands (1993)
72. K. L. Harvey, C. Zwaan: Solar Phys., **148**, 85 (1993)
73. J. Heyvaerts: Particle acceleration by solar flares. In: E.R. Priest (ed.) Solar Flare Magneto-hydrodynamics. Gordon and Beach, New York, p. 429 (1981)
74. G.D. Holman, M.E. Pesses: Astrophys. J., **267**, 837 (1983)
75. R.F. Howard: Ann. Rev. Astron. Astrophys., **34**, 75 (1996)
76. D. Hughes, M. Paczuski: Phys. Rev. Lett., **88**, 054302 (2002)
77. D. Hughes, M. Paczuski, R.O. Dendy, P. Halander, K.G. McClements: Phys. Rev. Lett., **90**, 131 (2003)
78. G.J. Hurford, R.A. Schwartz, S. Krucker, R.P. Lin, D.M. Smith, N. Vilmer: Astrophys. J. Lett., **595**, L77 (2003)
79. H. Isliker, A. Anastasiadis, D. Vassiliadis, L. Vlahos: Astron. Astrophys., **363**, 1134 (1998)
80. H. Isliker, A. Anastasiadis, L. Vlahos: Astron. Astrophys., **363**, 1134 (2000)
81. H. Isliker, A. Anastasiadis, L. Vlahos: Astron. Astrophys., **377**, 1068 (2001)
82. H. Isliker, L. Vlahos: Phys. Rev. E, **67**, 026413 (2003)
83. H. Isliker, L. Vlahos: unpublished result (2003)
84. H. Ji et al.: Phys. Rev. Lett., 92, 115001 (2004)
85. A.L. Kiplinger: Astrophys. J., **453**, 973 (1995)
86. J.G. Kirk: *Plasma Astrophysics*. Springer, Berlin, p. 225 (1994)
87. B. Kliem: Astrophys. J., **90**, 719, (1994)
88. B. Kliem, M. Karlicky, A.O. Benz: Astron. Astrophys., **360**, 715 (2000)
89. R.A. Kopp, G.W. Pneuman: Solar Phys., **50**, 85 (1976)
90. S. Krucker, R.P. Lin: Solar Phys., **210**, 229 (2002)
91. S. Krucker, M.D. Fivian, R.P. Lin: Adv. Space Res., **35(10)**, 1707 (2005)
92. S. Krucker, G.J. Hurford, R.P. Lin: Astrophys. J. Lett., **595**, L103 (2003)
93. S. Krucker, R.P. Lin: Astrophys. J. Lett., **542**, L61 (2000)
94. J. Kuijpers: Lect. Notes Phys., **469**, 101 (1996)
95. R.M. Kulsrud, A. Ferrari: Astrophys. Space Sci., **12**, 302 (1971)
96. P. Langevin: C.R. Acad. Sci. (Paris), **146**, 530 (1908)
97. T.N. Larosa, R.L. Moore: Astrophys. J., **418**, 912 (1993)
98. Y.T. Lau: Solar Phys., **148**, 301 (1993)
99. J.K. Lawrence, A.A. Ruzmaikin, A.C. Cadavid: Astrophys. J., **417**, 805 (1993)
100. J.K. Lawrence, A.C. Cadavid, A.A. Ruzmaikin: Astrophys. J., **465**, 425 (1996)
101. K.D. Leka, A. Skumanich: Solar Phys., **188**, 3 (1999)
102. A.J. Lichtenberg, M.A. Lieberman: *Regular and Stochastic motion*. Springer-Verlag, New York (1983)
103. R.P. Lin, the RHESSI Team: Proc. 10th European Solar Physics Meeting, Prague (ESA SP-506, 2002), p. 1035 (2002)
104. R.P. Lin, et al.: Astrophys. J. Lett., **595**, L69 (2003)
105. Y.E. Litvinenko: Solar Phys., **212**, 379 (2003)
106. D.W. Longcope: Living Rev. Solar Phys., **2**, 7 (2005)
107. D. W. Longcope, H.R. Strauss: Astrophys. J., **437**, 851 (1994)
108. E.T. Lu, R.J. Hamilton: Astrophys. J. Lett., **380**, L89. (1991)
109. E.T. Lu, R.J. Hamilton, J.M. McTiernan, K.R. Bromund: Astrophys. J., **412**, 841 (1993)
110. P.C.H. Martens: Astrophys. J. Lett., **330**, L131 (1988)
111. S. Masuda, T. Kosugi, H. Hara, S. Tsuneta, Y. Ogawara: Nature, **371**, 595 (1994)
112. A.N. McClynmont, L. Jiao, Z. Mikic: Solar Phys., **174**, 191 (1997)
113. S.W. McIntosh, P. Charbonneau, J.P. Norman, T.J. Bogdan, H.L. Liu: Phys. Rev. E, **65**, 46125 (2002)
114. K.P. Macpherson, A.L. MacKinnon: Astron. Astrophys., **350**, 1040 (1999)
115. D.B. Melrose: Particle Acceleration in Cosmic Plasmas. In: G.P. Zank, T.K. Gaisser (eds.) Institute of Physics, New York, p. 3 (1992)
116. N. Meunier: Astrophys. J., **515**, 801 (1999)
117. N. Meunier: Astron. Astrophys., **420**, 333 (2004)

118. L.J. Milano, P. Dmitruk, C.H. Mandrini, D.O. Gómez, P. Demoulin: Astrophys. J., **521**, 889 (1999)
119. J.A. Miller, D.A. Roberts: Astrophys. J., **452**, 912 (1995)
120. J.A. Miller, T.N. LaRosa, R.L. Moore: Astrophys. J., **461**, 445 (1996)
121. J.A. Miller, P.J. Cargill, et al.: J. Geophys. Res., **102**, 14631 (1997)
122. R. Metzler, J. Klafter: Phys. Rep., **339**, 1 (2000)
123. E.L. Moghaddam-Taaheri, L. Vlahos, H.L. Rowland, K. Papadopoulos: Phys. Fluids, **28**, 3356 (1985)
124. E.L. Moghaddam-Taaheri, C.K. Goertz: Astrophys. J., **352**, 361 (1990)
125. M.E.J. Newman, G.T. Barkema: *Monte Carlo Methods in Statistical Physics*. Oxford University Press, Oxford (2001)
126. C. Nodes, G.T. Birk, H. Leach, R. Schopper: Phys. Plasmas, **10**, 835 (2003)
127. A. Nordlund, K. Galsgaard: Solar and Stellar Magnetic Activity. In: G.M. Simnett, C.A. Allisandrakis, L. Vlahos (eds.) Solar and Heliospheric Plasma Physics. Springer Verlag, Berlin, (1996)
128. D.R. Nicholson: *Introduction to Plasma Theory*. John Wiley and Sons, New York (1983)
129. G. Nicolis, I. Prigogin: *Exploring Complexity: An Introduction*. Freeman Co., New York (1989)
130. M. Øieroset, R.P. Lin, T.D. Phan, D.E. Larson, S.D. Bale: Phys. Rev. Lett., **89**, 195001 (2002)
131. M. Onofri, L. Primavera, F. Malara, P. Veltri: Phys. Plasmas, **11**, 4837
132. M. Onofri, H. Isliker, L. Vlahos: Phys. Rev. Lett., **96**, 151102 (2006)
133. K. Papadopoulos: Rev. Geophys. Space Phys., **15**, 113 (1977)
134. E.N. Parker: Astrophys. J., **174**, 642 (1972)
135. E.N. Parker: Astrophys. J., **264**, 642 (1983)
136. E.N. Parker: Astrophys. J., **330**, 474 (1988)
137. C.E. Parnell, P.E. Jupp: Astrophys. J., **529**, 554 (2000)
138. V. Petrosian, T. Donaghy: Astrophys. J., **610**, 550 (1999)
139. V. Petrosian, H. Yan, A. Lazarian: Astrophys. J., **644**, 603 (2006)
140. H.E. Petschek: The Physics of Solar Flares. In: W.N. Hess (ed.), *Proc. AAS-NASA Symposium*, SP-50 p. 425 (1964)
141. E.R. Priest, T.G. Forbes: *Magnetic Reconnection: MHD theory and applications*. Cambridge University Press, Cambridge (2000)
142. E.R. Priest, T.G. Forbes: Astr Astrophys. Rev., **10**, 313 (2002)
143. E.R. Priest, G. Hornig, D.I. Pontin: J. Geophys. Res., **108**, 1285 (2003)
144. P.L. Prietchett: J. Geophys. Res., **106**, 3783 (2001)
145. J. Qiu, J. Lee, D.E. Gary, H. Wang: Astrophys. J., **565**, 1335 (2002)
146. H. Risken: *The Fokker-Planck Equation*, 2nd Ed., Springer, Berlin (1989)
147. I.I. Rousev, et al.: Astrophys. J. Lett., **588**, L45 (2003)
148. P. Saint-Hilaire, A.O. Benz: Solar Phys., **210**, 287 (2002)
149. T. Sakao, T. Kosugi, S. Masuda: *Observational Plasma Astrophysics: Five Years of Yohkoh and Beyond*. In: T. Watanabe, T. Kosugi, A.C. Sterling (eds.) ASSL, Kluwer Academic Publishers, Boston, MA, **229**, 273 (1998)
150. P.E. Seiden, D.G. Wentzel: Astrophys. J., **460**, 522 (1996)
151. R. Schopper, G.T. Birk, H. Lesch: Phys. Plasmas, **6**, 4318 (1999)
152. K. Shibata et al.: Astrophys. J. Lett., **451**, L83 (1995)
153. K. Shibata, S. Tanima: Earth Planets Space, **53**, 473 (2001)
154. D.M. Smith, G.H. Share, R.J. Murphy, R.A. Schwartz, A.Y. Shih, R.P. Lin: Astrophys. J. Lett., **595**, L81 (2003)
155. L. Sui, G.D. Holman: Astrophys. J. Lett., **596**, L251 (2003)
156. L. Sui, G.D. Holman, B.R. Dennis, S. Krucker, R.A. Schwartz, K. Tolbert: Solar Phys., **210**, 245 (2002)
157. V.S. Titov, E.R. Priest, P. Démoulin: Astron. Astrophys., **276**, 564 (1993)

158. R. Turkmani, L. Vlahos, P.J. Cargill, K. Galsgaard, H. Isliker: Astrophys. J. Lett., **620**, L59 (2005)
159. R. Turkmani, P.J. Cargill, K. Galsgaard, L. Vlahos, H. Isliker: Astron. Astrophys., **449**, 749 (2006)
160. B.A. Tverskoi: Soviet Phys.-JEPT Lett., **25**, 317 (1967)
161. A.M. Veronig, J.C. Brown, B.R. Dennis, R.A. Schwartz, L. Sui, A.K. Tolbert: Astrophys. J., **621**, 482 (2005)
162. B. Viticchié, D. DelMoro, F. Berrilli: Astrophys. J., **652**, 1734 (2006)
163. L. Vlahos, et al.: Energetic Phenomena on the Sun. In: M., Kundu, B., Woodgate, (eds.) NASA Conference Publication 2439 (1986)
164. L. Vlahos: Statistical Description of Transport in Plasmas, Astro- and Nuclear Physics. In: J. Misquich, G. Pelletier, P. Schuck, (eds.) Nova Science Publishers Inc., New York (1993)
165. L. Vlahos: Space Sci. Rev., **68**, 39 (1994)
166. L. Vlahos, M. Georgoulis, R. Kluiving, P. Paschos: Astron. Astrophys., **299**, 897 (1995)
167. L. Vlahos: Radio Emission from the Stars and the Sun. In: A.R. Taylor, J.M., Paredes, (eds.) *ASP Conference Series*, 93, ASP press, San Francisco, p. 355. (1996)
168. L. Vlahos: ESA Publications, **SP-505**, 105 (2002)
169. L. Vlahos, T. Fragos, H. Isliker, M. Gergoulis: Astrophys. J. Lett., **575**, L87. (2002)
170. L. Vlahos, M. Georgoulis: Astrophys. J. Lett., **603**, L61 (2004)
171. L. Vlahos, H. Isliker, F. Lepreti: Astrophys. J., **608**, 540 (2005)
172. Vlahos L.: The high Energy solar Corona: Waves, Eruptions, Particles. In: L. Klein, A.L. MacKinon, (eds.) Lect. Notes Phys., p. 15. (2007)
173. D.G. Wentzel, P.E. Seiden: Astrophys. J., **390**, 280 (1992)
174. T. Yokoyama, K. Shibata: Astrophys. J., **549**, 1160 (2001)
175. G.M. Zaslavsky: Phys. Report, **731**, 461 (2002)
176. Zeiler, et al.: J. Geophys. Res., **107**, 1230 (2002)
177. L.M. Zelenyi, A.V. Milanov, G. Zimabrdo: Multiscale Magnetic Structures in the Daistant Tail. In: A. Nishida, D.N. Baker, S.N. Cowley (eds.) American Geophysical Union, Washington DC, p. 37 (1998)
178. V. Zharkova, M. Gordovsky: Astrophys. J., **604**, 884 (2005)
179. V. Zharkova, M. Gordovsky: MNRAS, **356**, 1107 (2005)
180. H. Zirin: *Astrophysics of the Sun*. Cambridge University Press, Cambridge (1988)

Chapter 6
Diagnostics of the Solar Wind Plasma

K. Issautier

6.1 Introduction

The solar wind is a fully ionized plasma, coming from the outer atmosphere of the Sun, the so-called solar corona, which expands as a supersonic flow into the interplanetary medium [54]. The first observations indicating that the Sun might be emitting a wind were made by Biermann in 1946 of comet tails [1], which are observed to point away from the Sun. Comets usually exhibit two tails: a dust tail driven by the radiation pressure and a plasma tail, which points in slightly different directions pushed by the "solar corpuscular radiation" of the Sun. In 1958, E.N. Parker explained theoretically this "particle radiation" using a simple fluid model [54], showing that the solar atmosphere is not in hydrostatic equilibrium but must expand into the interplanetary medium as a wind. The existence of this solar wind was debated until it was indeed confirmed by spacecraft Lunik 2 and 3 [15] and continuously observed by Mariner 2 [52]. The Parker theory is discussed fully in Chap. 7 (Velli).

The solar wind forms a bubble in the interstellar medium, the heliosphere. The wind becomes subsonic after crossing the heliospheric shock. After an intermediate region of subsonic flow through the heliosheath (by analogy to the terrestrial magnetosheath) the solar wind reaches an interface, the so-called "heliopause," where the solar wind pressure and the interstellar pressure balance each other [16]. The nature of this boundary is far from being well known since its assumed position, around 100–200 AU from the Sun according to some theoretical models [29], has not yet been reached by spacecraft. On December 16, 2004, Voyager 1 crossed indeed the termination shock at 94 AU [7, 18].

Because of its proximity, the Sun is an ideal laboratory since in situ and/or remote sensing observations can give important details about our star, in particular, and some clues to understanding winds from other stars. Dupree in 1996 discussed

K. Issautier (✉)
Observatoire de Paris, LESIA, UMR 8109 CNRS, 92195 Meudon, France
karine.issautier@obspm.fr

Issautier, K.: *Diagnostics of the Solar Wind Plasma*. Lect. Notes Phys. **778**, 223–246 (2009)
DOI 10.1007/978-3-642-00210-6_6   © Springer-Verlag Berlin Heidelberg 2009

methods to detect stellar wind from cool stars, as giants or supergiants [8], and gave characteristics of these winds on a color–luminosity diagram (see Fig. 8 of [8]). Typically, the mass loss rate of cool stars such as the Sun, i.e., the mass lost per second, is around 2.10^{-14} M_s/year for our Sun, whereas other stars may lose much more material as a wind, three or four orders of magnitude more than the Sun.

After four decades of extensive observations, the origin and acceleration of the solar wind are still not fully understood, but is almost certainly linked to the coronal temperature in excess of 1 MK. Coronagraph observations show that the corona exhibits a variety of visible structures with different spatial and temporal scales. Polar plumes rising from the polar regions of the Sun suggest a global magnetic field, while polar coronal holes seen in soft X-rays as dark regions on images from the Yohkoh spacecraft, for example, are the origin of the fast solar wind.

This chapter addresses the issue of measuring solar wind properties. The inner corona can be imaged by remote sensing instruments, and the results are briefly described in Sect. 6.2, as well as in Chap. 7 (Velli). However, since the solar wind extends into interplanetary space, the possibility exists of making in situ measurements of the plasma and fields. A description of such techniques form the bulk of this chapter.

6.2 Measuring the Key Parameters in the Inner Heliosphere

To better understand the Sun, a major step would be to fly a probe as close as possible to our star. Such a "solar probe" has been discussed for many years, but the very high cost and risk are inhibiting factors. However, measuring the key parameters of the corona can be done with imaging instruments from spacecraft in the neighborhood of the Earth since part of solar energy is lost by radiative processes. In addition, to characterize the features of the solar wind, in situ probes are very useful.

6.2.1 The Corona

The SKYLAB mission [19] gave coronal hole temperatures of up to nearly 1 MK, but these data were limited due to the poor spectral resolution and to the low intensities in coronal holes. The SOlar and Heliospheric Observatory (SOHO) was able to determine a reliable electron temperature profile above a large polar coronal hole and in quiet coronal regions. Using the two SOHO spectrometers CDS (coronal diagnostics spectrometer) and SUMER (solar ultraviolet measurements of emitted radiation), electron temperatures have been measured as a function of height above the limb in a polar coronal hole and in quiet regions of the Sun, from the line ratio 1032A/173A in oxygen VI (see [6]).

Temperatures of around 0.8 MK are found close to the limb, rising to a maximum of less than 1 MK at 1.15 R_s, then falling to around 0.4 MK at 1.3 R_s. However, [30] deduced values of the order 1.5 MK at 1.5 R_s, which are significantly different from the SOHO results. Besides, observations from the solar wind ion composition spectrometer (SWICS) instrument on Ulysses, during the south polar pass, have been used to derive the "freezing in" temperatures in the solar wind. The charge–state ratio of O^{7+}/O^{6+} is a proxy for the coronal temperature [3, 66]. Summarized profiles of temperature are given in the review paper of [4].

Looking at the Sun with a line of sight perpendicular to the general magnetic field will give information on mass motions on the basis of the Doppler dimming of O VI 1032/1037 resonance lines, which affects the absorption profile of the ray, the intensity of the emission profile being related to the radial velocity of the plasma. In 2000, Giordanno et al. examined the profile of the outflow speed of the oxygen ions [12], using the ultraviolet coronograph spectrometer (UVCS) in the corona from 1.5 to 4 R_s, which reaches 300 km/s above 2.1 R_s. This is evidence for ion acceleration primarily near 1.6–2.1 R_s in the fast solar wind. The authors also showed that beyond 1.8 R_s the velocity distribution of the oxygen is highly anisotropic, which may be evidence that the heating process is operating preferentially in the direction perpendicular to the magnetic field.

Plots of temperature and density measurements obtained in the corona and in the solar wind are shown in Fig. 6.1. One can see quite a good agreement between the density measurements made by SOHO in the corona and in the solar wind. We can also note that the electron temperature variation forms a plateau low in the corona and, higher up, decreases to match the solar wind data at 1 AU.

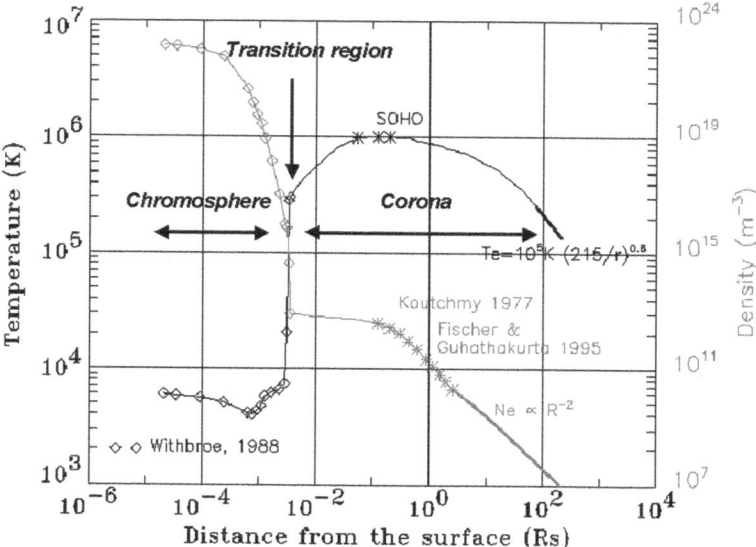

Fig. 6.1 Profiles of the electron density and temperature observed from remote sensing measurements in the transition region and corona

6.2.2 The Solar Wind

After the pioneering mission of Mariner 2 in the solar wind four decades ago [52], many other successful missions like Helios, Ulysses, Wind, ACE provided extensive measurements of in situ solar wind parameters, which are important to improve our understanding of this medium and to relate the interplanetary observations to the solar ones. Accurate measurements of the electron density and temperature are key elements for improving our knowledge of energy transport in the solar wind. Electrostatic analyzers gave the first clear proof of the existence of the wind. Particle properties are deduced from measurements of the electron and ion velocity distribution functions with high-time resolution, generally combining electrostatic analyzers, and particle detectors. From these observations, moments are derived ([51, 10]). However, these quantities are polluted at low energies by photoelectrons emitted by the spacecraft body due to solar UV light and the spacecraft potential, which itself depends on the electron density and temperature of the plasma requiring difficult corrections [14].

Although these effects can cause difficulties in the measurement of the electron plasma distributions,[1] electrostatic analyzers yield the basic knowledge of the solar wind plasma properties. In situ observations of solar wind electrons show that their distribution contains three components: the core, the halo, and the strahl. Core electrons, i.e., electrons with a speed smaller than the thermal speed, are characterized by a bi-Maxwellian distribution with an anisotropy at 1 AU of $T_{\parallel}/T_{\perp} \sim 1\text{--}1.5$. Halo electrons form a suprathermal tail of the total electron distribution. At the orbit of the Earth, the density ratio of the halo electrons to those in the core is $N_{halo}/N_{core} \sim 0.05$, while their temperature ratio is $T_{halo}/T_{core} \sim 6$. The third population, which is not always present, is the strahl. It is a sharply field-aligned beam, where the electron energy is comparable or higher to that of the halo electrons [11].

Another technique makes use of a wire dipole antenna coupled to a sensitive radio receiver; it is a passive plasma wave experiment. When an antenna is immersed in a plasma, it receives electrostatic waves which excite the plasma. This quasi-thermal noise (QTN) is a function of the velocity distributions of the particles, and thus mainly yields accurate values of the electron density and core temperature. This type of experiment has been successfully operated over the last decades on various spacecraft flying through different media: ISEE-3/ICE [31, 40] (solar wind and cometary tail), AMPTE–IRM [20, 55, 32] (Earth's magnetosphere), ULYSSES [67, 42, 43, 46, 33, 22] and references therein (interplanetary medium, Io Torus and Jovian magnetosphere), WIND [2, 62](solar wind), CASSINI [17, 49] (interplanetary medium, Jovian and Saturnian magnetospheres), and NOZOMI [36] (Mars environment). The main advantage of the method is that it is nearly independent of thespacecraft potential and photoelectron perturbations since the volume sensed

---

[1] Ion measurements are far less sensitive to photoelectron and spacecraft potential effects due to the significantly higher ion kinetic energies compared to photoelectron and spacecraft energies.

by the antenna is much larger than the spacecraft volume [44]. This method is complementary to particle analyzer technique and can be used to cross-check other plasma sensors. Through thermal noise spectroscopy, it provides reliable and accurate in situ measurement of plasma parameters, such as the density and the bulk electron temperature.

Basic principles of thermal noise spectroscopy and particle analyzers are reminded in the next section. Each instrumentation itself is detailed in addition to its main advantages and drawbacks.

6.3 Measuring the Key Parameters of the Solar Wind

6.3.1 Particle Analyzers

The quality of the plasma moments is affected by the quality of the distribution measurements. We will examine the general principles of particles detectors, with emphasis on the advantages and limitations of the instruments used. Detailed description of particle analyzer technique is given in the tutorial paper of [9].

6.3.1.1 Principles

A generic instrument for measuring plasma velocity distributions consists of a velocity space filter system, a detector, and a counter. The velocity space filter controls access to the detector, so that it is only those charged particles from a pre-selected region (volume element) in velocity space which may reach the detector at a given instant in time. The detector is able to respond to an arriving charged particle by generating an electrical signal. The counter records the electrical signals and their time of arrival. The distribution is characterized by making measurements for a set of different regions in velocity space.

The velocity space filter must be able to select particles arriving from a restricted velocity space volume. A three-dimensional velocity space distribution is constructed by making measurements which sample the full 4π solid angle at a desired resolution, for each of a set of energy values which cover a required energy range, again at a desired resolution. Particles that enter the analyzer from directions with different polar angles are transmitted to different parts of the detector and may be distinguished from one another according to where they strike the detector. The detector is usually subdivided into a number of zones (or sectors), and particles arriving in a given zone are grouped together by the counter.

In certain specific cases, it is possible to identify different species by examining the energy per charge distributions, i.e., E/q. An example is the case of the solar wind proton and alpha particle populations, which have about the same bulk speed but different masses. The kinetic energy of bulk motion of the two populations differs by the ratio of their masses, i.e., 4/1. The spread of particle energies relative to the kinetic energy of bulk motion is characterized by the thermal speed of the population, and for the solar wind populations the thermal speeds are small compared to the difference in their kinetic energy of bulk motion. Thus the two populations occupy distinct and non-overlapping ranges of E/q.

Under conditions where a particle distribution is expected to be gyrotropically symmetric, a measurement of a two-dimensional pitch angle distribution is sufficient to characterize the particle distribution. The three-dimensional distribution is reconstructed by simply rotating the pitch angle distribution about the magnetic field. The assumption of gyrotropic symmetry is considered to be often valid for plasma electron distributions, although not usually for ion distributions. Note that gyrotropy is a property of the distribution function in a frame in which there are no cross-field drifts, e.g., the species rest frame, which does not usually coincide with the spacecraft/instrument rest frame. Electron drifts are typically much smaller than the electron thermal speeds for hot plasma distributions, so that the measurement frame is a good approximation to the species rest frame. Thus for electron measurements, a two-dimensional distribution containing the magnetic field direction can simply be selected from a three-dimensional measurement and treated as an electron "pitch angle distribution." Indeed, it is quite common practice for electron measuring instruments, in particular, to transmit pitch angle distributions in preference to full three-dimensional distributions (there is a substantial reduction in the telemetry required per distribution). Ion drifts are not usually negligible compared to ion thermal speeds for hot plasma distributions. Ion pitch angle distributions are usually reconstructed during ground-based data analysis working with three-dimensional distributions.

6.3.1.2 Measurements

The plasma moments are calculated from integration of the electron and ion velocity distributions, providing independent parameters for the protons, the alpha particles, and the core and halo electron components. Note that no assumptions on the velocity distributions are made. In particular, the integration of the electron core distribution is computed starting slightly above the spacecraft potential and continuing to the core/halo breakpoint energy while the integration of the halo distribution starts slightly above the core/halo breakpoint energy to the highest detector energy. Note that for the halo integration the halo electrons below the core–halo breakpoint energy are neglected [10, 57].

A typical example of velocity distribution function is shown in Fig. 6.2, measured by Wind spacecraft at 1 AU in the fast solar wind. On the reduced function (bottom of the figure), in the plane $[v_{\parallel}, v_{\perp}]$ (with respect to the magnetic field direction), one can see the core and halo feature of the velocity distribution. Note that the velocity distribution is far from a Maxwellian at high energy. A strahl component is present (pointing anti-sunward) showing an important heat flux.

Fig. 6.2 The *upper panel* shows a typical example of the velocity distribution function parallel and perpendicular to the magnetic field obtained from electron analyzers on the WIND spacecraft at 1 AU. The *lower panel* shows reduced electron velocity distribution functions associated with this distribution. From [61]

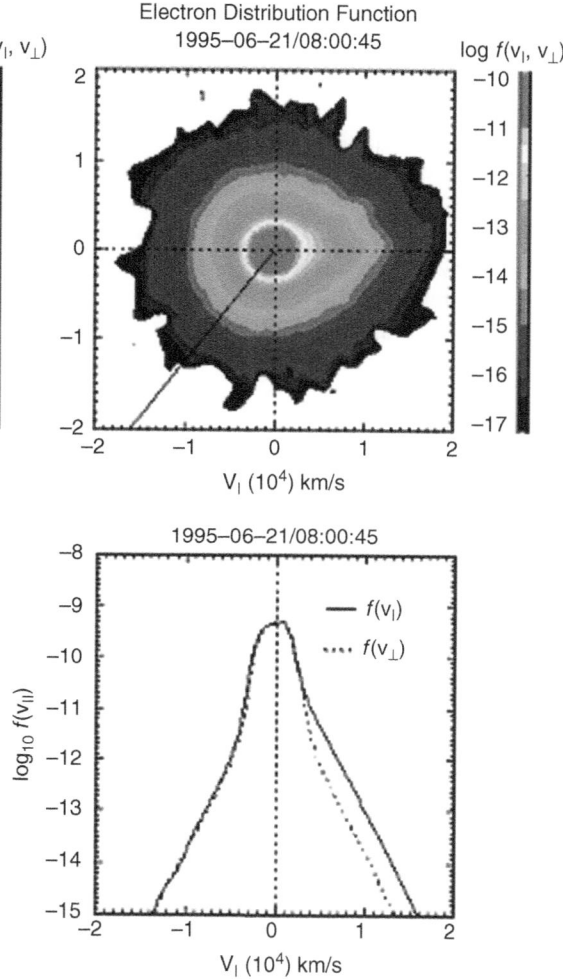

6.3.1.3 Advantages/Drawbacks

One cause of error is counts which are not due to plasma particles. These "background" counts can have several sources, including high-energy penetrating particles, detector dark counts, solar UV-induced counts, photoelectrons, or electronic noise. Penetrating particle background counts are only important where there is a high flux of energetic particles, for example, during solar energetic particle events. UV-induced counts are only seen when the detector is facing the Sun. To minimize errors, the magnitude of the various backgrounds must be estimated or empirically determined and subtracted before moments summation, which is not an easy task.

Photoelectrons are a particularly severe problem because they can dominate the count rates of electron instruments at low energies. In the solar wind, the photoelectrons emitted by the spacecraft body are usually dominant compared to the ambient electrons one is trying to measure. Thus the spacecraft charges positively, reaching a potential ϕ with respect to the plasma. As a consequence, the photoelectrons with energies smaller than $e\phi$ return to the surface and may be collected by the detector. Only those emitted with an energy larger than $e\phi$ can escape into the plasma.

However, the spacecraft electric potential is much more delicate to estimate since this potential depends itself on the values of the density and temperature that one wished to measure. Several techniques can be used to determine the spacecraft electric potential (see, for example, [59, 64, 62] respectively for Helios, Ulysses, and Wind instruments). To retrieve the correct solar wind electron distribution functions from the measurements, one has first to distinguish between spacecraft photoelectrons and solar wind electrons. This is performed by removing the data corresponding to energy channels lower than a given threshold energy. The choice of this energy depends on the different energy channel distributions specific of each spacecraft experiment. Then the distributions have to be corrected for spacecraft charging effects due to the emission of photoelectrons as well as with respect to currents from solar wind electrons and protons.

An alternative is to use the solar wind electron density and temperature from another experiment insensitive to spacecraft potential as a reference. Maksimovic et al. [34] proposed using the solar wind electron density obtained from the spectroscopy of quasi-thermal noise (QTN) around the electron plasma frequency, which allows accurate in situ electron diagnostics of space plasmas. In this way, [62] improved the determination of electron parameters in the WIND particle analyzer experiment and corrected their measured electron velocity distributions.

In spite of the spacecraft charging and low-energy cutoff effects, the electron electrostatic analyzers remain an essential means for investigating the actual shape and the fine details of the electron distribution functions in the solar wind, and in space plasmas generally.

6.3.2 QTN Method

6.3.2.1 Principle

The quasi-thermal noise (QTN) diagnostics of space plasmas is based on the generalization of the simple problem of an antenna immersed in black body radiation. Whereas the analysis of these electromagnetic fluctuations yields the black body temperature, an antenna in an equilibrium plasma is mainly excited by thermal plasma waves; in a weakly magnetized plasma these are Langmuir waves, which enable one to measure the local plasma density and temperature. Although space plasmas are not in equilibrium, they are often stable, so that the electrostatic fluctuations measured by an electric antenna are determined by the velocity distributions of the particles. The analysis of the spectrum of these fluctuations enables one to measure in situ the electron density and temperature and other parameters [44]. This method has been generalized to magnetized plasmas, in which it allowed one to measure also the magnetic field magnitude with a great accuracy [42], and to dusty plasmas [45].

6.3.2.2 Measurements

When a passive electric antenna is immersed in a stable plasma, the thermal motion of the ambient particles produces electrostatic fluctuations, which are completely determined by the particle velocity distributions. A sensitive and well-calibrated receiver connected to a wire dipole antenna can measure these electric field oscillations, and turn them into spectra. The analysis of the voltage power spectrum enables one to measure in situ plasma parameters [41]. It yields an accurate determination of the electron density since the electron thermal motions excite Langmuir waves, which produce a spectral peak just above the plasma frequency (see Fig. 6.3). Moreover, the electrons passing around the antenna induce voltage pulses on it, producing above a noise level which decreases as the observing frequency increases and a plateau just below this frequency.

Figure 6.3 shows a typical example of the power spectrum measured in the solar wind on Ulysses, using the URAP radio receiver. The main feature of the spectrum is a peak located in the vicinity of the plasma frequency, which is the characteristic frequency of the plasma. From the cutoff position, it is possible to determine the electron density since $f_p \propto \sqrt{n_e}$. The shape and the level of the spectrum yields the moments of the electron distributions, i.e., the halo and core electron density ratio and the halo and core electron temperature ratio. At low frequencies, the spectrum is due to the electrostatic fluctuations of the protons. Since the proton thermal speed is much smaller than the plasma bulk velocity, the proton noise spectrum is Doppler-shifted by the solar wind speed. Issautier et al. extended the method

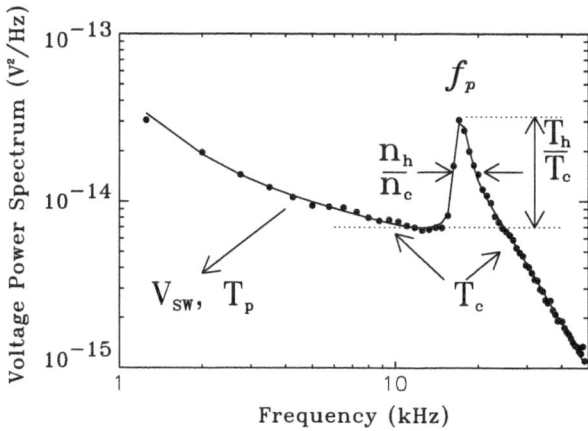

Fig. 6.3 Typical example of the thermal noise power spectrum measured in the solar wind from ULYSSES data measured by the URAP (Unified Radio and Plasma Wave) instrument. See text for details. From [25]

[21, 24] using this lowband frequency part to measure the solar wind speed and proton temperature [22]. The solid line is the theoretical spectrum which best fits the data (dots). The deduced plasma parameters (not shown on this figure), i.e., the electron density N_e, the core temperature T_c, and bulk speed V_{sw}, in addition to suprathermal parameters deduced from the fine structure of the peak [23, 44] are also routinely determined. The method is accurate to typically 1–2% for N_e, and to 10–20% for T_c, and has no free parameter. The fit rms error is usually only about 5%. Note that an accurate determination of T_c and other plasma parameters requires a sensitive and well-calibrated receiver, while N_e can be measured from the cutoff frequency of the spectrum.

To summarize, in practice, a plasma diagnostic is performed by (1) assuming a model of the velocity distribution (we describe the distribution of the electrons by the superposition of a cold (c) and a hot (h) Maxwellian [10] and that of the protons by one drifting Maxwellian), (2) calculating the theoretical spectrum produced by these distributions, and (3) deducing the parameters of the model by fitting the theory to the data.

Figure 6.4 shows an example of dynamical spectrogram (top) acquired by WIND/TNR experiment. A color bar chart is indicated on the right. On the bottom panel, the corresponding plasma parameters are plotted routinely, mainly the electron density (from the QTN fitting and the neuronal network) and the core temperature (from the fitting procedure).

6.3.2.3 Which Materials and Geometry Are Currently Used?

Three main hardware elements are necessary to perform thermal noise spectroscopy in space: a pair of long and thin wire booms (or alternatively by double sphere antennas), electric field preamplifiers, and a sensitive receiver. First, let us remind why we need long and thin antenna booms to detect quasi-thermal noise. As said before, the thermal noise spectrum peaks just above the plasma frequency. The type of induced electrostatic waves depends on the ratio between the plasma frequency and the electron gyrofrequency F_{eg} defined by F_{eg} [Hz] = 28 B [nT], where B is the static field. In the solar wind, when $f_p \gg F_{eg}$, the thermal motion of the electrons excites Langmuir waves. The electrostatic waves generated have a (frequency-dependent) wavelength longer than the Debye length λ_D where λ_D [m] $\approx 7(T_e$ [eV] / N_e [cm^{-3}])$^{1/2}$.

As a consequence, the antenna length L needs to exceed a few λ_D to detect the Langmuir wave peak and cutoff. In this case, the precision is of the order a few percent for the total electron density and better than 15% for the core temperature [44]. A precision of 15% cannot be reached for the suprathermal population if L is only of the order of a few Debye lengths, especially for a large T_h/T_c ratio. Indeed, very good spectral resolution is necessary to evaluate, with the same accuracy, the suprathermal electron density and temperature [5]. A further increase of L can solve this problem because the longer the antenna, the sharper the Langmuir wave peak [67].

The radius of the antenna, a, also needs to fulfill requirements imposed by the physical interaction between the plasma and the antenna. Indeed, the particles whose

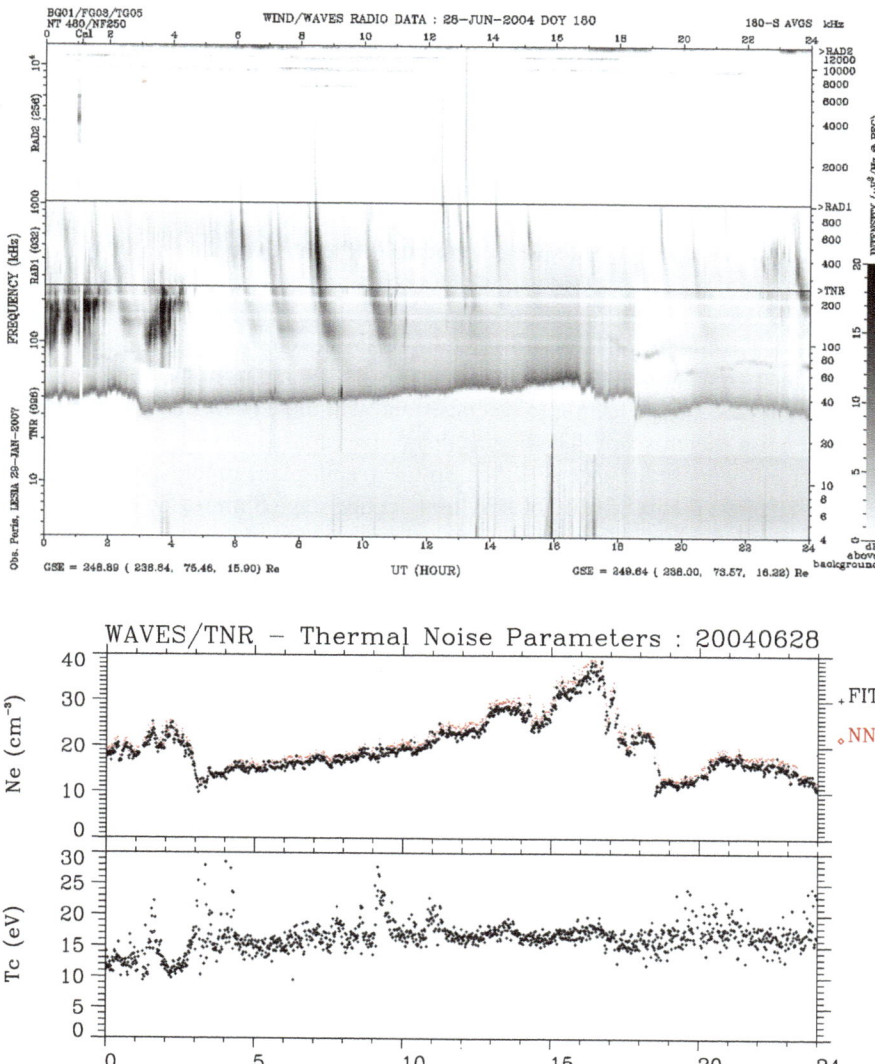

Fig. 6.4 Radio astronomy receiver spectrogram of the thermal noise receiver (TNR) instrument on the WIND spacecraft. The electrostatic Langmuir waves emissions are seen around 40 kHz. From it, the QTN analysis gives the total electron density of the plasma and the fittings of all the spectra give also the electron core temperature (*bottom panel*). Due to the large dynamic of the intensity of the wave emissions in the interplanetary medium and the magnetosphere, the dynamic range of the radio receiver has to cover several decades (e.g., low for electrostatic Langmuir emissions and high for electromagnetic solar type III burst radio emissions). On Wind, two multichannel receivers cover the frequency range 4–256 kHz in five overlapping and logarithmically space frequency bands. Each of them is divided into either 32 or 16 logarithmically spaced channels

trajectory intercepts the antenna surface stop on it or are reflected and/or induce a secondary emission; in addition, the surface emits photoelectrons. The noise generated by this physical phenomenon is generally modeled by simply adding a shot noise component V_{SN}^2 [56]. In the solar wind, the shot noise spectral density measured by a wire dipole perpendicular to the solar direction has been analytically derived [40] as $V_{SN}^2 \propto (f_p/f)^2 T_c (a/L)[ln(a/L_D)]^2$ for frequency $f \ll f_p$, $L \gg L_D$ and the solar wind speed V_{SW} tends to 0.

This equation clearly shows that the radius of the wire antenna needs to be thin enough to minimize the shot noise which occurs at frequencies below the plasma frequency. For example, the radius of each of the two wire booms of the Ulysses spacecraft, 1.1 mm, fulfills the above-mentioned necessary requirement to estimate the proton thermal noise at low frequencies. Finally, the radius of the boom has to be smaller than the Debye length, to be consistent with most of the theory involved in thermal noise spectroscopy. The booms themselves consist of two 35 m monopoles diametrically opposed (hence a 72.5 m tip-to-tip spin-plane dipole antenna). In the case of Ulysses, they are made of beryllium–copper tape 5 mm wide and 0.04 mm thick (hence a 5×0.04, cross section, equivalent to a tube of 2.2 mm diameter).

On Ulysses, the electric field preamplifiers are located at the foot of each antenna mechanism. For thermal noise spectroscopy, the preamplifiers have to cover a large frequency range, typically several decades (e.g., 1 kHz to 1 MHz). In order to extract other parameters than the plasma frequency, quasithermal noise has to be estimated with a good precision. In particular, noise induced by the preamplifiers needs to be kept as low as possible.

6.3.2.4 Main Advantages

Since 1979, the thermal noise spectroscopy technique has been applied to a wide variety of media (see detailed references in Sect. 3.1): the solar wind, the Io torus, cometary tail environments, and various planetary environments (Earth, Jupiter, Saturn). In these media, the robustness of thermal noise spectroscopy for obtaining the electron density is in particular due to the fact that thermal noise spectroscopy is independent of gain calibration. Moreover, when the fluctuations of the electron density in the solar wind are too fast to be measured by any particle instrument, thermal noise spectroscopy is still able to yield the density. It is worth underlining the fact that the measure of the electron density through thermal noise spectroscopy does not require sensitive preamplifiers and receivers. Only other plasma parameters like the core electron temperature and the solar wind bulk velocity do. The core electron temperature can also be reliably and routinely extracted, but only if a sensitive and well-calibrated receiver is used. Finally, unlike electrostatic analyzers, thermal noise spectroscopy is in general relatively immune to spacecraft photoelectrons and charging effects to estimate the density and the core electron temperature. Because of its reliability and accuracy, it also serves to calibrate other plasma sensors.

6.3.2.5 Main Drawbacks

Thermal noise spectroscopy is not well suited to estimating the density and the temperature of suprathermal electrons since these measurements strongly depend on the shape of the peak, hence on the frequency resolution of the receiver. Moreover, like other passive measurements, the determination of the core plasma parameters is disturbed in the presence of strong plasma instabilities, and natural radio emissions (a solar type III burst, for example). Thermal noise spectroscopy is noise sensitive. Indeed, thermal noise spectroscopy may be sensitive to interferences originating from other instruments on the same spacecraft platform (which occurs on most platforms at least for certain modes of operation), or from even moderate level of natural emissions generated non-locally. When a solar type III burst occurs just above the plasma frequency (electromagnetic waves do not propagate below the plasma frequency in an unmagnetized plasma), it perturbs the thermal noise spectrum above the electron plasma frequency. Hence, the core temperature cannot be deduced from the hole spectrum. However, recent improvement on thermal noise spectroscopy gives the opportunity to yield this parameter, in this case, using only the low-frequency range (below f_p) of the spectrum [24].

6.3.2.6 Diagnostics in Magnetized Plasmas

An extension of the method to magnetized plasmas was first used in the Jovian magnetosphere when Ulysses crossed the Io plasma torus [42]. When the electron gyrofrequency F_{eg} is no longer negligible compared to f_p, the wave spectrum is modified [46], since the electron thermal motion excites Bernstein waves. Figure 6.5 shows a spectrum acquired by WIND in the Earth's plasmasphere. The

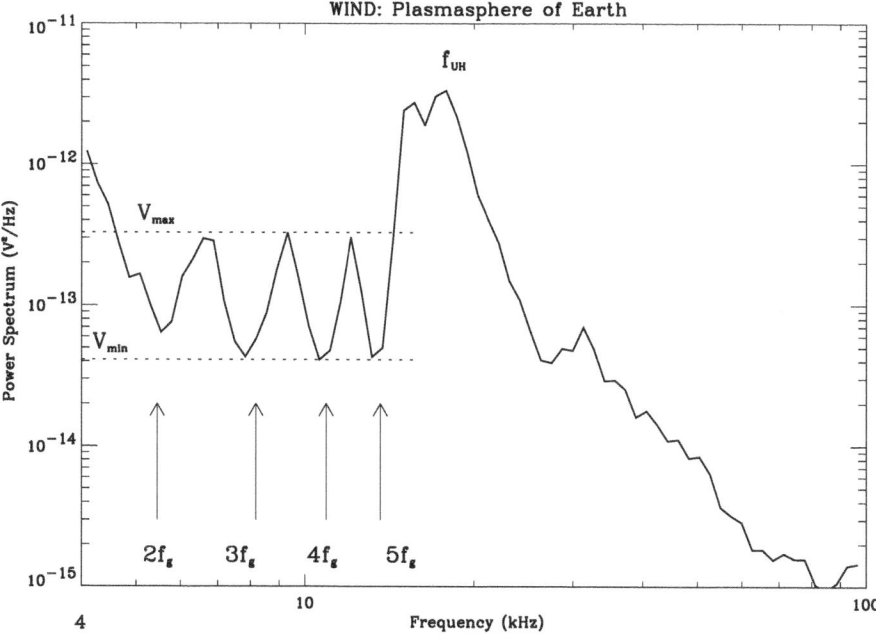

Fig. 6.5 Voltage power spectrum acquired by Wind in the Earth's plasmasphere (*solid line*). The levels V_{min} and V_{max} yield an estimate of the temperatures: in this case, $T_c \simeq 9.5 \times 10^4$ K and $T_h \simeq 6.6 \times 10^5$ K; the determination of $F_{eg} \simeq 2.7$ kHz and $f_{UH} \simeq 15.6$ kHz yields $n_e \simeq 2.9$ cm^{-3}

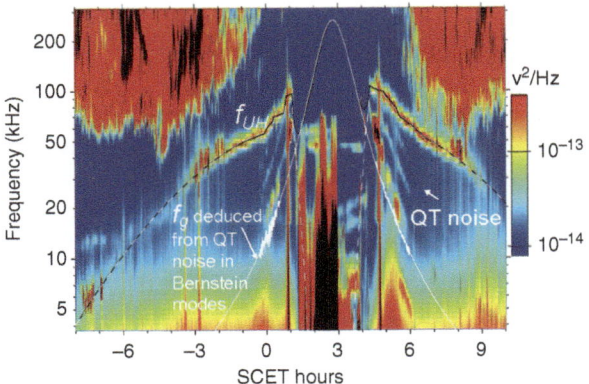

Fig. 6.6 Voltage power spectrum acquired by Cassini in the Saturn magnetosphere during the closest approach in 2004. From [49]

spectrum peaks at the upper-hybrid frequency f_{UH} and exhibits weak bands with well-defined minima at gyroharmonics below f_{UH}. Such a spectrum allows one to determine (i) the electron density from the detection of f_{UH} [47], (ii) an estimate of the cold [42] and hot [65] electron temperatures, T_c and T_h, from the minimum and maximum levels of the gyroharmonics bands, respectively, and (iii) the magnetic field magnitude from the detection of the spectral minima at gyroharmonics, which is in agreement with the magnetometer data to within a few percent. Recently, on Cassini, Moncuquet et al. [49] studied the Kronian magnetosphere and deduced the density and temperature of the electrons using the QTN method, in addition to the magnetic field. An example of the dynamic spectrum acquired during the closest approach of Cassini to Saturn is seen in Fig. 6.6.

6.3.2.7 Examples in Other Environments

The thermal noise spectroscopy method was first used on a large scale with the radio experiment on the spacecraft ISEE-3/ICE when it crossed the tail of comet Giacobini–Zinner [40]. The experiment yielded the profiles of cometary electron density and temperature during the encounter. These results are showed in Fig. 6.7. These are unique since the ICE electrostatic electron analyzer could not detect adequately the cold cometary electrons in the plasma sheet ($n = 670$ cm^{-3}, $T = 1.3 \times 10^4$K) as the effects of spacecraft potential and photoelectrons could not be properly eliminated.

6.4 Comparison of Solar Wind Plasma Parameters

A detailed comparison between solar wind plasma parameters from particle analyzers and quasi-thermal noise (QTN) method on Ulysses is given by Issautier et al. [24]. Similarly, the statistical study made by WIND using 3DP particle analyzer and TNR radio experiments is reviewed by Salem et al. [62]. Below, as an example, we briefly focus on the comparison of the electron density and core temperature during Ulysses fast latitudinal scan in 1994–1995 during solar minimum of activity.

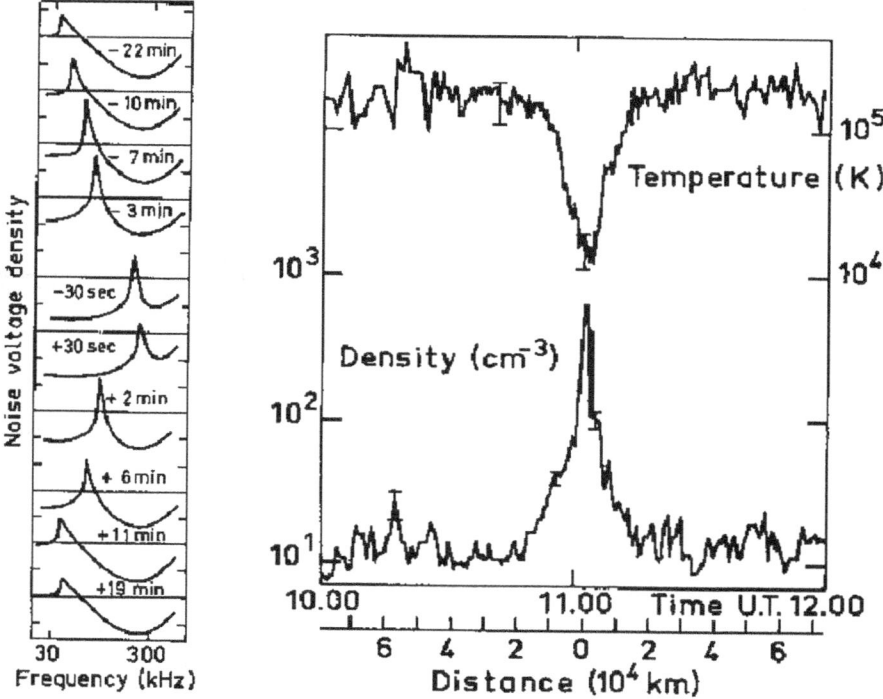

Fig. 6.7 Series of thermal noise spectra measured by the radio experiment aboard ICE during the crossing of the tail of comet Giacobini–Zinner (*left side*). The corresponding profiles of electron density and temperature are displayed as a function of the distance from the tail axis at 7800 km from the nucleus. Adapted from [40]

6.4.1 Electron Density

Figure 6.8 shows a scatter plot of the electron density n_e deduced from QTN spectroscopy, using the URAP radio data on Ulysses, and $n_{e(SWOOPS)}$ acquired by SWOOPS particle analyzers. For the Ulysses fast latitudinal scan, the QTN analysis gives about 173,000 data points whereas the SWOOPS instrument yields around 166,000 measurements since its resolution rate is lower. From the histogram, one can see that by comparing both data sets there is a systematic offset of 50% compared to the electron spectrometer. Indeed, as we said before, it is due to spacecraft effects which compromise the measurements. However, a detailed comparison [24] of the electron density n_e from QTN and the total ion density n_i from ion spectrometer with $n_i \equiv n_p + 2n_{alpha}$ (n_p and n_{alpha} are the proton and alpha number densities, respectively) shows a very good agreement. The average offset between the two data sets differs by less than 5%. In addition, n_i and n_e (from QTN) exhibit very similar variations as a function of time [23]. An example of comparing electron density from SWOOPS experiment and the QTN method, and of total ion density is showed in Fig. 6.9. One can see that the electron density from QTN is superimposed on

Fig. 6.8 (**a**) Scatter plot of the electron density measurements from SWOOPS and URAP experiment on Ulysses during the fast latitude scan in 1994–1995. (**b**) Histogram of the difference between the two data. Note that we did not sort out or select the slow or fast solar wind observations since we want here to obtain statistical results including both kinds of wind. There is an offset of 50% compared to electron spectrometer

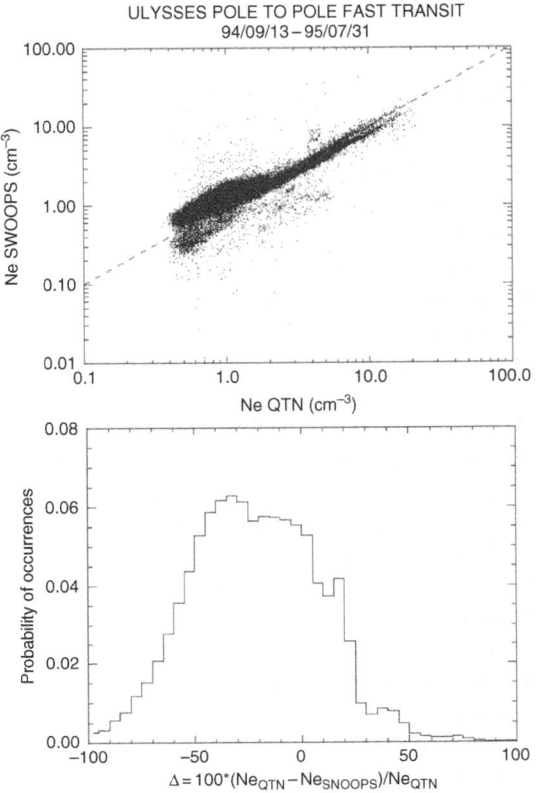

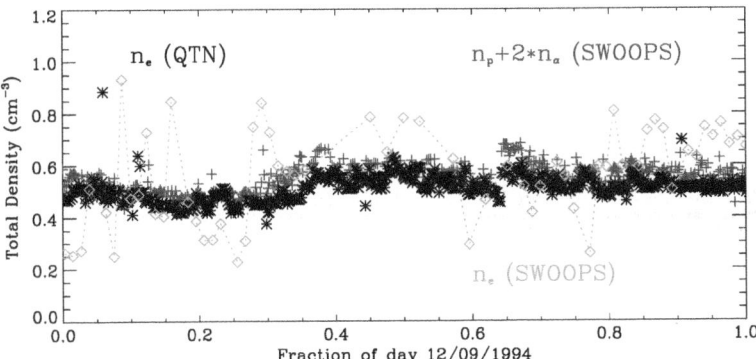

Fig. 6.9 Comparison of solar wind density measurements on Ulysses, by the QTN technique and particle analyzers

that given by ion spectrometer within a few percent while there is large difference compared to the electron spectrometer. This instrument is polluted by spacecraft effects and needs cross-calibration, using QTN method, for example.

6.4.2 Core Temperature

A statistical study on the core temperature T_c measured by both types of instruments during the fast latitudinal scan of Ulysses shows a systematic offset of 5–10% depending on the solar wind flow speed (see Fig. 6.10). The solid line in Fig. 6.10 is the result of the fit between the data and a Gaussian model.

It is noteworthy that the superposition of both T_c data sets [23] reveals that for low-core temperature values, i.e., 7.5×10^4 K, particle analyzer systematically provides lower values than the QTN determinations. This occurs primarily for high-speed solar wind coming from polar coronal holes. On the other hand, for the highest temperatures, particle analyzer finds T_c values higher than QTN determinations. This mainly occurs near ecliptic plane for slow wind flow. Differences in the T_c determinations may be the result of the complicated spacecraft potential and sheath, and/or mixing of photoelectrons into the core electron population.

Figure 6.11 shows an example of the core electron temperature, derived by both techniques, from a stream interface crossing by Ulysses during its fast latitudinal scan. Note that the URAP experiment has four times the temporal resolution of SWOOPS at this time. One can see the disagreement in T_c for that period, T_c from SWOOPS being roughly constant at about 1.2×10^5 K, while T_c from QTN is around 7.5×10^4K before the stream interface, jumps to 1.1×10^5K at the discontinuity and then falls back to its initial value. The temperature enhancement from QTN analysis

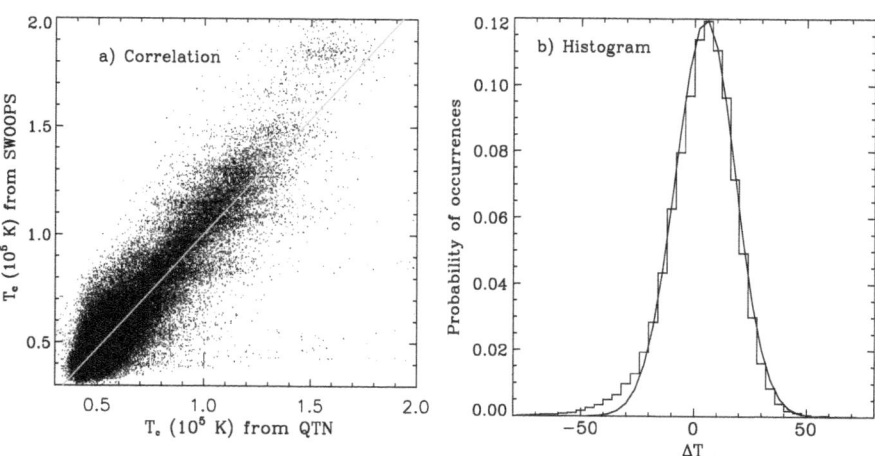

Fig. 6.10 (**a**) Scatter plot of the core temperature measurements from SWOOPS and URAP experiment on Ulysses during the fast latitude scan in 1994–1995. (**b**) Histogram of the difference between the two data. The *solid line* is the Gaussian model which best fit the data. (From [24])

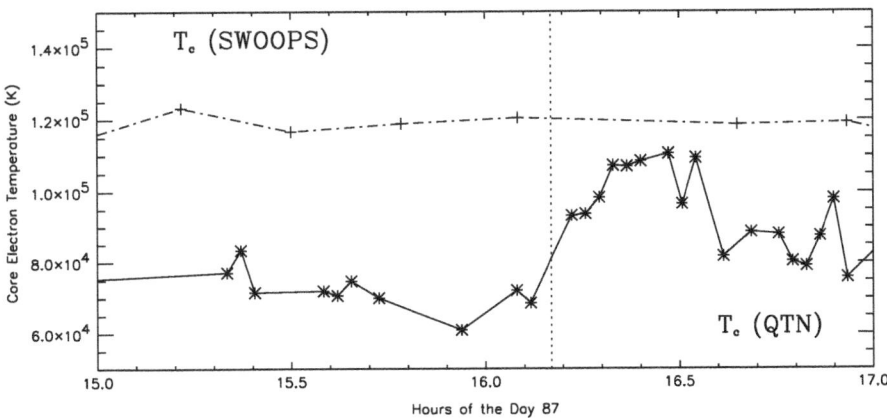

Fig. 6.11 Comparison of the core electron temperature obtained on Ulysses by SWOOPS and URAP experiments during the stream interface crossing on March 28, 1995. From [24]

at the interface is about 1.4 the initial T_c value, in agreement to the result obtained by [13] using IMP data at 1 AU.

Finally, we can summarize both techniques as follow. Particle analyzers provides reliable ion diagnostics in addition to suprathermal (halo) electron parameters and yields fundamental information on particle velocity distributions, whereas the quasi-thermal noise method accurately provides total electron density and core electron temperature.

6.5 Large-Scale Variations of the Heliosphere

6.5.1 Basic Results Over Solar Cycle: Ulysses in Three Dimension

Since its launch in October 1990, the Ulysses spacecraft has given us excellent opportunities to study the large-scale structure of the solar wind over a wide range of heliographic distances and latitudes. During its first pole-to-pole transit in 1994–1995, achieved in 10 months near the 1996 solar activity minimum, Ulysses confirmed the rather simple bimodal speed structure of the heliosphere [53]. Indeed, in situ measurements provided the first direct proof of the steady-state fast solar wind, with a speed around 750 km/s [58] and a low density of ~3 cm⁻³ [23] originating from large polar coronal holes poleward 22°S and 21°N. In contrast, at lower latitudes, Ulysses encountered alternating streams of high and low speed flows due to the crossings of the warped heliospheric current sheet (HCS) [37] and/or density compression regions.

In 2000–2001, Ulysses explored again the solar wind from pole-to-pole during the rising phase of solar cycle 23 near maximum. During that period the state of the corona was dramatically more complex than at minimum, and revealed different regimes of wind, from slow and intermediate flows, i.e., streamers, to sporadic fast wind from small coronal holes, extending to all heliographic latitudes. Indeed, compared to solar minimum the HCS is much more tilted and warped; it is no longer centered on the equatorial plane [38] and extends to higher latitudes up to 70°. With increasing solar activity, SOHO/LASCO coronograph images reveal the presence of numerous coronal mass ejections. In addition, there are an increased number of

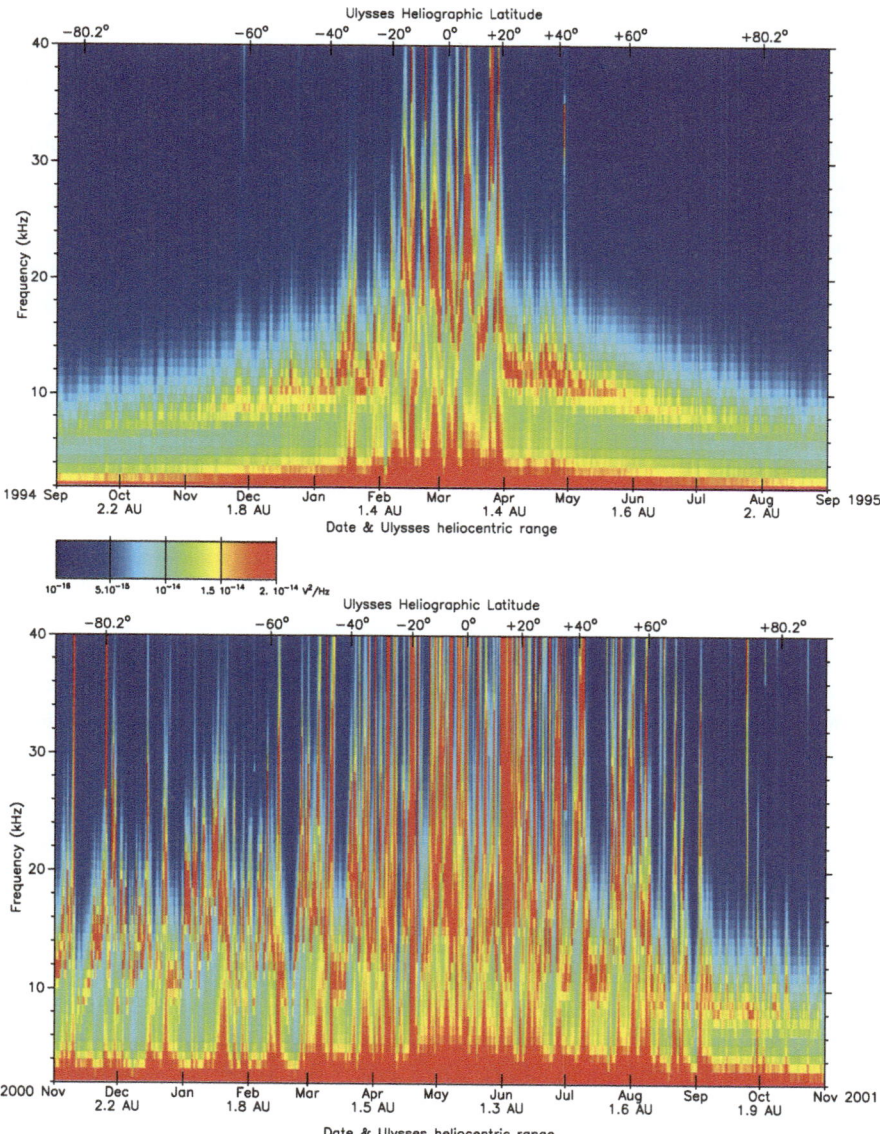

Fig. 6.12 Pole-to-pole radio spectrograms from ULYSSES/URAP observations obtained near solar minimum (*top*) and maximum (*bottom*). The color bar chart indicates the coded intensity of the signal. From [27]

active regions on the solar surface. With the transition of solar activity from minimum to maximum, large polar coronal holes are replaced by small equatorward coronal holes [68, 39].

Figure 6.12 shows radio spectrograms (intensity coded by a color scale as a function of frequency and time) acquired by the Ulysses/URAP experiment during the fast latitude scan near solar minimum (top) and maximum (bottom). The intense fluctuating line is the plasma line f_p of the solar wind. In the top panel one can see that near the 1996 solar minimum, there are two distinct regions in latitude. Between 22°S and 21°N heliographic latitudes, Ulysses was embedded in the solar wind from the streamer belt and crossed the HCS several times. In contrast, poleward of 40°, for several months Ulysses observed the steady-state and stationary fast solar wind emanating from well-developed polar coronal holes in both the southern and northern hemispheres [37].

As the solar cycle rises to its maximum, the bottom panel of Fig. 6.12 shows the radio spectrogram where the plasma line varies a lot at all heliographic latitudes, and appears to be the extension of that observed in the equatorial plane at solar minimum (Fig. 6.12 top panel). At activity maximum, the solar wind is indeed dominated by a mixture of fast and slow solar wind flows, together with numerous CMEs, shocks, and other transient events. However, a relatively quiet and steady-state period of fast wind could be seen at the end of the northern polar pass, namely from September to November 2001 in Fig. 6.12). Ulysses is then embedded in a polar coronal hole extending beyond 72°N, where the wind speed was observed to be nearly constant at 750 km/s [39].

The accurate electron diagnostics using the QTN method give new opportunity to understand the three-dimensional structure of the solar wind over a full solar cycle [22, 26, 27]. For polar coronal holes, Issautier et al. [22] obtained the radial profiles for the electron density and core temperature. In addition, their histograms normalized to 1 AU using these deduced radial profiles revealed a roughly normal distribution, corresponding to the high-speed wind, less dense and cooler than the slow solar wind. A north/south asymmetry was found between the two hemispheres [26, 27].

6.5.2 Basic Results in the Ecliptic Range: Wind

Since its launch in November 1994 WIND spacecraft has frequently observed the in-ecliptic solar wind upstream of the Earth's bow shock. The WIND/WAVES Thermal Noise Receiver (TNR) was specially designed to measure the in situ plasma thermal noise spectra. Applied to the TNR data and associated with a neural network, the quasi-thermal noise method yield routinely (every 4.5 s) the electron density and core temperature of the solar wind. Recently, the study of [28] showed more than 9 years of the solar wind plasma parameters obtained from the Wind spacecraft, encompassing almost the whole solar cycle 23. The large sample of N_e and T_c data points (∼2 millions) allows one to study the dependence of the solar wind structure on solar activity and with different regimes of the solar wind. The corresponding distributions have been compared with those of Ulysses for common periods in order to better understand the origin of the different types of winds over a full solar cycle [28, 63]. Figure 6.13 reveals a complex distribution for both N_e and T_c pointing to a mixture of different regimes of wind.

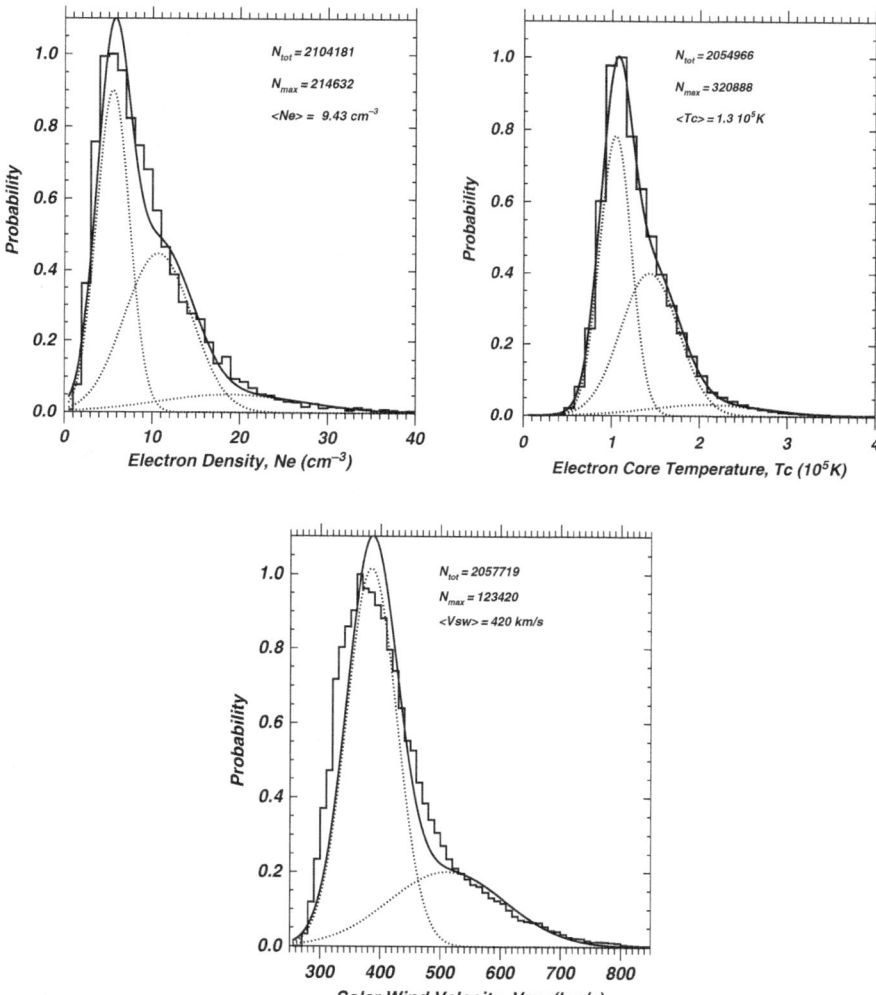

Fig. 6.13 Histograms of N_e, T_c, and V_{sw} obtained from November 1994 to December 2003. We indicate the total number of data used to make the histogram (N_{tot}), the number of points at the peak (N_{max}), and the mean of the parameter over the whole period. The three-Gaussian fit to the observed histograms is shown by a *continuous line* and individual Gaussians by *dotted lines*. The V_{sw} histogram is fitted with two Gaussians corresponding to the fast and slow wind, respectively. (From [28])

On average, the solar wind is cooler (by 20%) and denser (by 15%) at solar minimum than at solar maximum [28, 27, 63]. Note that large-scale variations combining IMP-8, Ulysses, and Voyager 2 proton density and speed measurements are reported by Richardson and Wang [60] from 1988 to 1998. The study of [28] confirms an anti-correlation of the electron density with the sunspot number during solar cycle 23 for the most dilute population (as seen in Fig. 6.13). In addition, a positive correlation of the core temperature is obtained with the sunspot number. The mean value of T_c is about 1.23×10^4 K near the 1996 solar minimum, whereas it is around 1.44×10^4 K near the 2001 solar maximum. The strongest correlation is obtained for the hottest population of the solar wind (seen in Fig. 6.13). This latter population includes overdense disturbed wind events, which are indeed more frequent during solar maximum due to solar activity.

6.6 Summary and Perspectives

The quasi-thermal noise method has been and is still successfully applied to various ionized environments as the tail of a comet [40], the solar wind [22], and the magnetospheres of planets [42, 46, 47]. It is implemented routinely on Ulysses and Wind spacecraft to measure the electron density and temperature in the solar wind, and produced unique measurements of the Io plasma torus aboard Ulysses, which led to a new understanding of the Io torus structure and stability [43, 48]. It is also used on Wind to investigate the Earth's magnetosphere during its perigees as well as on IMAGE. Recently, [49] also measured the electron parameters in the magnetosphere of Saturn with the RPWS experiment on Cassini. In 2006–2007, a third flyby of the solar poles by Ulysses occurred near the solar activity minimum. Accurate electron diagnostics using this method enable one to understand the three-dimensional structure of the solar wind at this stage of the solar cycle.

Quasi-thermal noise measurements serve as a reference for other techniques, and has been used to calibrate and cross-check other sensors [24, 34, 33, 62]. Indeed, the technique is weakly sensitive to spacecraft potential and photoelectron perturbations, since it senses a large plasma volume, at least of the order of the local Debye length cube. Finally, the method is proposed on several future missions such as Solar Orbiter [35] in the solar wind and is selected on Bepi-Colombo/MMO which is dedicated to investigate the Mercury magnetosphere [50].

References

1. L. Biermann: *Kurze original milteilungen*. Naturwissenschaften, Berlin, pp. 34-87 (1946)
2. J.L. Bougeret, M.L. Kaiser, P.J. Kellogg, R. Manning, K. Goetz, S.J. Monson, N. Monge, L. Friel, C.A. Meetre, C. Perche, L. Sitruk, S. Hoang: Space Sci. Rev., **71**, 231 (1995)
3. A. Bürgi, J. Geiss: Solar Phys., **103**, 347 (1986)
4. S.R. Cranmer: Space Science Rev., **101**, 229 (2002)
5. Y. Château, N. Meyer-Vernet: J. Geophys. Res., **96**, 5825 (1991)

6. C. David, A.H. Gabriel, F. Bely-Dubau, A. Fludra, P. Lemaire, K. Wilhelm: Astron. Astrophys., **336**, L90 (1998)
7. R.B. Decker, S.M. Krimigis, E.C. Roelof, M.E. Hill, T.P. Armstrong, G. Gloeckler, D.C. Hamilton, L.J. Lanzerotti: Science, **309**, 5743, 2020 (2005)
8. A.K. Dupree: Solar Wind Eight. In: D. Winterhalter, J.T. Gosling, S.R. Habbal, W.S. Kurth, M. Neugebauer (eds.) AIP Conference Proceedings 382, Woodbury, New-York (1996) p. 66
9. A.N. Fazakerley, S.J. Schwartz, G. Paschmann: Analysis Methods for Multi-spacecraft data. In: G. Paschmann, P.W. Daly (eds.) ISSI Scientific Report, p. 91 (1998)
10. W.C. Feldman, J.R. Asbridge, S.J. Bame, M.D. Montgomery, (eds.) S.P. Gary: J. Geophys. Res., **80**, 4181 (1975)
11. W.C. Feldman, E. Marsch: Cosmic Winds and the Heliophere. In: J.R. Jokipii, C.P. Sonett, M.S. Giampapa (eds.) Kinetic Phenomena in the Solar Wind. University of Arizona Press, Tuscon, pp. 617 (1997)
12. S. Giordano, E. Antonucci, M.A. Dodero: Adv. Space Res., **25**, 1927 (2000)
13. J.T. Gosling, J.R. Asbridge, S.J. Bame, W.C Feldman: J. Geophys. Res., **83**, 1401 (1978)
14. R.C.L. Grard: J. Geophys. Res., **78**, 2885 (1973)
15. K.I. Gringauz, V.V. Bezrukikh, V.D. Ozerov, R.E. Rybshinskiy: Soviet Phys. *Docklady*, English version, **5**, 361 (1960)
16. S. Grzedzielski, D.E. Page: *Physics of the Outer Heliosphere*. Pergamon Press, UK (1990)
17. D.A. Gurnett, et al.: Space Sci. Rev., **114**, 395 (2004)
18. D.A. Gurnett, W.S. Kurth: Science, **309**, 2025 (2005)
19. M.C.E. Huber, et al.: Astrophys. J., **194**, L115 (1974)
20. B. Häusler, et al.: IEEE Trans. Geosci. Remote Sens., **GE-23**, 267 (1985)
21. K. Issautier, N. Meyer-Vernet, M. Moncuquet, S. Hoang: Geophys. Res. Lett., **13**, 1649 (1996)
22. K. Issautier, N. Meyer-Vernet, M. Moncuquet, S. Hoang: J. Geophys. Res., **103**, 1969 (1998)
23. K. Issautier, N. Meyer-Vernet, M. Moncuquet, S. Hoang: J. Geophys. Res., **104**, 6691 (1999)
24. K. Issautier, R.M. Skoug, J.T. Gosling, S.P. Gary, D.J. McComas: J. Geophys. Res., **106**(15), 665 (2001)
25. K. Issautier, M. Moncuquet, N. Meyer-Vernet, H. Hoang, R. Manning: Astrophys. Space Sci., **277**, 309 (2001)
26. K. Issautier, M. Moncuquet, S. Hoang: Solar Wind Ten. In: M. Velli (ed.) (AIP Conference Proceedings) **679**, 59 (2003)
27. K. Issautier, M. Moncuquet, S. Hoang: Solar Phys., **221**, 351 (2004)
28. K. Issautier, et al., Adv. Space Res., **35**, 2141 (2005)
29. V. Izmodenov, G. Gloeckler, Y. Malama: Geophys. Res. Lett., **30**(10), (2003)
30. Y.-K. Ko, L.A. Fisk, J. Geiss, G. Gloeckler, M. Guhathakurta: Solar Phys., **171**, 345 (1997)
31. R. Knoll, G. Epstein, S. Hoang, G. Huntzinger, J.L. Steinberg, J. Fainberg, F. Grena, R.G. Stone, S.R. Mosier, IEEE Trans. Geosci. Electron., **GE-16**, 199 (1978)
32. E.J. Lund, J. LaBelle, R.A. Treuman: J. Geophys. Res., **99**(23), 651 (1994)
33. M. Maksimovic, et al.: Geophys. Res. Lett., **100**(19), 881 (1995)
34. M. Maksimovic, et al.: Geophys. Res. Lett., **25**, 1265 (1998)
35. M. Maksimovic, K. Issautier, N. Meyer-Vernet, C. Perche, M. Moncuquet, M.I. Zouganelis, S.D. Bale, N. Vilmer, J.-L. Bougeret: Adv. in Space Res., **26**, 1471 (2005)
36. H. Matsumoto, et al.: Earth Planets Space **50**(3), 223 (1998)
37. D.J. McComas, et al.: J. Geophys. Res., **105**(10), 419 (2000)
38. D.J. McComas, R. Goldstein, J.T. Gosling and R.M. Skoug: Space Sci. Rev., **97**, 189 (2001)
39. D.J. McComas et al.: Geophys. Res. Lett., **29**(9), 4-1, (2002)
40. N. Meyer-Vernet, P. Couturier, S. Hoang, C. Perche, J.L. Steinberg, J. Fainberg, C. Meetre: Science, **232**, 370 (1986)
41. N. Meyer-Vernet, C. Perche: J. Geophys. Res., **94**, 2405 (1989)
42. N. Meyer-Vernet, S. Hoang, M. Moncuquet: J. Geophys. Res., **98**, 21163 (1993)
43. N. Meyer-Vernet, et al.: Icarus **116**, 202 (1995)

44. N. Meyer-Vernet, S. Hoang, K. Issautier, M. Maksimovic, R. Manning, M. Moncuquet, R.G. Stone: Measurement Techniques in Space Plasmas: Fields. In: Borovsky et al. (ed.) Geophys. Monograph Series, **103** 1998, pp. 205–210
45. N. Meyer-Vernet, et al.: ESA SP **-476** (2001)
46. M. Moncuquet, N. Meyer-Vernet, S. Hoang, J. Geophys. Res., **100**, 21,697 (1995)
47. M. Moncuquet, N. Meyer-Vernet, S. Hoang, R.J. Forsyth, P. Canu, J. Geophys. Res., **102**, 2373 (1997)
48. M. Moncuquet, et al.: J. Geophys. Res., **107**, 1260 (2002)
49. M. Moncuquet, A. Lecacheux, N. Meyer-Vernet, B. Cecconi, W.S. Kurth, et al.: Geophys. Res. Lett., **32**, 1029 (2005)
50. M. Moncuquet et al.: Adv. Space Res., **38**, 680 (2006)
51. M.D. Montgomery, S.J. Bame, A.J. Hundhausen: J. Geophys. Res., **73**, 4999 (1968)
52. M. Neugebauer, C.W. Snyder: J. Geophys. Res., **71**, 4469 (1966)
53. M. Neugebauer: The Heliosphere Near Solar Minimum: The Ulysses Perspective. A. Balogh, R. Marsden, E.J. Smith (eds.) (Springer-Praxis, UK) p. 43, (2001)
54. E.N. Parker: Astrophys. J., **128**, 664 (1958)
55. G. Paschmann, et al.: IEEE Trans. Geosci. Remote Sens., **GE-23**, 262 (1985)
56. M. Petit: Ann. Telecommun., **30**, 351 (1975)
57. J.L. Phillips: Adv. Space Res., **13**(6), 47 (1993)
58. J.L. Phillips, et al.: Geophys. Res. Lett., **22**, 3301 (1995)
59. W.G. Pilipp, H. Miggenrieder, M.D. Montgomery, K.- H. Muhlhauser, H. Rosenbauer, R. Schwenn: J. Geophys. Res., **92**, 1075 (1987)
60. J. Richardson, C. Wang: Geophys. Res. Lett., **26**, 561 (1999)
61. C. Salem: Ondes, turbulence et phénomènes dissipatifs dans le vent solaire à partir des observations de la sonde WIND. PhD Thesis, University P7, France (2000)
62. C. Salem, et al.: J. Geophys. Res., **106**, 21,710 (2001)
63. C. Salem, et al.: Adv. Space Res., **42**(4), 491 (2003)
64. E.E. Scime, J.L. Phillips, S.J. Bame: J. Geophys. Res, **99**(14), 769 (1994)
65. D. Sentman: J. Geophys. Res., **87**, 1455 (1982)
66. R. von Steiger, et al.: J. Geophys. Res., **105**(27), 217 (2000)
67. R.G. Stone et al.: Astron. Astrophys. Suppl., **92**, 291 (1992)
68. Y.-M. Wang, N.R. Sheeley: J. Geophys. Res., **99**, 6597 (1994)

Chapter 7
Physical Processes in the Solar Wind

M. Velli

7.1 The Solar Wind

The solar wind is a supersonic flow of particles emanating from the solar corona. It expands continuously to form the heliosphere, a bubble in the interstellar medium where solar particles and electromagnetic fields dominate. The existence of this flow has been established for over 40 years now, and abundant data has accumulated concerning its average properties as well as intermittent energetic manifestations which impact the Earth's magnetosphere (causing geomagnetic storms and aurorae). Still, many of the phenomena occurring in the heliosphere, from plasma heating to particle acceleration, from bursty energetic manifestations involving large-scale energy storage and triggering of instabilities, to cascade and dissipation mechanisms in turbulence, present fundamental questions of modern nonlinear physics that are only partially understood. In this chapter I will focus on why and how the solar wind arises, why it becomes supersonic, and on some current questions in the physics of the acceleration process, including wave propagation and turbulence development and decay.

Although knowledge of a solar influence at the Earth's orbit dates back to Carrington's observations in the second half of the nineteenth century that aurorae often occurred several hours after white light solar flares, the first direct indication of a continuous outflow of fast particles from the Sun came from Biermann's investigation in the 1950s of the shape of the cometary ion tails, from which he deduced an average outflow speed of around 475 km/s. In 1957 Chapman showed how a static conductive corona starting at 10^6 K at the Sun should maintain a high density out to large distances (in fact, after an initial decrease, the density should increase again), and in 1958 Parker [36] argued, on the basis of the unreasonably high pressures that static and breeze solutions yielded at large distances, that "probably it is not possible for the solar corona, or, indeed, perhaps the atmosphere of any star, to be in complete hydrostatic equilibrium out to large distances." He then proceeded to

M. Velli (✉)

Dipartimento di Astronomia, Università di Firenze, Largo Enrico, Fermi, 5 50125, Firenze, Italy, and Jet Propulsion Laboratory, California Institute of Technology, Pasadena, CA, USA
velli@arcetri.astro.it

Velli, M.: *Physical Processes in the Solar Wind*. Lect. Notes Phys. **778**, 247–268 (2009)
DOI 10.1007/978-3-642-00210-6_7   © Springer-Verlag Berlin Heidelberg 2009

show that a viable solution yielding negligible pressures at infinity consisted of a flow accelerating continuously and becoming supersonic at large distances.

The subsequent Parker – Chamberlain debate on supersonic/subsonic evaporation was cut short by the in situ measurement of a steady, supersonic wind with a flux density of a few times 10^8 cm^{-2} s^{-1} by the Luna 2 (1959) and Explorer (1961) spacecraft [18, 41], in what appeared to be a clear confirmation of Parker's model of an expanding corona. Subsequent in situ measurements via satellites, notably Helios I and II in the inner heliosphere, ACE and WIND at 1 astronomical unit (AU), the Voyagers in the outer heliosphere, and Ulysses via its polar passes, as well as the continuous coronal and solar wind monitoring from SOHO, have given us a much more detailed picture of the solar wind, its structure and dynamics. [Readers are referred to Chap. 6 (Issautier) for a description of in situ measurement techniques.]

Though this picture differs in many respects from the original wind ideas of Parker, and suggests more complex dynamical processes at the heart of the solar corona and heliosphere as a whole, the starting point in any understanding of the wind remains the hydrodynamics of the spherical expansion of hot gases under the gravity of a central star.

7.2 Hydrodynamics of a Featureless Solar Wind Expansion: Why the Solar Wind Is Supersonic

Consider hydrostatic balance for a spherically symmetric ionized atmosphere with gravity

$$\frac{\partial p}{\partial r} = -m_p n \frac{g}{r^2}, \tag{7.1}$$

where g/R_0 is the gravitational acceleration at the solar surface (R_0 is the solar radius) and we have normalized distances to the solar radius, $r = R/R_0$. Also m_p is the proton mass, n the number density so that the mass density $\rho = m_p n$. Recalling that $p = 2n\kappa T$, integration yields

$$n\kappa T = n_0 \kappa T_0 \exp\left(-\int_1^r dr \, \frac{m_p}{2\kappa T} \frac{g}{r^2}\right). \tag{7.2}$$

Convergence of the integral in the exponential then implies that a static spherically symmetric extended atmosphere with a temperature profile decreasing with distance less rapidly than $1/r$ requires a finite pressure at infinity to be confined, the same being true if the atmosphere "evaporates" as a subsonic flow, or breeze.

However, Mestel (quoted in [39]), first remarked that it would not take a large fall in coronal temperature for the pressure of the local interstellar medium (ISM) to be sufficient to suppress the solar wind entirely. Indeed, the pressure of the ISM, $p_{ISM} \simeq 1.24 \, 10^{-12}$ dyne/cm^2 would suffice to confine a 4×10^5 K static corona with base density 10^9 cm^{-3}.

So although correct, the argument for a supersonic wind does not appear to be as strong on the basis of pressure arguments only. In reality, the dependence of spherically symmetric, non-rotating atmospheres with flows on changes of the external conditions is much more subtle. It is useful and instructive to discuss the stability of spherically symmetric flows, from winds to accretion and back, together.

For the sake of analytical simplicity, we will consider here only isothermal flows, i.e., flows for which the temperature may be considered constant out to large radial distances. The strong electron thermal conduction makes this approximation an appropriate one for a solar-type corona, although, as we shall see, a significant ion to electron temperature imbalance appears to be present in the fast wind emanating from coronal holes.

7.2.1 Stationary Isothermal Flows: Breezes, Winds, Accretion

The equations of motion for one-dimensional, spherically symmetric, stationary isothermal flow neglecting self-gravity may be written in the form:

$$\frac{\partial}{\partial r}\left(\rho v r^2\right) = 0, \quad p = c_s^2 \rho, \tag{7.3}$$

$$v\frac{\partial v}{\partial r} = -\frac{1}{\rho}\frac{\partial p}{\partial r} - \frac{g}{r^2}, \tag{7.4}$$

where v is the velocity and c_s is the (constant) sound speed. For a static atmosphere, the pressure profile is given by $p = p_0 \exp(-g/c_s^2 + g/rc_s^2)$ which, as discussed above, implies a non-vanishing asymptotic value for the pressure at large distances: $p_\infty^s = p_0 e^{-g/c_s^2}$. In terms of the Mach number $M = v/c_s$, the hydrodynamic equations may be written as (a prime denoting radial derivatives throughout this section)

$$\left(M - \frac{1}{M}\right) M' = \frac{2}{r} - \frac{g}{r^2 c_s^2}, \tag{7.5}$$

which may be integrated, and expressed in two equivalent ways

$$\frac{1}{2}\left(M^2 - M_0^2\right) - \log\left(\frac{M}{M_0}\right) = 2\log r + \frac{g}{rc_s^2} - \frac{g}{c_s^2}, \tag{7.6}$$

$$\frac{1}{2}\left(M^2 - M_0^2\right) + \log\left(\frac{p}{p_0}\right) = \frac{g}{rc_s^2} - \frac{g}{c_s^2}, \tag{7.7}$$

where M_0 is the base Mach number. The second form is essentially the conservation of energy flux, where for an isothermal atmosphere the enthalpy is expressed as $\log p$ instead of $\gamma p/(\gamma - 1)$. By definition in this case, attention is limited to positive energy flows. A general and very interesting discussion of the behavior of flows with a more realistic energy equation in the cases of positive energy, as well as vanishing total energy and various intermediate limits, may be found in [39].

Equation (7.5) has a singular point at the sonic point, $r = g/2c_s^2$, $M = 1$. Solutions to the above equations may be represented in the (M, r) phase plane illustrated in Fig. (7.1), which following the symmetry of (7.5) is symmetric in the sign of M. The diagram is divided into four parts labeled I – IV (eight when considering positive and negative M) by the two critical (transonic) solutions which cross at the sonic point $r = g/2c_s^2$, $M = 1$. Single-valued continuous flow profiles $M(r)$ which are subsonic for all r, the breezes, lie below both transonic solution curves (region I).

Among flows which are subsonic at the atmospheric base, the accelerating transonic one has the special property that the density and pressure tend to zero at large distances: because of the small but finite values of the pressure of the ambient "external" medium, a terminal shock transition, connecting to the lower branch of the double-valued solutions filling region II will in general be present [20].

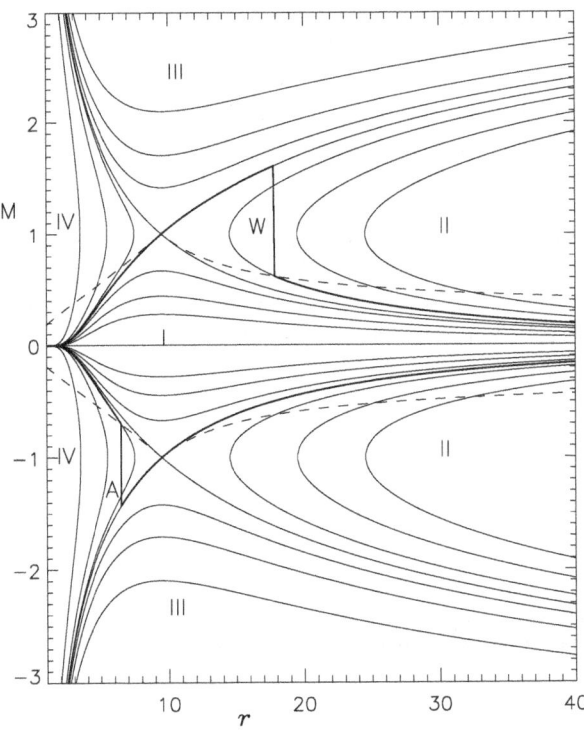

Fig. 7.1 The (M,r) phase plane. The continuous curves are appropriate both for positive and negative M. The *dashed line* intersection with double-valued curves defines the shock position for winds (region II, curve W) or accretion flows (region IV, curve A)

The jump conditions across such a shock are found from conservation of mass and momentum across the shock, which read (superscripts $-,+$ denote the solution immediately upstream and downstream from the shock, respectively)

$$\rho^- M^- = \rho^+ M^+, \quad (mass), \tag{7.8}$$

$$\rho^+ (M^+)^2 + \rho^+ = \rho^- (M^-)^2 + \rho^-, \quad (momentum), \tag{7.9}$$

from which one finds immediately that $M^+ M^- = 1$. This gives a way to graphically construct shock transitions; it is sufficient to plot the curves corresponding to $1/M$ for the transonic solutions (dashed lines in Fig. 7.1), and to connect the transonic solution with the double-valued curve in region II where the dashed line intersects them: such a solution is given by curve W.

The downward transonic solution in itself is not a possible solution for outflows, since a continuous transition from supersonic to subsonic flows is unstable. However, it plays the same role as the Parker wind solution plays for inward-directed accretion flows: for negative M solutions, the same construction leads to accretion shocks in the flow [30], this time in the region labeled IV, and one such transition is shown in Fig. 7.1 (curve A).

For given base values of the pressure, the position of the shock is uniquely determined by the pressure of the interstellar medium, and the distance from the critical point to the shock decreases as the interstellar medium pressure increases: conservation of mass across the shock immediately gives the asymptotic pressure in terms of the upstream Mach number M^- as

$$p_\infty = p_0 M^- \exp\left(M_*^2 - 2\frac{g}{c_s^2} - M^{-2} + \frac{1}{M^{-2}} \right) / 2, \tag{7.10}$$

where M_* is the base Mach number of the upward transonic solution. p_∞ is a monotonically decreasing function of M^-, which is itself, obviously, a monotonically *increasing* function of the shock position r_s, so that increasing p_∞ decreases r_s. When p_∞ reaches a value $p_\infty^c = p_0 \exp(M_*^2/2 - g/c_s^2)$, the shock distance $r_s = r_c = g/2c_s^2$ and the discontinuity in the flow velocity reduces to a discontinuity in the derivative of the profile. This is the fastest possible (or critical) breeze, made up of the section of upward transonic solution below r_c and the section of downward transonic solution beyond r_c.

For the breeze solutions, with a base Mach number M_0 such that $M_* > M_0 \geq 0$, the asymptotic pressure is easily calculated to be

$$p_\infty = p_0 \exp\left(M_0^2/2 - g/c_s^2 \right) \geq p_\infty^s. \tag{7.11}$$

It follows that the pressure required to confine a breeze *increases* with increasing base Mach number and is greater, if only slightly, than that of a static atmosphere. The limiting value of p_∞ is again p_∞^c. For a given base pressure and asymptotic pressures $p_\infty^c > p_\infty \geq p_\infty^s$ it, therefore, appears that **two** possible stationary out-flow solutions exist, a supersonic shocked wind and a subsonic breeze. When such conditions occur, it is frequently the case that one of the solutions may be unstable, i.e., any small perturbation may lead the flow to evolve away from stationarity.

To discuss the stability of stationary flows, we introduce time-dependent small perturbations (sound waves) and linearize the equations of motion around the stationary state. We will apply boundary conditions which allow the configuration to evolve from one stationary solution to another with the same density and pressure boundary values: this means the perturbing sound waves must leave the pressure (and density) unperturbed at the atmospheric base and infinity.

It is convenient to introduce characteristic variables $y^\pm = \hat{m} \pm \hat{p}$, where $\hat{p} = \tilde{p}/p$ is the adimensional normalized pressure perturbation and $\hat{m}$ is the Mach number (velocity) fluctuation. In these variables an outward (inward) propagating sound wave has $y^- = 0$ ($y^+ = 0$). In our stability analysis, the condition requiring no pressure perturbations at the inner and outer boundary becomes simply that $y^+ = y^-$ at these radii.

Assuming a time dependence $y^\pm = y^\pm(r) \exp(-i\omega + \gamma)t$ the linearized equations become

$$(M \pm 1) y^{\pm\prime} - i(\omega + i\gamma)y^\pm + \frac{1}{2}\left(y^\pm + y^\mp\right) \frac{M'}{M}(M \mp 1) = 0. \qquad (7.12)$$

In a uniform flow one has stable propagating waves with dispersion relation $\omega^\pm = (M \pm 1)k$, where k is a radial wave number. In the presence of a nonuniform but stable flow, (7.12) describes wave propagation and reflection, and a conserved flux, the wave action flux, exists (in a static medium, the wave energy flux is conserved; when there are mass motions, this is replaced by the wave action flux: see Sect. 7.4 for a discussion of wave energy and [48] and references therein for a discussion of wave action conservation). When $\gamma \neq 0$ the wave action evolution equation becomes

$$\left[\frac{(M+1)^2}{M}|y^+|^2 - \frac{(M-1)^2}{M}|y^-|^2\right]' + 2\frac{\gamma}{M}\left[(M+1)|y^+|^2 - (M-1)|y^-|^2\right] = 0,$$
$$(7.13)$$

the first square bracket being proportional to the wave action.

Notice that for $|M| < 1$ the term in the second square bracket is positive definite. Integrating this equation between 1 and r and imposing the boundary condition $y^+ = y^-$ at the extremes, we find the following estimate for γ:

$$\gamma = \frac{2\left(|y^+|_0^2 - |y^+|_r^2\right)}{\int_1^r dr \, M^{-1}\left[(M+1)|y^+|^2 - (M-1)|y^-|^2\right]}, \qquad (7.14)$$

where $|y^+|_0^2$, $|y^+|_r^2$ are the fluctuation amplitudes at the atmospheric base and r, respectively. It follows then that if the perturbation amplitude is non-vanishing at the base but tends to zero at great distances, the flow is unstable.

Now for $\omega = 0$ and large r the asymptotics of the solutions to (7.12) for breezes may be found by first noticing that at large distances the breeze solutions (from (7.7)) behave as

$$M \sim \frac{M_0}{r^2} \qquad \frac{M'}{M} \sim -\frac{2}{r}.$$

(7.15)

As a consequence, the amplitude of the perturbed outward and inward propagating sound waves are found to have, in terms of the eigenvalue γ, leading order asymptotic solutions

$$y^+ \sim \frac{e^{\mp \gamma r}}{r} \left(1 \pm \frac{1}{2\gamma r} \right), \qquad y^- \sim \pm \frac{e^{\mp \gamma r}}{2\gamma r^2}.$$

(7.16)

The difference between the two solutions, therefore, may now be written simply as

$$y^+ - y^- \sim \frac{e^{\mp \gamma r}}{r},$$

(7.17)

and the boundary condition of vanishing perturbed pressure at infinity, $y^+ = y^-$ as $r \to \infty$ translates into the fact that γ should be positive, if the first solution is the correct one, or negative, if the second one is correct.

In both cases the amplitudes of the fluctuations tend to zero at great distances: this means that the numerator of (7.14) is positive. Consistency then requires γ to be positive, and breeze solutions must therefore be *unstable*.

The growth rate is largest for high values of the base Mach numbers but both the static atmosphere ($M_0 = 0$) and the critical breeze ($M_0 = M_*$) are marginally stable. In the latter case, as in the shocked wind solutions, the perturbation equations also become singular at the sonic point, because the phase speed of the inward propagating wave vanishes there: an additional regularity condition must be imposed in the stationary equations, effectively isolating the region below the sonic point from the region beyond it. The presence of this additional boundary condition is the mathematical reason behind the stability of flows with a continuous subsonic/supersonic transition.

The breeze instability is driven by the unfavorable stratification (7.11). As the flow decelerates in the spherical geometry, pressure and mass tend to accumulate and actually increase with distance from the Sun. So imagine a static atmosphere, and let the pressure at infinity increase: an *inflow*, not an outflow, is the intuitively expected result. And these flows should be stable, which follows immediately from the analysis presented above; the denominator in (7.14) changes sign, so the only consistent way to satisfy the boundary conditions is to choose the second solution in (7.16), implying a negative value for γ.

In fact, the stationary equations are symmetrical in M, while the perturbation equations are invariant under a change in sign of both M and γ. So the Parker "breeze" solutions are irrelevant to outflow, but relevant to accretion.

7.3 A Wind-Accretion Hysteresis Cycle

Given that breezes are unstable, the pressure range $p_\infty^c > p_\infty \geq p_\infty^s$ presents strange properties, in that there is a stable outflow consisting in a supersonic wind with a terminal shock, even though the stratification of pressure would seem to favor accretion. This is evidence for bi-stability in the theory of spherically symmetric flows, carrying with the possibility of hysteresis cycles in the transitions from accretion to winds and vice versa. Consider what happens if the pressure difference between the coronal base and the distant medium varies, starting from a supersonic shock wind.

As p_∞ increases, the shock moves inward, decreasing in amplitude in the process, up to a limit where there is a final "shocked wind" consisting in what was previously called the "critical breeze," with a discontinuity only in the derivative of the flow speed and not in the flow itself. Once this critical breeze is reached, there is no neighboring outflow solution capable of sustaining a higher pressure at infinity, because the breeze solutions, which are unstable anyhow, would require a smaller pressure at infinity.

The only possibility for the flow is to collapse into its symmetrical $(M \to -M)$ critical breeze accretion profile, which is also marginally stable. As the pressure is increased further, an accretion shock is formed *below* the sonic point, connecting the symmetrical of the downward transonic solution to one of the double-valued curves in region IV (as shown by the curve labeled A in Fig. 7.1). For $p_\infty > p_\infty^c$ there is a unique shocked accretion flow [30], and the shock position moves inward from the critical point as the pressure is increased beyond p_∞^c (if the pressure is too high, the shock may collapse onto the star).

Consider now what happens if, starting from a shocked accretion flow, the pressure at the surface increases or alternatively the pressure of the ISM decreases. The shock moves outward, but this time, as p_∞ decreases below p_∞^c, the flow can evolve with continuity into subsonic accretion. As p_∞ decreases further, the accretion–breeze velocities decrease, but when p_∞ decreases below p_∞^s, the flow must accelerate again into a supersonic shocked wind.

This scenario, predicted in [49], had gone unobserved previously. Parker himself had proved, erroneously, that all breezes were stable, and (in a private conversation with the author) mentioned he had not known of the Bondi accretion solution when he first thought about the solar wind problem. In addition, some numerical simulations of spherically symmetric flows had been interpreted as showing how the transitions in the Parker/Bondi diagram were continuous [27], i.e., that all solutions could be obtained continuously with small changes in the boundary conditions, a misconception which is still widespread. Even in those simulations, however, there was some indication that something was not quite right, in that for certain changes of the pressure at the outer boundary the transition from breezes through subsonic accretion to accretion shocks occurred somewhat abruptly.

The Velli hysteresis cycle theory explains this feature: the stratification produced by breezes, though globally unstable, is not locally unstable everywhere; for example, below the critical point the pressure in breezes decreases with height more rapidly than in the static case. Inspection of (7.7) actually shows that this is true out to the radius r_s where the Mach number of the flow has decreased to the same level as the base Mach number, which may be calculated by imposing $M = M_0$ in (7.7), i.e.,

$$2 \log r_s + g/r_s - g = 0.$$

This equation is independent of the base Mach number, which also means that at this height the pressure is the same for all breezes, while below this radius, the pressure at a given height is a monotonically *decreasing* function of base Mach number.

As the boundary conditions are imposed at closer and closer distances r_b, the growth rate of the instability is reduced, and marginal stability is obtained when $r_b = r_s$. Imposing boundary conditions below this radius stabilizes the breezes, but consequently destabilizes subsonic accretion, as is shown in Fig. 7.2b, where the

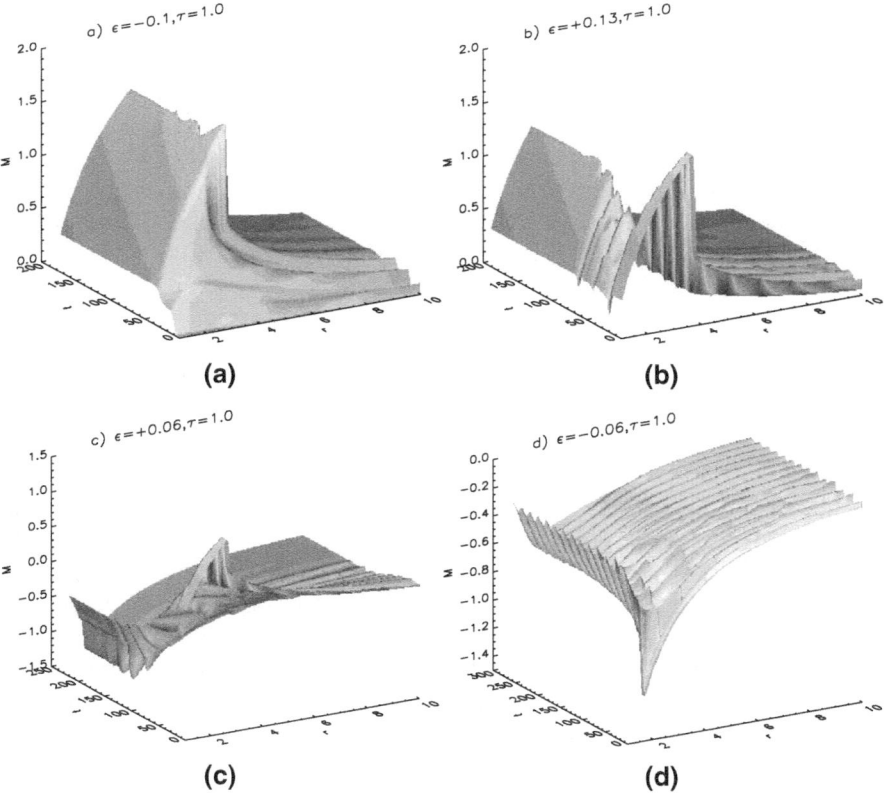

Fig. 7.2 The Velli cycle : (**a**) from the static atmosphere to shocked wind (the pressure p_∞ is decreased just below the static value) (**b**) the pressure p_∞ is increased to above the static value but just below the critical value p_∞^c (**c**) a further small increase in pressure above p_∞ causes the wind to collapse into shocked accretion (**d**) return to the conditions in (**b**) yields a stable accretion breeze

maximal growth rate for breezes (continuous line) and accretion (dashed line) as a function of r_b is plotted. For large values of g this marginal stability radius depends exponentially on g as $r_s \simeq \exp(g/2) - g/2$.

When the boundary is at $r_b < r_s$ the equilibrium flows still present an hysteresis-type cycle in terms of the enthalpy jump between r_b and the coronal base, but in a reversed order with respect to that previously described: supersonic accretion is blown into supersonic winds as the base pressure is increased beyond a critical value, while an outflow breeze phase exists before the collapse to accretion as the pressure at the outer boundary is increased beyond the value appropriate to a static atmosphere.

A numerical simulation to test the original hysteresis cycle prediction was carried out in [7]. The numerical domain was large enough so that the outer pressure boundary condition was imposed in the regime of unstable outflow breezes. The results are summarized in Fig. 7.2a–d. First (Fig. 7.2a), the static atmosphere was taken as the initial condition and the pressure at the outer boundary was lowered by a small amount. The solution rapidly expands and develops into a shocked wind. Then (Fig. 7.2b), the pressure was increased, pushing the shock inward close to the critical point. Subsequently, (Fig. 7.2c), the pressure was further increased slightly, collapsing the wind to supersonic shocked accretion. Finally (Fig. 7.2d), the outer pressure was decreased again to the value it had before the shocked wind collapsed, but this time a stable accretion breeze resulted.

In generalizing to polytropic or other more realistic equations for the energy, some attention is necessary since the density and pressure may fall to zero at a finite distance, and transonic flows do not exist for all polytropic indices ($\gamma \leq 3/2$ below the sonic point is a necessary condition [37]). With these caveats, the discussion of the isothermal case is easily generalized. The energy equation now becomes (as the sound speed varies, c_{s0} is its base value)

$$v^2/2 + c_s^2/(\gamma - 1) - g/r = v_0^2/2 + c_{s0}^2/(\gamma - 1) - g.$$

For breezes the asymptotic behavior $v \sim 1/r^2$ still holds, so that in fact we may write

$$c_{s\infty}^2/(\gamma - 1) = v_0^2/2 + c_{s0}^2/(\gamma - 1) - g,$$

which shows that the temperature at great distances from the central object increases with the base Mach number, upto the value which, for a given base density and pressure, gives a transonic flow [20]. Conservation of energy across the shock then implies that independently of the asymptotic pressure, $c_{s\infty}$ is always the same. It is still true that given the base density and pressure, for a range of pressures at great distances between that of the static atmosphere and that of the critical breeze there are two solutions, an unstable breeze (or stable accretion) and a shocked wind, but now the thermodynamic state of the distant medium is different, the breeze having a higher density and lower temperature.

7.4 Solar Wind Energetics and Empirical Solar Wind Models

As mentioned in the Introduction, and detailed in other chapters of this book, the solar wind is a far cry from the simple, spherically symmetric solar wind discussed in
the previous section. The wind varies over the solar cycle, but maintains its simplest
configuration during solar activity minimum, when large polar coronal holes dominate the outer solar atmosphere away from the equator, with a helmet streamer belt
straddling the magnetic equator. Correspondingly, the solar wind presents a steady
high-speed wind, averaging above 750 km/s, away from the neutral sheet, with a
much slower wind of about 400 km/s in the ecliptic plane. Winds of 750 km/s would
emanate from a 2.5 MK degree isothermal corona, while a 400 km/s wind would
arise from an isothermal corona below 1 MK. Understanding the energetics of the
solar wind expansion requires going beyond the spherically symmetric isothermal
Parker wind. In particular, an important role in determining the solar wind speed
is played by the *expansion factors* of flux tubes, as well as the mechanical energy
deposited in the wind and its location (height in the corona). Also fundamental are
the coronal energy loss processes, namely, radiative losses, thermal conduction back
down to the transition region, and of course the quantity of interest and the energy
lost into the solar wind.

Consider the expansion of a plasma from the Sun along a small field-aligned flux
tube, in a stationary state. The energy flux along the tube, which must be conserved
provided there are no losses across the flux tube boundaries, may be written as [19]

$$F_0 = F_m + F_q + F_{rad} + F_{sw}, \tag{7.18}$$

where F_m represents the mechanical energy flux (leading to coronal heating and
wind acceleration), F_{rad} the radiative losses integrated from the base to the given
distance along the flux tube, F_q the conductive flux, and

$$F_{sw} = |\dot{M}| \left(\frac{1}{2} V^2 + \frac{\gamma p}{(\gamma - 1)\rho} - \frac{GM_\odot}{R} \right) \tag{7.19}$$

is the energy flux in the solar wind, comprised of the enthalpy flux, the kinetic
energy flux carried away by the wind, and the gravitational potential energy flux.
Fixing the flux tube base somewhere near the top of the chromosphere we may
write:

$$F_0 = F_{m,0} - |\dot{M}| \frac{GM_\odot}{r} = F_{m,0} - |\dot{M}| \frac{1}{2} V_g^2, \tag{7.20}$$

where at the base the integrated radiative loss vanishes by definition, and the conductive flux is also negligible at the chromospheric base (because the heat conducted
down from the corona is mostly radiated away above the chromosphere; generally
speaking conduction and radiation may be neglected separately, though they play
an important role in the detailed variations of solar wind speed and mass flux).

Also, the temperature and flow speed of the solar wind plasma are so small that they may be neglected compared to the potential energy flux, the product of the solar wind mass flux, $|\dot{M}|$, and the square of the escape speed from the solar surface ($V_g = 618$ km/s). Far from the Sun heat conduction may be neglected, the enthalpy, gravitational potential, and mechanical energy fluxes are negligible compared to the solar wind kinetic energy, so that

$$F_0 = F_{rad,\infty} + |\dot{M}|\frac{1}{2}V_\infty^2, \qquad (7.21)$$

where V_∞ is the asymptotic flow speed in the wind. It is then possible to write down an expression relating the total mechanical input flux, the solar wind mass loss, and the total amount of energy radiated away:

$$V_\infty^2 = 2\frac{F_{m,0} - F_{rad,\infty}}{|\dot{M}|} - V_g^2. \qquad (7.22)$$

Traditionally, this equation has been used to understand how the mass flux depends on the mechanical energy flux $F_{m,0}$ which presumably heats and accelerates the wind. Also, the dependence of the asymptotic speed on flux tube expansion rates is hidden in this expression. Taking into account the fact that the fast solar wind speed is of the same order as the escape speed from the Sun, the mass flux and mechanical energy flux appear to be proportional. The radiative loss, though smaller than the other terms (typically less than 15%), regulates the transition region pressure, playing a fundamental buffer role in keeping the mass flux nearly constant (in the Parker isothermal theory, because the density is essentially static up to the critical point, there is an exponential dependence of the mass flux on temperature).

A basic fluid model of the fast solar wind which fits remote sensing data of density and velocity close to the Sun as well as in situ density, magnetic field, temperature and their gradients at 1 AU may be constructed assuming an average temperature profile peaking at about 3 million degrees (average of electron and ion temperature) at 3 solar radii, and falling as $r^{-0.8}$ at 1 AU. A super-radially expanding flux tube geometry from the coronal hole base is considered (a classical form originally derived by [26, 32]) with a maximal overexpansion compared to radial of about 7.25, the expansion occurring within 2 solar radii. A supplementary pressure gradient which boosts the wind to its asymptotic speed of about 750 km/s is given by an Alfvén wave flux (more about which in the subsequent sections) with an amplitude of about 20 km/s at the coronal base. Profiles for the temperature, wind speed, Alfvén speed, Alfvénic Mach number, plasma β, and relative magnetic and velocity fluctuations in the Alfvén waves are shown in Fig. 7.3. The mechanical energy flux required to drive such a standard fast solar wind is about 10^5 erg/cm^2/s, increasing or decreasing with the overall expansion of the wind compared to spherical.

As detailed in another chapter, the fast solar wind appears to originate from coronal holes, while the slow solar wind originates from areas within or adjacent to the quiet Sun and the magnetic activity belt. The fast wind, therefore, appears to

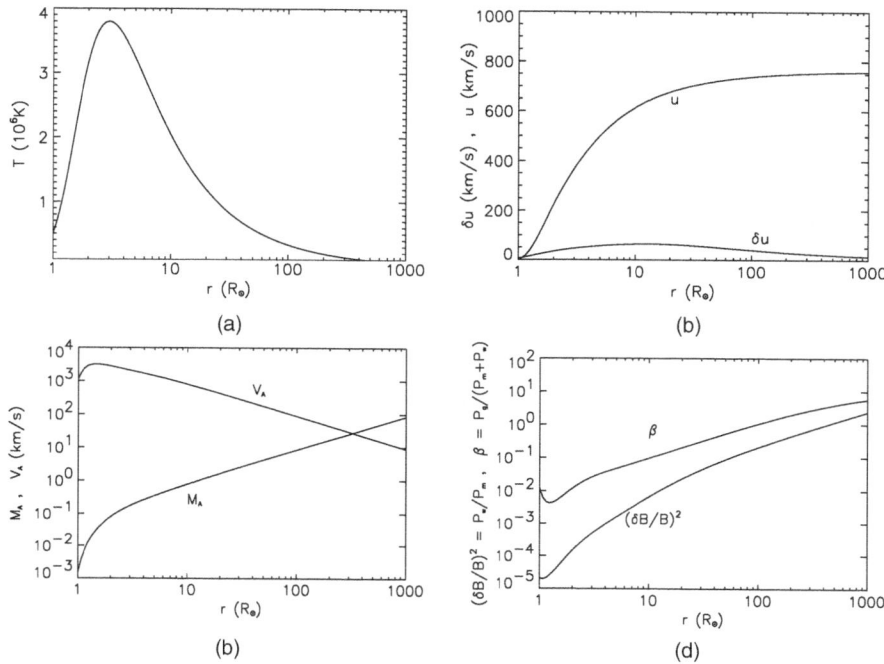

Fig. 7.3 A fast solar wind model. The temperature profile (**a**) peaks at 3 $R_\odot$ and then falls with radius as $T \sim r^-$. The coronal hole is assumed to expand with the parametrization given by [32]. Alfvén critical point is in $r = 20\,R_\odot$ (**c**) while Alfvén waves (**b**, lower curve, **d** lower curve) supply an additional acceleration, while plasma *beta* (**d**) and densities fit typical 1 AU values

originate from regions where the electron temperature is low, while the slow wind comes from hotter regions of the corona. This is confirmed by the well-known anti-correlation between solar wind speed and "freezing-in" temperature of the different ionization states of oxygen and magnesium as observed in the solar wind, e.g., by the Ulysses spacecraft [12]. Fisk [10] proposed a solar wind model aimed at explaining the anti-correlation, based on reconnection between emerging and low-lying loops in the solar corona and the open magnetic field as the source of energy, while the release of matter confined in the previously closed loops is the source of mass. The basic equation is given by (7.22). Assuming that the loop which reconnects is in thermodynamic equilibrium, its total mass is proportional to base density and temperature, and if the mass flux is some proportion of the mass in the loop, (7.22) yields an anti-correlation because mass flux is in the denominator on the RHS. The total mechanical flux in the Fisk model turns out to be proportional to the total open magnetic flux in the heliosphere:

$$V_\infty^2 = \left(\frac{B_{loop,i}}{\rho_{loop,i}} \right) \left(\frac{\int \mathbf{B}_{open} \cdot d\,\mathbf{h}}{\mu_0 r_0} \right) \left(\frac{V_g^2}{c_s^2} \right) G - V_g^2, \qquad (7.23)$$

where $B_{loop,i}$ and $\rho_{loop,i}$ are the magnetic field and density in the initially closed loop, while G is a function which depends weakly on loop length and temperature. In this model, radiative losses have been neglected, and the mass flux has been prescribed as a fixed fraction of loop mass. Also, there is the question of how to incorporate the effect of an intrinsically dynamic phenomenon in an equation derived within the assumptions of stationary state. In other words, other phenomena could be occurring during reconnection leading to losses of mass, momentum, and energy across the boundaries of a hypothetical flux tube. The model does, however, appear to fit the observed anti-correlation of freezing in temperature to wind speed [13].

Subsequent papers based on similar ideas take radiation also into account [42], leading to a different scaling law which we do not reproduce here. In this case, however, the stationary equilibrium properties of the closed coronal loop are even more fundamental to the calculation. They may be easily altered by the dynamical processes intervening during reconnection between the closed loop – open field that is postulated to be the source of the wind.

In conclusion, attempts to include the effects of dynamical coronal heating mechanism in which magnetic field lines are continuously opened and closed to supply the solar wind still requires more theoretical work to be placed on a firm foundation. On the observational side, it remains to be seen whether observations of the lower corona, perhaps with the multi-point view afforded by the STEREO spacecraft, or the higher time cadence of Hinode and SDO, will show the required interchange reconnection events.

7.5 Alfvénic Fluctuations and the Solar Wind

The discussion turns now to Alfvén waves and Alfvénic turbulence and their possible role in solar wind acceleration and dynamics. For a detailed introduction to the theory of turbulence, with applications to magnetized plasmas, see Chap. 3 (Carbone and Pouquet) which also addresses the important issue of intermittency. Only specific aspects pertaining to the solar wind will be discussed here.

Fluctuations in the high-speed solar wind streams with periods below a few hours down to minutes and less are found to be dominated by what is known as Alfvénic turbulence. This is a well-developed turbulence spectrum which has all the properties of a flux of large amplitude, constant magnetic field magnitude Alfvén waves propagating away from the Sun. The properties of such fluctuations have been summarized in [17, 46] as far as Helios observations are concerned, while the observations within the high-speed flow at polar latitudes by the Ulysses spacecraft are described in [22] and the review paper [21].

Alfvénic fluctuations were first identified in the trailing edges of high-speed solar wind streams by Belcher and Davis [3]. Denoting the magnetic fluctuations and velocity fluctuations by **b** and **v**, respectively, and defining the Elsasser variables $\mathbf{z}^{\pm} = \mathbf{v} \mp \text{sign}(\mathbf{B})\mathbf{b}/\sqrt{\mu_0 \rho}$ (we have incorporated changes in the sign of the average field in the definition of Alfvén waves), we may characterize these fluctuations by the relations $\delta |B|^2 << |\mathbf{b}|^2$, i.e., small total magnetic intensity fluctuations; $|\mathbf{z}^+| >> |\mathbf{z}^-|$, i.e., outward propagating waves dominate: $|\delta\rho/\rho|^2 << |\mathbf{v}/c_s|^2 = M_T^2$, where c_s is the sound speed and M_T the turbulent Mach number.

On the other hand, in other periods (in the slow solar wind, or often at solar activity maximum) the fluctuation properties change such that all the inequalities $(<<, >>)$ above become $\simeq$. With few exceptions, at least at solar minimum, solar wind turbulence varies continuously between the Alfvénic state (in the polar wind and in trailing edges of high-speed streams in the ecliptic plane) and the standard state (slow wind at magnetic sector crossings).

The shape of the velocity and magnetic field spectra observed in situ are strongly suggestive of a nonlinear cascade, where one expects to find energy at all possible wave vectors. In the analysis of solar wind fluctuations this fully developed turbulence point of view was first adopted by Coleman [5].

Homogeneous, incompressible MHD turbulence predicts Alfvénic turbulence as the asymptotic outcome when initial conditions have $u \simeq \delta b/\sqrt{\mu_0\rho}$ (u being the absolute value of velocity fluctuations), while the observed evolution with heliocentric distance is such that Alfvénic turbulence decays toward "standard": the power index of the transverse magnetic field spectrum is typically $\alpha \simeq -1$ for lower frequencies close to the Sun, decreasing to the Kolmogorov value $\alpha \simeq -1.67$ at higher frequencies. At lower frequencies a dominance of kinetic energy over magnetic energy (f $\leq 10^{-4}$ Hz) is observed, due to the presence of the large-scale slow–fast wind shear flow, while in the Alfvénic domain there is a slight dominance of magnetic over kinetic energies.

The bend in the spectrum moves to lower frequencies with increasing distance from the Sun, the evolution being somewhat faster within high-speed streams in the ecliptic plane and slower in the polar wind. Together with the evolution in the shape of the spectrum, the specific energy in the fluctuations also varies with distance from the Sun, in a way which is roughly consistent, $e \sim r^{-1}$ (r being heliocentric distance, normalized to the solar radius) with the conservation of wave action at the lowest frequencies ($\simeq 10^{-3}$ Hz). In the decay toward standard turbulence, the waves tend to become more mixed (the dominance of one component over the other is lost). So there is an apparent paradox: incompressible MHD turbulence which starts out with some predominance of outward waves ends up with only the outward waves left (as will be shown below), while the turbulence observed in situ evolves in the opposite direction.

Over the past 15 years many models have been developed to describe the evolution of the Alfvénic spectrum, starting all the way from the photospheric motions at the base of corona holes (e.g. [35, 50, 6]). Such models have included nonlinear evolution due to the in situ generation of inward modes in the solar wind (such modes are necessary in incompressible MHD to have nonlinear interactions) and the interaction of the waves with the large-scale magnetic field and velocity shears in the current sheet and between fast and slow streams.

As mentioned above, the existence in the solar wind of a well established, scale-invariant spectrum made up of non-interacting waves is problematical, as remarked in [8, 9]: given a solar source of outgoing waves, the observed spectrum should reflect its properties, as well as the filtering of the intervening medium, and hence gaps in the spectrum or peaks at some typical generation frequencies (or harmonics) should appear, which is not observed in the data.

However, the standard situation is that of a mixture of both types of Alfvén waves, with moderate dominance of outward propagating waves, so that the non-linear couplings are not vanishingly small. For example, on the average, the ratio z^+/z^- calculated between 10^{-4} and 10^{-2} Hz was about 0.6 during the first 3 months of the Helios mission (at heliocentric distance between 0.3 and 1 AU and at solar minimum).

Consider first this quasi-symmetric regime ($z^+ \sim z^-$), which one may attempt indeed to describe by "standard" MHD theory. Just as in any wind tunnel, one cannot expect to observe fully developed turbulence too close to tunnel entry; some time must be allowed to let the nonlinear effects develop significantly. This time is the turnover time τ_{nl}, which for an eddy of size l depends on the rms energy contained around this scale, about u^2 if u is the velocity fluctuation amplitude within the eddy. Then $\tau_{nl} \sim l/u = (U/u)T$, if $T = l/U$ is the typical timescale measured in the spacecraft frame. At a given heliocentric distance R, the transport time by the average flow is $\tau_{tr} = R/U$. A third timescale, $\tau_{ad} \sim 1/(\nabla \cdot \mathbf{U}) = -(1/\rho)D\rho/Dt$, describes the rate of change of the plasma specific volume $1/\rho$ associated with the geometry of the expansion, and does not depend on the spatial scale of the eddies.

The adiabatic and transport times are of the same order of magnitude in the super-sonic region of the wind (they are identical for a spherical expansion with constant speed). For example, at R = 0.3 AU, the distance of closest approach of the Sun by the Helios spacecrafts, $\tau_{tr} = R/U = 35$ h; in the Alfvénic range of periods 1 h > T > 3 min, $u/U \simeq 0.05$, and thus $\tau_{nl} < \tau_{tr} \sim t_{ad}$. If we follow a plasma parcel which is convected with the wind, nonlinear effects will strongly affect only the part of the spectrum for which the average turnover time is smaller than the transport time. Therefore the width of the inertial range, if one does exist, must depend on the radial distance. Except if the spectrum is very steep, the nonlinear time decreases with the scale; hence there will always be a critical scale $L(R) \sim (u/U)R$ below which nonlinear effects dominate [47].

As heliocentric distance R increases, the adiabatic time also increases, and so does L(R), as long as u does not decrease as fast as R^{-1}, i.e., as long as the turbulent specific energy does not decrease as fast as R^{-2}. The nonlinear interactions are thus free to redistribute the energy among the degrees of freedom available between the scale L(R) and a dissipation scale $l_d \sim l_g$.

A fully developed turbulent state is expected when several orders of magnitude separate the two scales ($L(R) \gg l_d$). According to the Kolmogorov theory of 1941 [25], two properties characterize such a state. First, the energy dissipation rate is independent from the viscosity of the fluid, i.e., it reaches a finite value in the limit of zero viscosity. Since the nonlinear interactions respect the conservation of energy, the dissipation rate must be given by the energy injection rate $\Pi^\circ = \epsilon$, which either comes from an external source, or from the largest, energy-containing eddies (here at scale L(R)).

Second, the energy is not transferred directly from the largest scale down to the dissipation scale, but instead is transferred via successive interactions between smaller and smaller (but at each step comparable) wave numbers, hence the name "energy cascade". During this cascade, eddies of a given size l break into smaller eddies, but are regenerated by the larger eddies, and so on; since all scales are in energetic equilibrium, the energy dissipation rate of eddies of size l, $\Pi(l)$, is independent of the scale l, i.e., and is equal to ϵ: $\Pi(l) = \Pi^{\circ} = \epsilon$. In the fluid case, this, together with the assumption of nonlinear interactions dominated by modes local in wave number space, leads to the Kolmogorov law for the spectrum $E_k \sim \epsilon^{2/3} k^{-5/3}$. In the case of a conducting fluid imbedded within a uniform magnetic field, self-distortion of fluid eddies is replaced by weaker interactions between propagating Alfvén waves. The propagation introduces an additional timescale, the Alfvén time $\tau_A \sim 1/V_a$, so the effective energy transfer is no longer equal to the eddy turnover time. Indeed, noting that, to lowest order, the nonlinear terms couple linear solutions (i.e., Alfvén wave packets) propagating in opposite directions, the coherent interaction time is reduced to τ_A [24], which is smaller than the eddy turnover time by the factor $\delta B/B_0$. This leads to $E_k \sim (\epsilon V_a)^{1/2} k^{-3/2}$, which is the Iroshnikov–Kraichnan spectrum.

The preceding arguments assume implicitly that both field amplitudes z^+ and z^- are comparable. From the MHD equations, one sees that the turnover times for z^+ and z^- eddies in reality depends on the amplitude of the other field:

$$\tau_{nl}^{\pm} = l/z^{\mp}. \tag{7.24}$$

Nonlinear interactions conserve the energies $E^{\pm}$ separately in both fields leading to the possibility of separate energy cascades, via distinct fluxes Π^+ and Π^-. If one assumes that the IK decorrelation effect holds, then equal Π^+ and Π^- fluxes are obtained:

$$\Pi_k^+ = \Pi_k^- = k^3 E_k^+ E_k^-/V_a, \tag{7.25}$$

and when both amplitudes are comparable, one obtains the IK spectrum.

If, however, one mode should dominate over the other, the equal dissipation of both fluxes should lead to an ever increasing imbalance, and asymptotically to the presence of waves propagating only in one direction, in a frozen, exact nonlinear state: the process called "dynamical alignment."

During this evolution, (7.25) implies that the sum of spectral indices for the waves propagating in either directions, i.e, m^+, m^- with $E_k^+ \sim k^{-m^+}$, $E_k^- \sim k^{-m^-}$, should be equal to 3 [15, 16]:

$$m^+ + m^- = 3. \tag{7.26}$$

Dynamical alignment was proven to occur via 2D numerical simulations in both incompressible and compressible MHD [38], but has not been observed in the solar wind, where the spectrum tends to evolve in the opposite way. The most probable reason lies in the anisotropic properties of Alfvén wave propagation. Although the Alfvén effect exists, it is limited to one spatial direction, given by the mean magnetic field, or by the combined effects of the largest eddies, while fluctuations with a dominant wave number in the plane perpendicular to this direction are only moderately influenced by this effect.

In the approximation of a strong, dominant, average axial field, where reduced magnetohydrodynamics (RMHD) applies [44], the Alfvén effect is so dominant along the direction of the mean field that nonlinear interactions are totally quenched in that direction, leading to a system of equations amenable to the 2D incompressible MHD equations in the perpendicular planes, coupled by linear propagation of Alfvén waves along the mean field. In this case, one may determine the expected spectral anisotropy by assuming that the spectrum in the parallel direction reflects a balance between nonlinear evolution within any plane othogonal to the magnetic field, and correlation between different planes due to Alfvén wave propagation from plane to plane. In other words, one expects that in equilibrium a scale develops parallel to the magnetic field such that the linear wave propagation along the field lines at that scale, is equal to the nonlinear time at that same scale in the perpendicular directions. Introducing the subscripts $\perp$ and $\parallel$, this balance may be expressed as

$$\tau_A = (k_\parallel V_a)^{-1} \simeq \tau_{NL\perp} = (k_\perp u_k)^{-1}. \tag{7.27}$$

If one now assumes that within the planes, the Alfvén effect due to the magnetic field of large eddies in the planes may be neglected, the velocity field at the scale $k_\perp$ satisfies $u_k \sim \epsilon^{1/3} k_\perp{}^{-1/3}$ so that

$$k_\parallel(k_\perp) \sim \epsilon^{1/3} k_\perp{}^{2/3}/V_a. \tag{7.28}$$

This relation defines a region in the $k_\parallel, k_\perp$ plane where the RMHD approximation is valid [34]. In astrophysical applications of full incompressible MHD, this spectral domain was identified in [14]. If, however, one takes into account the Alfvén effect even in the perpendicular direction, one may define a regime of "weak turbulence" where three-wave interactions dominate the nonlinear cascade. In this case a simple extension of the dimensional analysis presented here shows that one obtains a perpendicular spectrum in the form $E_{k_\perp} \approx k_\perp^{-2}$ [33, 11]. This follows by recognizing that the cascade occurs mainly in the perpendicular direction, but the Alfvén effect which leads to the lengthening of the effective interaction time, $\tau^* \sim (\tau_{nl}/\tau_A)\tau_{nl} \sim (B_0/\delta B)\tau_{nl}$, contains in the τ_A term only the parallel term $\tau_A \sim l_\parallel/V_a$.

Although the solar wind is neither incompressible, nor isotropic, or homogeneous, the observed slopes are close to those inferred by the above arguments. While Coleman [5] argued in favor of an Iroshnikov–Kraichnan spectrum ($-3/2$ slope), most of the observational evidence seems to be for spectral slopes in the solar wind turbulence very near the Kolmogorov slope, as soon as the heliocentric distance is larger than about 1 AU (see, for example, [2]). The present state of numerical simulations does not help to clarify the situation: even when the conditions are favorable (an incompressible, homogeneous fluid with large-scale magnetic fields), it is difficult, because of the limited resolution available, to measure spectral slopes accurately enough, and thus to determine which of the Kolmogorov or Iroshnikov–Kraichnan phenomenology is valid [4, 31, 28].

If one assumes the interactions are coherent as in Kolmogorov's theory [29], i.e., assume the transfer times $\tau^{*\pm}$ to be equal to the turnover times for both Elsasser fields, then the transfer rates for outward and inward modes are distinct:

$$\Pi_k^{\pm} = k^{5/2} E_k^+ E_k^- / (E_k^{\mp})^{1/2} = \epsilon^{\pm}. \tag{7.29}$$

In this case, the constraint of constant fluxes leads to the Kolmogorov spectra for both fields, whatever the asymmetry in the fluxes $\epsilon^{\pm}$.

As seen in the paragraphs on linear evolution of waves in the solar wind, expansion effects appear as supplementary terms in the equation, which involve the "average" Alfvén and advection velocities V_a and U and their radial derivatives. The definition of the timescale T over which the full equations are averaged in order to separate the large-scale wind expansion from the small-scale fluctuations is somewhat arbitrary. The basic requirement is that T^{-1} should be less than the lowest frequency considered. The amplitude can then be split into two parts, an average $<z>$ and a fluctuating z. Upon substraction of the time average from the original MHD equations and assuming incompressible fluctuations, one obtains now the full equations [51]:

$$\frac{\partial z^{\pm}}{\partial t} + (V_g^{\pm}) \cdot \nabla z^{\pm} + z^{\mp} \cdot \nabla (V_g^{\mp}) + \frac{1}{2}(z^{\pm} - z^{\mp})\left(\frac{1}{2}\nabla \cdot U \pm \nabla \cdot V_a\right), \tag{7.30}$$

$$= -\frac{1}{\rho}\nabla p^T - \left(z^{\mp} \cdot \nabla z^{\pm} - <z^{\mp} \cdot \nabla z^{\pm}>\right),$$

where $V_g^{\pm} = U \mp V_a$ is the linear group velocity . An equation for the evolution of the quadratic quantities, such as the energy spectrum, may be obtained by multiplying (7.31) by $z^{\pm}(x', t')$, averaging and then Fourier transforming with respect to the "fast" variables $x - x', t - t'$. Assuming isotropy and time stationarity of the spectral quantities, one obtains equations for the spectra which depend both on wave number k (fast variable) and distance r (slow variable) [46]:

$$\nabla \cdot (V_g^{\pm} E_k^{\pm}(r)) + E_k^{\pm}(r)\frac{1}{2}\nabla \cdot U + M_k^{\pm}(r) = NL\left(\Pi_k^{\pm}\right), \tag{7.31}$$

where the derivatives on the left-hand side denote derivatives with respect to the slow variable r. In (7.31) the first term is the wave energy flux, the second term is the work done by the waves in accelerating the bulk flow. The third term $M^{\pm}$ represents the linear coupling between the two wave species, due to the large-scale inhomogeneities:

$$\mathbf{M}^{\pm}(\mathbf{k}, \mathbf{r}) = -\mathbf{z}_{\mathbf{k}}^{+}\mathbf{z}_{\mathbf{k}}^{-} :: \left(\nabla \mathbf{V}_{g}^{\pm} - \frac{1}{2}\delta_{ij}\left(\nabla \cdot \mathbf{V}_{a} \pm \frac{1}{2}\nabla \cdot \mathbf{U} \right) \right) \qquad (7.32)$$

and effectively couples the $\pm$ equations with that for the residual energy $E^r = \mathbf{z}^+ \cdot \mathbf{z}^- = (u^2 - b^2)$. Its order of magnitude may be estimated as $\approx E^r/\tau_{ad}$. In the linear case of transverse Alfvén waves, $\mathbf{M}^{\pm}$ and the nonlinear terms may be neglected (the distinction between energy spectral densities and energies at a given scale may in this case be dropped without altering the equations), and (7.31, 7.32) reduces to the equation of conservation of the adiabatic invariant $S^{\pm} = E^{\pm}/(k.V_a)$:

$$\nabla \cdot (\mathbf{V}_g^{\pm} S^{\pm}) = 0.$$

In the solar wind the Alfvén speed may be neglected compared to the wind speed, so that we obtain from (7.32) [23]:

$$\nabla \cdot (\mathbf{U}\rho E^{\pm}) + \rho E^{\pm}\nabla \cdot \mathbf{U}/2 = 0,$$

which in a spherical expansion at constant speed gives $E^{\pm} \propto r^{-1}$ (recall that $E^{\pm}$ is the specific energy; the result usually quoted is $\rho E^{\pm} \propto r^{-3}$). Both spectra E^+ and E^- thus decrease with distance in a self-similar way, i.e., without changing their shape. The observations confirm that this simple linear description is valid at low frequencies between $(2 \times 10^{-4}, 2 \times 10^{-3})$ Hz, while at larger frequencies they show that total turbulent energy decays more rapidly than simply predicted by the adiabatic change, suggesting again that turbulent dissipation (and thus nonlinear interactions) are at work [1, 43].

These considerations apply only in the inner heliosphere. Indeed, at larger distances from the Sun, the energy decay rate appears to be roughly frequency independent [2]. The radial dependence, however, is $\propto r^{-3.5}$, which is slightly steeper than the WKB dependence. In order to explain the changes in spectral shapes in the inner heliosphere, [47, 45] developed a model in which the nonlinear flux on the right-hand side of (7.31) is calculated following dimensional analysis: isotropy in $\mathbf{k}$ is necessarily assumed, so that the flux depends on the wave vector amplitude only.

The model describes the two regimes on either side of the critical scale $L(r)$ at which $\tau_{nl} \approx \tau_{ad}$. At small wave vector the nonlinear effects are negligibly small, the only modifications are limited to the WKB decay $u^2 \propto r^{-1}$; at larger wave vector they dominate so that locally an equilibrium spectrum in $k^{-5/3}$ is maintained (or $k^{-3/2}$, depending on the form (a) or (b) of the nonlinear flux [11]) the WKB effects acting to fix the energy level at a given frequency. In this way, when starting with a flat (k^{-1}) spectrum as similar to what is observed at 0.3 AU, one obtains the steepening toward equilibrium spectrum, first at high frequencies, and later on at low frequencies.

These models approximate the ratio E^+/E^- by an observational constant. However, both E^+ and E^- obey coupled evolution equations, the coupling occurring through expansion and nonlinear effects. Hence there is no reason to expect that the ratio remain constant. Then one may speculate on whether the situation found in the

Alfvénic periods, where $z^- \ll z^+$, is the result of the evolution of the turbulence, or is a distinct property of high-speed winds.

More recent phenomenological models aimed at understanding the full evolution of an Alfvénic spectrum from the base of the corona out into the solar wind have been developed by Verdini and Velli, and Cranmer et al. [50, 6]. Simulations show that, in a highly stratified atmosphere, the nonlinear interactions of Alfvén waves launched from the photosphere are able to generate and sustain an incompressible turbulent cascade, which displays the observed Alfvénicity. The efficiency of turbulence in transporting energy to the dissipative scales is, however, still unclear. The spectral slope at different coronal heights evolves with distance, subject to expansion and driving effects, affecting the radial dependence of dissipation. The initial spectrum of Alfvén waves in the photosphere cannot be constrained by in situ data collected in the far solar wind, since local processes contribute to its shaping there.

The whole area of the fluctuations driving the solar wind is being revolutionized by the Hinode observations (http://solarb.msfc.nasa.gov/) of fluctuations in the chromosphere and corona, which appear to show large amplitude (20–25 km/s amplitudes) in the chromosphere.

The upcoming years could be crucial in understanding the origins and driving of turbulence in the solar wind.

References

1. B. Bavassano, M. Dobrowolny, F. Mariani, N. F. Ness: J. Geophys. Res., **87**, 3617 (1982)
2. B. Bavassano, E.J. Smith: J. Geophys. Res., **91**, 1706 (1986)
3. J.W. Belcher, L. Davis: J. Geophys. Res., **76**, 3534 (1971)
4. D. Biskamp, H. Welter: Phys. Fluids, **B1**, 1964 (1989)
5. P.J. Coleman: Astrophys. J., **153**, 371 (1968)
6. S.R. Cranmer, A.A. van Ballegooijen, R.J. Edgar: Astrophys. J. Supp., **171**, 520 (2007)
7. L. Del Zanna, M. Velli, P. Londrillo: Astron. Astrophys., **330**, L13 (1998)
8. M. Dobrowolny, A. Mangeney, P.-L. Veltri: Phys. Rev. Lett., **45**, 144 (1980)
9. M. Dobrowolny, A. Mangeney, P.-L. Veltri: Astron. Astrophys. (A & A), **83**, 26 (1980)
10. L.A. Fisk: J. Geophys. Res., **108**, 10.1029/2002JA009284 (2003)
11. S. Galtier, S.V. Nazarenko, A.C. Newell, A. Pouquet: Astrophys. J., **564**, L49 (2002)
12. J. Geiss, G. Gloeckler, R. von Steiger: Space Sci. Rev., **72**, 49 (1995)
13. G. Gloeckler, T.H. Zurbuchen, J. Geiss: J. Geophys. Res., **108**, 10.1029/2002JA009286 (2003)
14. P. Golreich, S. Sridhar: Astrophys. J., **438**, 763 (1995)
15. R. Grappin, U. Frisch, J. Léorat, A. Pouquet: Astron. Astrophys. **105**, 6 (1982)
16. R. Grappin, J. Leorat, A. Pouquet: Astron. Astrophys., **126**, 51 (1983)
17. R. Grappin, A. Mangeney, E. Marsch: J. Geophys. Res., **95**, 8197, (1990)
18. K.-I. Gringauz, V.-V. Bezrukikh, V.-D. Ozerov, R.-E. Rybchinskiy: Soviet Phys. (English transl.), **5**, 361–364, (1960)
19. V.H. Hansteen, E. Leer, T.E. Holzer: Astrophys. J., **482**, 498 (1997)
20. T.E. Holzer, W.I. Axford: Ann. Rev. Astron. Astrophys., **8**, 30, (1970)
21. T.S. Horbury, B.T. Tsurutani: In: The Heliosphere Near Solar Minimum. In: A. Balogh, R.G. Marsden, E.J. Smith (Eds.), Springer-Praxis, UK (2001).
22. T.S. Horbury, A. Balogh, R.J. Forsyth, E.J. Smith: J. Geophys. Res., **101**, 405 (1996)
23. S.A. Jacques: Astrophys. J. **215**, 942, (1977)

24. R.H. Kraichnan: Phys. Fluids, **8**, 1385 (1965)
25. A. Kolmogorov: Soviet Phys. Doklady, **30**, 301 (1941)
26. R.A. Kopp, T.E. Holzer: Solar Phys., **49**, 43 (1976)
27. P. Korevaar: Astron. Astrophys., **226**, 209 (1989)
28. J. Mason, F. Cattaneo, S. Boldyrev: Phys. Rev. Lett., **97**, 255002 (2006)
29. W.H. Matthaeus, M.L. Goldstein, D. Montgomery: Phys. Rev. Lett., **51**, 1484 (1983)
30. W.H. McCrea: Astrophys. J., **124**, 461 (1956)
31. W.C. Muller, R. Grappin: Phys. Rev. Letts., **95**, 114502 (2005)
32. R.H. Munro, B.V. Jackson: Astrophys. J., **213**, 874 (1977)
33. C.S. Ng, A. Bhattacharjee: Phys. Plasmas, **4**, 605 (1997)
34. S. Oughton, P. Dmitruk, W.H. Matthaeus: In: Turbulence and Magnetic Fields in Astrophysics. In: E. Falgarone, T. Passot (eds.) Springer-Verlag, Berlin Vol. 614 (LNP) pp. 28–55, (2003)
35. S. Oughton, P. Dmitruk, W.H. Matthaeus: Solar Wind 11/SOHO 16, Connecting Sun and Heliosphere. B. Fleck, T.H. Zurbuchen, H. Lacoste (eds.) ESA SP-592, Noordwijk, (2005)
36. E.N. Parker: Astrophys. J., **128**, 664 (1958)
37. E.N. Parker: Interplanetary Dynamical Processes. Interscience, New York (1963)
38. J.M. Picone, R.B. Dahlburg, *Phys. of Fluids* **B 3**, 29 (1991)
39. P.H. Roberts, A.M. Soward: Proc. Roy. Soc. London A, **328**, 185 (1972)
40. D.A. Roberts, M.L. Goldstein, L.W. Klein, W.H. Matthaeus: J. Geophys. Res., **92**, 12023 (1987)
41. F. Scherb: Space Res. **4**, 797 (1964)
42. N.A. Schwadron, D.J. McComas: Astrophys. J., **599**, 1395 (2003)
43. R. Schwenn: in Solar Wind 5, Nasa Conf. Publ., CP2280, 489 (1983)
44. H.R. Strauss: Phys. Fluids, **19**, 134 (1976)
45. C.Y. Tu: J. Geophys. Res., **93**, 7 (1988)
46. C.Y. Tu, E. Marsch, K.M. Thieme: J. Geophys. Res., **94**, 11739 (1989)
47. C.Y. Tu, K. Pu, J. Wei: Geophys. Res., **89**, 9695 (1984)
48. M. Velli: Astron. Astrophys., **270**, 304 (1993)
49. M. Velli: Astrophys. J., **432**, L55 (1994)
50. A. Verdini, M. Velli: Astrophys. J., **662**, 669 (2007)
51. Y.C. Whang: J. Geophys. Res., **85**, 2285 (1980)

Chapter 8
Physics of Stellar Coronae

M. Güdel

8.1 Introduction

For the plasma physicist, the solar corona offers an outstanding example of a space plasma, and surely one that deserves a lifetime of study. Not only can we observe the solar corona on scales of a few hundred kilometers and monitor its changes in the course of seconds to minutes but we also have a wide range of detailed diagnostics at our disposal that provide immediate access to the prevalent physical processes.

Yet, solar physics offers a rich field of unsolved plasma-physics problems. How is the coronal plasma continuously heated to $>10^6$ K? How and where are high-energy particles episodically accelerated? What is the internal dynamics of plasma in magnetic loops? What is the initial trigger of a coronal flare? How does the corona link to the solar wind, and how and where is the latter accelerated? How is plasma transported into the solar corona?

Why, then, study stellar coronae that remain spatially unresolved in X-rays and are only marginally resolved at radio wavelengths, objects that require exposure times of several hours before approximate measurements of the ensemble of plasma structures can be obtained?

There are many reasons. In the context of the *solar–stellar connection,* stellar X-ray astronomy has introduced a range of stellar rotation periods, gravities, masses, and ages into the debate on the magnetic dynamo. Coronal magnetic structures and heating mechanisms may change as these parameters are varied. Parameter studies could provide valuable insight for constraining relevant theories. Different topologies and sizes of magnetic field structures lead to different wind mass-loss rates, and this will regulate the stellar spin-down rates differently.

Including stars into the big picture of coronal research has also widened our view of coronal plasma physics. While solar coronal plasma resides typically at $(1–5) \times 10^6$ K with temporary excursions to ≈ 20 MK during large flares, much higher temperatures are found on some active stars, with steady plasma temperatures

M. Güdel (✉)

Institute of Astronomy, ETH Zurich, 8093 Zurich, Switzerland,
`guedel@astro.phys.ethz.ch`

Güdel, M.: *Physics of Stellar Coronae.* Lect. Notes Phys. **778**, 269–325 (2009)
DOI 10.1007/978-3-642-00210-6_8

Table 8.1 Symbols and units used throughout the text

Symbol, acronym	Explanation
R_* or R	Stellar radius [cm]
$R_\odot$	Solar radius [7×10^{10} cm]
c	Speed of light [3×10^{10} cm s^{-1}]
k	Boltzmann constant [1.38×10^{-16} erg cm^3 K^{-1}]
m_e	Mass of electron [9.1×10^{-28} g]
d	Distance [pc]
p	Pressure [dyne cm^{-2}]
L	Coronal loop semi-length [cm]
T	Coronal electron temperature [K]
T_b	Radio brightness temperature [K]
n, N	Non-thermal electron density (N: integrated in energy)
N	Rate of flares
n_e	Electron density [cm^{-3}]
n_H	Hydrogen density [cm^{-3}]
B	Magnetic field strength [G]
f	Surface filling factor [%]
Γ	Loop area expansion factor (apex to base)
E	Energy [erg]
L_R	Radio luminosity [erg s^{-1} Hz^{-1}]
L_X	X-ray luminosity [erg s^{-1}]
L_{bol}	Stellar bolometric luminosity [erg s^{-1}]
$\Lambda = \Lambda_0 T^\chi$	Cooling function [erg s^{-1} cm^3]
τ	(Decay) timescales, also: optical depth
ν	Radio frequency
ν_p, ω_p	Plasma frequency, angular plasma frequency
ν_c, Ω_c	Gyrofrequency, angular gyrofrequency
κ	Absorption coefficient
ε	Kinetic energy of electron
γ	Lorentz factor
L_B	Magnetic scale height
HRD	Hertzsprung–Russell diagram
EM	Emission measure
Q, DEM	Differential emission measure
EMD	(Discretized, binned) emission measure distribution

of several tens of MK and flare peaks beyond 100 MK. Energy release in stellar flares involves up to 10^5 times more thermal energy than in solar flares, and pressures arise that are not encountered in the solar corona.

This chapter provides a "stellar astronomer's view" of magnetic coronae. It cannot replace the knowledge of detailed physical processes in the solar corona, nor do we expect to find entirely new physical concepts from the study of stellar coronae without guidance from solar physics. However, as I hope to show in the following sections, stellar astronomy has provided some unexpected and systematic trends that may well help understand systematics of coronal behavior across a large range of stellar parameters. In this sense, the goal of this chapter is to present an overview of the basic, observed stellar coronal phenomena and their interpretation by analogy (where appropriate) to solar physics, rather than the derivation of the basic plasma physical mechanisms themselves. For a deeper understanding of the latter, I must refer the reader to appropriate textbooks and lectures on solar coronal physics. Table 8.1 lists symbols and units used throughout the present text.

8.2 Stellar Coronae – Defining the Theme

We consider coronae to be an ensemble of closed magnetic structures above the stellar photosphere and chromosphere together with their plasma content, regardless of whether the latter is thermal bulk plasma or a non-thermal population of accelerated particles. Although often narrowed down to some specific energy ranges, coronal emission is intrinsically a multi-wavelength phenomenon revealing itself from the meter-wave radio range to gamma rays. The most important wavelength regions from which we have learned *diagnostically* about *stellar* coronae include the radio (decimetric to centimetric) range and the X-ray domain. The former is sensitive to accelerated electrons in magnetic fields, and has provided the only direct means of imaging stellar coronal structure, through very long baseline interferometry.

The soft X-ray (0.1–10 keV) diagnostics have been instrumental in our understanding of physical processes in the hot, magnetically trapped coronal plasma, and the recent advent of high-resolution X-ray spectroscopy with the *Chandra* and *XMM-Newton* X-ray observatories is now measuring the physical parameters of coronal plasma directly. The adjacent extreme ultraviolet (EUV) range contains diagnostics relevant for the same temperature range as X-rays.

8.3 The Coronal Hertzsprung–Russell Diagram

Before discussing specific physical problems in stellar coronal physics, I will briefly review the phenomenology of stellar radio (here: 1–20 GHz) and X-ray emission, and summarize stellar classes that are prolific coronal emitters. Figure 8.1 presents Hertzsprung–Russell diagrams (HRD) of detected X-ray (left) and radio (right) stars. They show all the basic features that we know from an optical HRD. Although the samples used for these figures are in no way "complete" (in volume or brightness), the main sequence is clearly evident, and so is the giant branch. The cool half of the subgiant and giant areas is dominated by the X-ray and radio-strong samples of RS CVn and Algol-type close binaries. The top right part of the diagram, comprising cool giants, is almost devoid of X-ray detections (although well populated by radio emitters). The so-called corona vs. wind dividing line (dashed in Fig. 8.1a; after [65]) separates coronal giants and supergiants to its left from stars with massive winds to its right. It is unknown whether the wind giants possess magnetically structured coronae at the base of their winds – the X-rays may simply be absorbed by the overlying wind material. Additionally, a very prominent population of (presumably) coronal radio and X-ray sources just above the main sequence is made up of various classes of pre-main sequence stars, such as classical and weak-lined T Tauri stars.

Moving toward A-type stars on the HRD, one expects, and finds, a significant drop of coronal emission owing to the absence of magnetic dynamo action in these

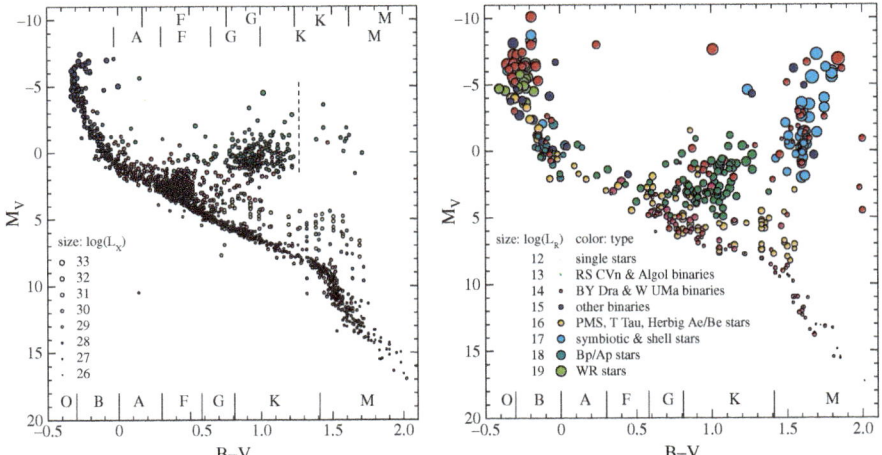

Fig. 8.1 *Left*: Hertzsprung–Russell diagram based on about 2000 X-ray detected stars extracted from survey catalogs (see [37] for references). The size of the circles characterizes $\log L_X$ as indicated in the panel at *lower left*. The ranges for the spectral classes are given at the *top* (*upper row* for supergiants, *lower row* for giants) and at the *bottom* of the figure (for main-sequence stars). *Right*: Similar, but for 440 radio stars detected between 1 and 10 GHz (after [36])

stars. However, this is also the region of the chemically peculiar Ap/Bp stars that possess strong magnetic fields and many of which are now known to be non-thermal radio sources as well. Finally, the very luminous radio and X-ray emissions from O and B stars are believed to originate in non-magnetic stellar winds – we will not discuss these stars further.

8.4 Non-flaring Radio Emission from Stellar Coronae

"Quiescent" (non-flaring) radio emission at levels of 10^{12}–10^{16} erg s^{-1} Hz^{-1} from magnetically active stars was entirely unanticipated but constitutes an important achievement in stellar radio astronomy: there simply is no solar counterpart! Quiescent emission can be defined by the absence of impulsive, rapidly variable flare-like events. Common characteristics of quiescent emission are (i) slow variations on timescales of hours and days, (ii) broadband radio spectra, (iii) brightness temperatures in excess of coronal temperatures measured in X-rays, and usually (iv) low polarization degrees.

8.4.1 Bremsstrahlung

The Sun emits steady, full-disk, optically thick thermal radio emission at chromospheric and transition region levels of a few 10^4 K. However, such emission cannot be detected with present-day facilities, except for radiation from the very nearest stars or giants subtending a large solid angle. Using the Rayleigh–Jeans approximation for the flux density S

$$S = \frac{2kT\tau v^2}{c^2} \frac{\pi R^2}{d^2} \approx 0.049 \left(\frac{T}{10^6 \text{ K}}\right) \left(\frac{v}{1 \text{ GHz}}\right)^2 \left(\frac{R}{R_\odot}\right)^2 \left(\frac{1 \text{ pc}}{d}\right)^2 \tau \text{ mJy} \quad (8.1)$$

(R = source radius, T = electron temperature, d = stellar distance, k = Boltzmann constant, v = observing frequency, τ = optical depth), we find for optically thick chromospheric emission (with $\tau = 1$, $T = 1.5 \times 10^4$ K, $v = 8.4$ GHz)

$$S \approx \frac{0.05}{d_{\text{pc}}^2} \left(\frac{R}{R_\odot}\right)^2 \text{ mJy}. \quad (8.2)$$

[Recall that 1 Jy is 10^{-26} W m^2 Hz^{-1}.] Optically thin free–free emission from the hot, X-ray-emitting plasma in coronal loops can be estimated as follows: the radio optical depth is

$$\tau = \int \kappa \, dl \approx \frac{0.16}{v^2 T^{3/2}} \int n_e^2 \, dl \quad (8.3)$$

and the X-ray volume emission measure EM, using a filling factor f in the approximation of a small coronal height,

$$\text{EM}_X = 2R^2 \pi f \int n_e^2 \, dl \quad (8.4)$$

(the factor of 2 accounts for the visible and invisible hemispheres, assuming that the EM is uniformly distributed). For plasma between 1 and 20 MK, the EM can be estimated from the X-ray luminosity L_X with a conversion factor $\Lambda(T)$ [71], with Λ larger for lower T:

$$L_X = \Lambda(T)\text{EM}_X \approx (1.5 - 6) \times 10^{-23}\text{EM}_X \quad [\text{erg s}^{-1}]. \quad (8.5)$$

Thus, we find (using $T_{\text{MK}} = T/[10^6 \text{ K}]$)

$$S = 2.6 \times 10^{-52} \frac{L_X}{\Lambda(T)T_{\text{MK}}^{1/2} f d_{\text{pc}}^2} \text{ mJy} \approx (4 - 17) \times 10^{-30} \frac{L_X}{T_{\text{MK}}^{1/2} f d_{\text{pc}}^2} \quad \text{mJy}. \quad (8.6)$$

Coronal bremsstrahlung contributions are presently out of reach for almost all stars.

8.4.2 Gyroresonance Emission

Because active stars show high coronal temperatures and large magnetic filling factors that prevent magnetic fields from strongly diverging with increasing height, the radio optical depth can become significant at coronal levels owing to gyroresonance absorption. This type of emission is observed above solar sunspots. The optical depth for the sth harmonic is [120]

$$
\begin{aligned}
\tau(s, \nu) &= \frac{\pi^3}{4} \frac{\nu_p L_B}{\nu_c} \frac{s^2(2s-2)!}{2^{2s-2}s![(s-1)!]^2} \left(\frac{s^2}{2\mu}\right)^{s-1} \\
&= 1.45 \frac{L_B}{R_*} \frac{R_*}{R_\odot} \frac{n_e}{\nu_{\mathrm{GHz}}} (1.3 \times 10^{-4}, 4.1 \times 10^{-6}, 1.8 \times 10^{-7}) \left(\frac{T}{20\ \mathrm{MK}}\right)^{s-1},
\end{aligned}
$$
$$(8.7)$$

where the three coefficients in the parenthesis are for the harmonics $s = 3, 4$, and 5, respectively, of the gyrofrequency, ν_{GHz} is the observing frequency in GHz, n_e is the electron density of the emitting hot plasma at temperature T, and $\mu = m_e c^2 / kT$. The magnetic scale height, L_B, is not precisely known but will be assumed to be $0.3 R_*$ (see arguments in [120]). Using typical X-ray derived temperatures ($[1\text{--}3] \times 10^7$ K) and X-ray emission measures $\mathrm{EM} = n_e^2 V$ (to estimate electron densities in the emitting volume), one finds that τ invariably reaches unity at $s = 3, 4$, or 5 for an observing frequency of 15 GHz while $\tau < 1$ for larger s. The highest harmonic that is still optically thick is relevant.

The measured flux from the optically thick layer is again given by Eq. (8.1). Observations have shown the following:

– The observed fluxes at 15 GHz from M dwarfs are not compatible with the above prediction if the emitting layer of hot plasma covers the entire stellar surface. The observed flux is typically much smaller. The microwave spectra are falling, which is not compatible with optically thick radiation ([120] – see Fig. 8.2).

– There are exceptions in which a rising spectrum from 5 to 15 GHz (Fig. 8.2) could be explained by gyroresonance emission, while the lower frequency spectrum cannot [42]. The same stars may show this feature only temporarily.

These results imply that the hot plasma in general is not coincident with the strong magnetic field regions in the corona. It must reside in lower B regions in between or above the cooler regions, possibly implying rather extended coronal structures seen in X-rays. This plasma could be induced by flares in which magnetic loops reconnect to larger structures in which the low-density gas will not produce appreciable gyroresonance emission [120]. The cooler plasma also usually observed in active stars might be trapped in the strong fields but its gyroresonance emission is also negligible.

Lower frequency radio emission cannot be due to gyroresonance emission: the radius of the optically thick layer would still be at $s = 3\text{--}5$, but it would then be located at $>3 R_*$ (for dMe stars). If we extrapolate the corresponding magnetic fields of more than 100 G down to photospheric levels, we would find photospheric field strengths much in excess of those observed [34].

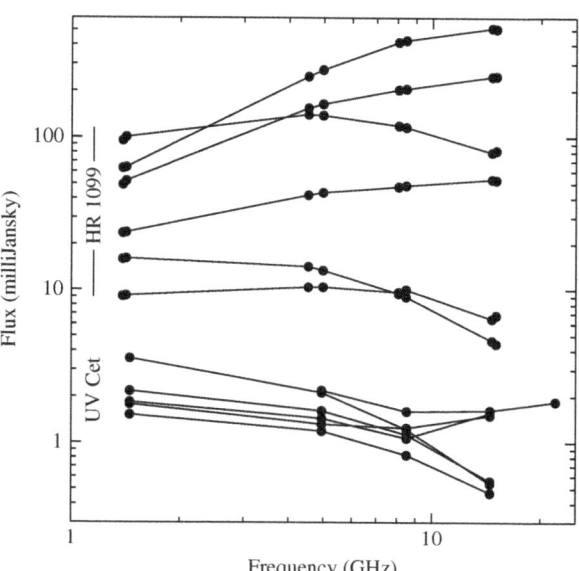

Fig. 8.2 Radio spectra of the RS CVn binary HR 1099 (*upper set*) and of the dMe dwarf UV Cet (*lower set*) at different flux levels. The gently bent spectra are indicative of gyrosynchrotron emission, and the high-frequency part of U-shaped spectra for UV Cet has been interpreted as a gyroresonance component (HR 1099 spectra: courtesy of S. M. White)

8.4.3 Gyrosynchrotron Emission

We could, however, allow for much higher T_{eff}. The optically thick layer would in that case shift to harmonics above 10, the range of gyrosynchrotron radiation. The optically thick source sizes are then more reasonable for M dwarfs, with $R \approx R_*$ [64], and the optically thin emission may still be strong enough for detection. However, for a thermal plasma, the spectral power drops like ν^{-8} at high frequencies, in contrast to observed microwave spectra that show $\nu^{-(0.3...1)}$ for magnetically active stars (Fig. 8.2).

Instead, the electron population could be *non-thermal*. Electron energy distributions in cosmic sources are often found to follow a power law, and this also holds for solar and stellar microwave flares [57]:

$$n(\gamma) = K(\gamma - 1)^{-\delta} \tag{8.8}$$

(γ = Lorentz factor). There is wide support for this model from estimates of the brightness temperature (e.g., based on the stellar radius or from resolved interferometric images). The important question then is how these coronae are continuously replenished with high-energy electrons.

The radio spectral time development implied by this model can be analytically calculated for the case of a short injection of electrons into the corona [21]. The outline of the derivation is as follows: For trapped electrons with non-thermal number density n, an equation of continuity applies in energy,

$$\frac{\partial n(\gamma, t)}{\partial t} + \frac{\partial}{\partial \gamma}\left[n(\gamma, t)\frac{d\gamma}{dt} \right] = 0, \tag{8.9}$$

with the solution

$$n(\gamma, t)d\gamma = n(\gamma_0)d\gamma_0, \tag{8.10}$$

where $n(\gamma_0)d\gamma_0$ is the initial distribution. We now need an expression for the energy loss rate of an electron, $d\gamma/dt$. In an extended stellar magnetosphere, the most relevant loss mechanisms are synchrotron loss and Coulomb-collisional loss, given by, respectively,

$$-\dot{\gamma}_{\text{coll}} = 5 \times 10^{-13} n_e \quad [\text{s}^{-1}], \qquad \tau_{\text{coll}} = 2 \times 10^{12} \frac{\gamma}{n_e} \quad [\text{s}] \tag{8.11}$$

$$-\dot{\gamma}_{\text{B}} = 1.5 \times 10^{-9} B^2 \gamma^2 \quad [\text{s}^{-1}], \qquad \tau_{\text{B}} = \frac{6.7 \times 10^8}{B^2 \gamma} \quad [\text{s}], \tag{8.12}$$

where n_e is the ambient thermal electron density: see, e.g., [85]; we have assumed a pitch angle of $\pi/3$. The total energy loss rate is thus

$$\frac{d\gamma}{dt} = \alpha + \beta\gamma^2, \tag{8.13}$$

with the appropriate coefficients from Eqs. (8.11) and (8.12). This equation has an analytical solution for $n(\gamma, t)$ if the initial distribution is a power law as given by Eq. (8.8), namely

$$n(\gamma, t) = K(1 + \tan^2 A\alpha t)A^\delta \frac{[A\gamma(1 + A\tan A\alpha t) - A + \tan A\alpha t]^{-\delta}}{(1 - A\gamma \tan A\alpha t)^{2-\delta}}, \tag{8.14}$$

where $A = (\beta/\alpha)^{1/2}$ and the initial power law has been bounded by $\gamma_{0,1} < \gamma_0$ (typically a low energy, e.g., $\gamma_{0,1} = 1.1$) and $\gamma_{0,2} > \gamma_0$ (typically a very high energy, formally including $\gamma_{0,2} = \infty$). The evolution of the initial power law boundaries can be computed; after a finite time, all non-thermal electrons have thermalized. Some characteristic results of these calculations are shown in Fig. 8.3.

Figure 8.3a shows the flattening with time of the electron energy distribution due to the collisional losses at lower energies, and a cutoff due to synchrotron losses at higher energies. Figure 8.3b illustrates the intensity of radiation after calculation of the emissivity and the absorption coefficient, together with an observed spectrum of the RS CVn binary UX Ari. In the early spectrum, the optically thick portion is well visible on the low-frequency side. As the electron energy decays, the spectrum becomes optically thin, and once the high-energy electron population is depleted, the spectrum begins to fall off steeply as there are no electrons left emitting at the observing frequency.

The time development of the emitted intensity as a function of magnetic field strength (Fig. 8.3c) shows that the initial radiation originates in fields of higher strength (the "core"), but the responsible electron population decays rapidly due to synchrotron losses. Eventually, radiation from the slowly decaying population in the weaker, more extended magnetic fields (the "halo") begins to dominate on longer timescales.

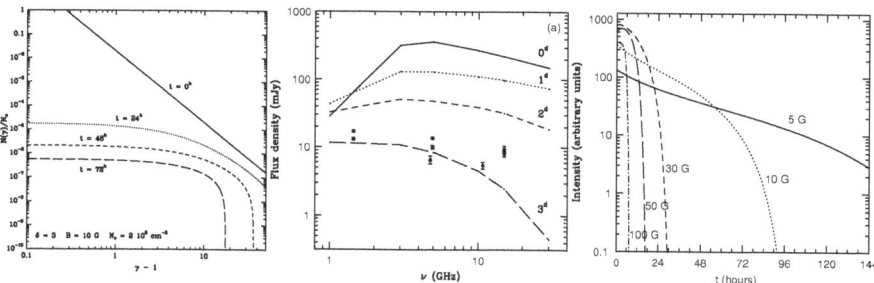

Fig. 8.3 Time-dependent spectral model of an RS CVn magnetosphere after injection of a power law electron population. *Left*: Evolution of the electron energy distribution due to collisional and synchrotron losses, starting from a power law. *Middle*: Evolution of the microwave spectrum with time, for a magnetic field of $B = 10$ G and an initial electron power law index of $\delta = 2$. The electron density in the thermal power law is $n_e = 2 \times 10^8$ cm^{-3}. *Right*: Time evolution of the radiation intensity at 5 GHz for different magnetic field strengths (from [21], courtesy of E. Franciosini)

A power law index of $\delta = 2$–3 (as also used above) appears to usually fit active-stellar microwave spectra such as those shown in Fig. 8.2 quite well [119, 108]. This suggests very hard electron distributions, similar to those seen in gradual solar flares [23]. But such distributions are observed during so-called quiescence – an indication that quiescent radio emission is due to a flare-like process? We will return to this hypothesis in later sections.

8.5 Thermal X-ray Emission from Stellar Coronae

8.5.1 High-Resolution X-ray Spectroscopy

The high-resolution X-ray spectrometers on *XMM-Newton* and *Chandra* cover a large range of spectral lines that are temperature- and partly density-sensitive. The spectra thus contain the features required for deriving X-ray emission measure distributions, abundances, coronal densities, and opacities.

Figure 8.4 shows examples of X-ray spectra. The stars cover the entire range of stellar activity: HR 1099 representing a very active RS CVn system, Capella an intermediately active binary, and Procyon an inactive F dwarf. The spectrum of HR 1099 reveals a considerable amount of continuum and comparatively weak lines, which is a consequence of the very hot plasma in this corona ($T \approx 5$–30 MK). Note also the unusually strong Ne IX/Fe XVII and Ne X/Fe XVII flux ratios if compared to the other stellar spectra (these ratios are due to an intrinsic compositional anomaly of the HR 1099 corona). The spectrum of Capella is dominated by Fe XVII and Fe XVIII lines which are preferentially formed in this corona's plasma at $T \approx 6$ MK (Fig. 8.5). Procyon, in contrast, shows essentially no continuum and only very weak lines of Fe. Its spectrum is dominated by the H- and He-like transitions of C, N, and O formed around 1–4 MK. The flux ratios between H- and He-like transitions are also convenient temperature indicators: The O VIII λ18.97/O VII λ21.6 flux ratio, for example, is very large for HR 1099 but drops below unity for Procyon.

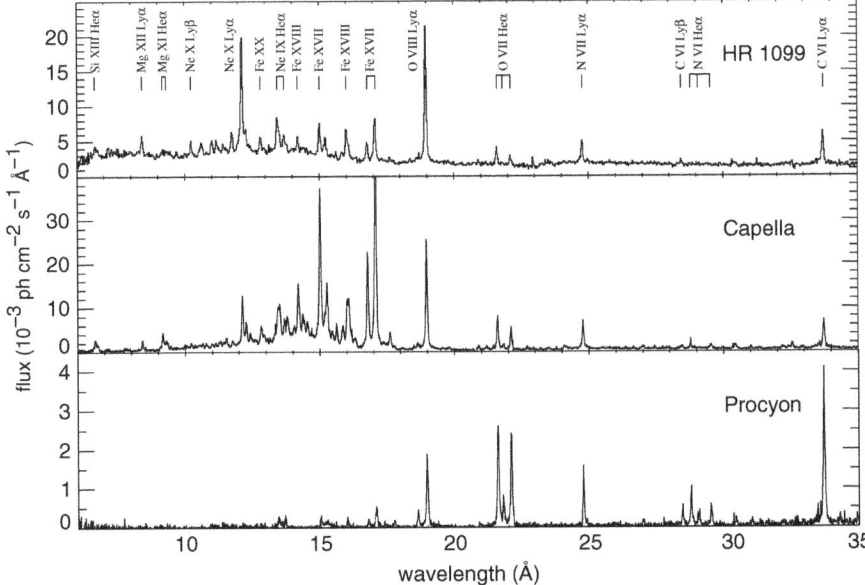

Fig. 8.4 Three high-resolution X-ray spectra of stars with largely differing activity levels: HR 1099, Capella, and Procyon. Data from *XMM-Newton* RGS

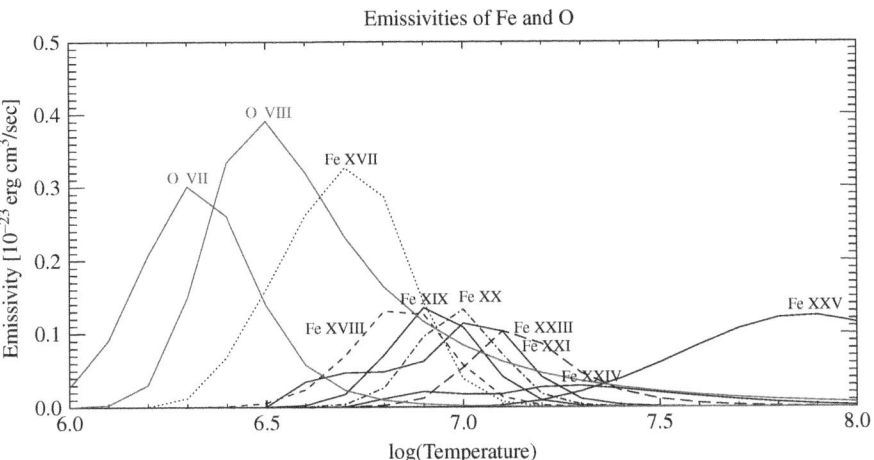

Fig. 8.5 Emissivities of several X-ray transitions, in particular of highly ionized iron and oxygen species (Fig. courtesy of A. Telleschi, after [105])

8.5.2 Thermal Coronal Components

The large range of temperatures measured in stellar coronae has been a challenge for theoretical interpretation. Whereas much of the solar coronal plasma can be well described by a component of a few million degrees, magnetically active stars have consistently shown a wide distribution of electron temperatures, reaching values as high as 50 MK outside obvious flares.

The flux ϕ_j observed in a spectral line from a given atomic transition can be written as

$$\phi_j = \frac{1}{4\pi d^2} \int AG_j(T)\frac{n_e n_H dV}{d\ln T}d\ln T, \qquad (8.15)$$

where d is the distance, $G_j(T)$ the "line cooling function" (luminosity per unit EM; Fig. 8.5) that contains the atomic physics of the transition as well as the ionization fraction for the ionization stage in question, and A the abundance of the element with respect to some basic tabulation used for G_j. For a fully ionized plasma with cosmic abundances, the hydrogen density $n_H \approx 0.85 n_e$. The expression

$$Q(T) = \frac{n_e n_H dV}{d\ln T} \qquad (8.16)$$

defines the *differential emission measure* (DEM). I will use this definition throughout but note that some authors define $Q'(T) = n_e n_H dV/dT$ which is smaller by one power of T. For a plane-parallel atmosphere with surface area S, Eq. (8.16) implies

$$Q(T) = n_e n_H SH(T), \qquad H(T) = \left|\frac{1}{T}\frac{dT}{ds}\right|^{-1}, \qquad (8.17)$$

where H is the *temperature scale height*.

8.5.3 Observational Results

"Discrete" (binned) emission measure distributions (EMDs) reflecting the full DEMs can be obtained by various inversion algorithms from the observed spectra. Most EMDs have generally been found to be singly or doubly peaked and confined on either side approximately by power laws [97]. Interestingly, EMDs are often very steep on the low-T side, and this is particularly true for the more active stars. For example, the slope of the stellar EMD in Fig. 8.6, (left) follows approximately $Q \propto T^3$ on the low-T side.

It is notable that the complete EMD shifts to higher temperatures with increasing stellar activity as seen in Fig. 8.6 (left and lower right) often leaving very little EM at modest temperatures and correspondingly weak spectral lines from ions of C, N, and O.

As a consequence, a relatively tight correlation between the characteristic coronal temperature (e.g., derived from a DEM) and the normalized coronal luminosity L_X/L_{bol} is found: *Stars at higher activity levels support hotter coronae*, with the most active stellar coronae reaching characteristic temperatures of several tens of MK. An example of solar analogs is shown together with the Sun itself during its activity maximum and minimum in Fig. 8.7. Here,

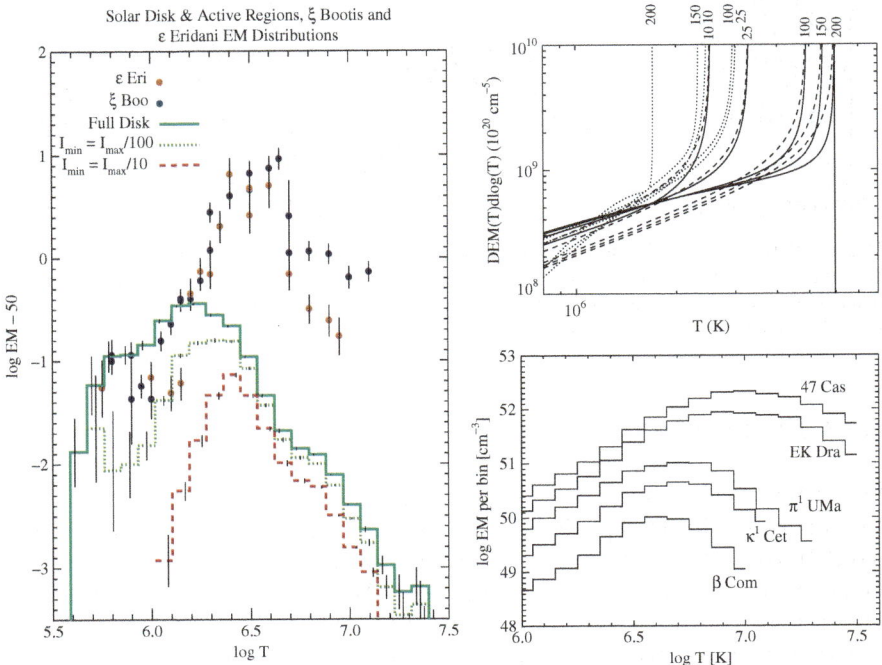

Fig. 8.6 *Left*: Emission measure distributions of two intermediately active stars (*bullets*) and the Sun (histograms; EM is in units of 10^{50} cm^{-3}). The histograms refer to full-disk solar EMDs derived from *Yohkoh* images at solar maximum, including also two versions for different lower cutoffs for the intensities in *Yohkoh* images (figure courtesy of J. Drake, after [28]). *Upper right*: Calculated differential emission measures of individual static loops. The *solid curves* refer to uniform heating along the loop and some fixed footpoint heating flux, for different loop half-lengths labeled above the figure panel in megameters. The *dashed curves* illustrate the analytical solutions presented by Rosner et al. [92] for uniform heating. The *dotted lines* show solutions assuming a heating scale height of 2×10^9 cm (figure courtesy of K. Schrijver, after [95]). *Lower right*: Examples of discrete stellar emission measure distributions of solar analogs (figure courtesy of A. Telleschi)

$$L_X \propto T^{4.5 \pm 0.3}, \tag{8.18}$$

$$EM \propto T^{5.4 \pm 0.6}, \tag{8.19}$$

where L_X denotes the total X-ray luminosity (but EM and T refer to the "hotter" component in standard 2-T fits to *ROSAT* data).

8.5.4 Interpretation of Differential Emission Measure

Equations (8.15) and (8.16) introduce the DEM as the basic interface between the stellar X-ray observation and the model interpretation of the thermal source. It con-

Fig. 8.7 Coronal temperature
vs. X-ray luminosity for solar
analogs. For details and
references, see Güdel [37]

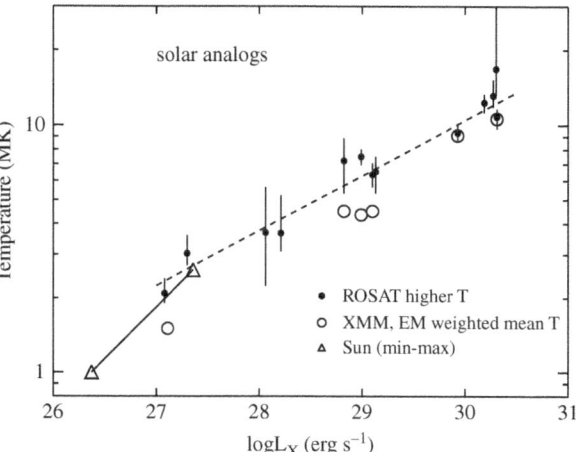

tains information on the plasma temperature and the density-weighted plasma mass
that emits X-rays at any given temperature. Although a DEM is often a highly de-
generate description of a complex real corona, it provides important constraints on
heating theories and on the range of coronal structures that it may describe. *Solar
DEMs* can, similar to the stellar cases, often be approximated by two power laws
$Q(T) \propto T^s$ on either side of their peaks (Fig. 8.6). The Sun has indeed given considerable guidance in physically interpreting the observed stellar DEMs.

8.5.4.1 The DEM of a Static Loop

The DEM of the plasma contained in a static magnetic loop follows from the hydrostatic equilibrium (see, e.g., derivation by Rosner et al. [92]). Under the conditions
of negligible gravity, i.e., constant pressure in the entire loop, and negligible thermal
conduction at the footpoints,

$$Q(T) \propto pT^{3/4-\chi/2+\alpha} \frac{1}{\left(1 - [T/T_a]^{2-\chi+\beta}\right)^{1/2}} \tag{8.20}$$

[19], where T_a is the loop apex temperature, and α and β are power law indices of,
respectively, the loop cross-sectional area S and the heating power q as a function of
T: $S(T) = S_0 T^\alpha$, $q(T) = q_0 T^\beta$, and χ is the exponent in the cooling function over
the relevant temperature range: $\Lambda(T) \propto T^\chi$. If T is not close to T_a and the loops
have constant cross section ($\alpha = 0$), we have

$$Q(T) \propto T^{3/4-\chi/2}, \tag{8.21}$$

i.e., under typical coronal conditions for non-flaring loops ($T < 10$ MK, $\chi \approx -0.5$),
the DEM slope is near unity. If strong thermal conduction is included at the

footpoints, then the slope changes to $+3/2$ if not too close to T_a [112]. The single-loop DEM sharply increases at $T \approx T_a$ (Fig. 8.6).

Loop expansion ($\alpha > 0$) obviously steepens the DEM. Increased heating at the loop footpoints (instead of uniform heating) makes the T range narrower and will also increase the slope of the DEM (see numerical calculations of various loop examples by Schrijver and Aschwanden [95] and Aschwanden and Schrijver [8], Fig. 8.6, upper right).

8.5.4.2 The DEM of Flaring Structures

Antiochos [5] discussed DEMs of solar flaring loops that cool by (i) static conduction (without flows), (ii) evaporative conduction (including flows), and (iii) radiation. The inferred DEMs scale, in the above order, as

$$Q_{cond} \propto T^{1.5}, \qquad Q_{evap} \propto T^{0.5}, \qquad Q_{rad} \propto T^{-\chi+1}. \qquad (8.22)$$

Since $\chi \approx 0 \pm 0.5$ in the range typically of interest for stellar flares (5−50 MK), all above DEMs are relatively flat (slope 1 ± 0.5). If multiple loops with equal slope but different peak T contribute, then the slope up to the first DEM peak can only become smaller. Non-constant loop cross sections have a very limited influence on the DEM slopes.

Stellar flare observations are often not of sufficient quality to derive temperature and EM characteristics for many different time bins. An interesting diagnostic was presented by Mewe et al. [72] who calculated the time-integrated (average) DEM of a flare that decays quasi-statically. They find

$$Q \propto T^{19/8} \qquad (8.23)$$

up to a maximum T that is equal to the temperature at the start of the decay phase.

Systems of episodically flaring loops were computed semi-analytically by Cargill [20], using analytic approximations for conductive and radiative decay phases of the flares. Here, the DEM is defined not by the internal loop structure but by the time evolution of a flaring plasma (assumed to be isothermal). Cargill argued that for radiative cooling, the (statistical) contribution of a flaring loop to the DEM is, to zeroth order, inversely proportional to the radiative decay time, which implies

$$Q(T) \propto T^{-\chi+1} \qquad (8.24)$$

up to a maximum T_m and a factor of $T^{1/2}$ less if subsonic draining of the cooling loop is allowed. Simulations with a uniform distribution of small flares within a limited energy range agree with these rough predictions, indicating a time-averaged DEM that is relatively flat below 10^6 K but steep ($Q[T] \propto T^4$) up to a few MK, a range in which the cooling function drops rapidly.

Let us next assume – in analogy to solar flares – that the occurrence rate of stellar flares is distributed in energy as a power law with an index α ($dN/dE \propto E^{-\alpha}$).

Then, an analytic expression can be derived for the time-averaged DEM of such a flare ensemble, i.e., a "flare-heated corona" [39]. We present a brief outline of the derivation.

Observationally, the flare peak temperature is correlated with the peak EM in both solar [32] and stellar flares [37]:

$$EM_0 = aT_0^b \quad [\text{cm}^{-3}], \tag{8.25}$$

with $b \approx 4.3 \pm 0.35$ in the range of $T = 10\text{--}100\,\text{MK}$ (see Fig. 8.20). The X-ray luminosity L_X of a plasma due to bremsstrahlung and line emission can be expressed as

$$L_X \approx \text{EM } \Lambda(T) = g \text{ EM } T^\chi, \tag{8.26}$$

with $\chi \approx -0.3$ over the above temperature range (for broadband X-ray losses).

From Eqs. (8.25) and (8.26) we obtain a relation between the flare peak temperature T_0 and its peak luminosity $L_{X,0}$,

$$T_0 = \left(\frac{L_{X,0}}{ag} \right)^{1/(b+\chi)}. \tag{8.27}$$

We will investigate the general case in which τ varies with the flare energy, namely

$$\tau = \tau_0 E^\beta, \tag{8.28}$$

where $\beta \geq 0$ is assumed and τ_0 is a constant adjusted to the larger detected flares) in which case

$$\frac{dN}{dL_{X,0}} = \frac{dN}{dE} \frac{dE}{dL_{X,0}} = k' L_{X,0}^{-(\alpha-\beta)/(1-\beta)}, \tag{8.29}$$

where the constant $k' = k\tau_0^{(1-\alpha)/(1-\beta)}/(1-\beta) > 0$ as long as $\beta < 1$ (which can be reasonably assumed). Since we neglect the short rise time of the flare, our flare light curves are described by their exponential decay at $t \geq 0$,

$$L_X(t) = L_{X,0}e^{-t/\tau}. \tag{8.30}$$

From hydrodynamic modeling, theory, and observations, it is known that during the flare decays $T \propto n_e^\zeta$, where n_e is the plasma density [91]. The parameter ζ is usually found between 0.5 and 2.

Integrating the above equations for all emission contributions from the decaying and cooling flare plasma over the entire distribution of flares leads to two expressions valid for different regimes:

$$Q(T) \propto \begin{cases} T^{2/\zeta} & \text{for } T < T_m \\ T^{-(b+\chi)(\alpha-2\beta)/(1-\beta)+2b+\chi} & \text{for } T > T_m \end{cases}, \tag{8.31}$$

where T_m (a free parameter) is the temperature of the DEM peak. It is controlled by the lower cutoff in the power law energy distribution of flares (for details, see [39]).

8.5.5 Discussion and Summary on the Temperature Structure

There is, as of yet, no clear explanation for the observed DEMs in active stars. The most notable feature is the often steep slope on the low-T side. The steepness is not easy to explain with standard static loop models, although strong footpoint heating may be a way out. Another explanation are magnetic loops with an expanding cross section from the base to the apex. The larger amount of plasma at high temperatures near the apex evidently steepens the DEM. Solar imaging does not prefer this type of loop, however.

Alternatively, the steep slope may be evidence for continual flaring; Eq. (8.31) predicts slopes between 1 and 4, similar to what is often found in magnetically active stars. There is other evidence that this model has some merit – we will encounter it again in the subsequent sections.

It is also well established that more luminous stars (of a given size) reveal hotter coronae (Fig. 8.7). Again, the cause of this relation is unclear. Perhaps increased magnetic activity leads to more numerous interactions between adjacent magnetic field structures. The heating efficiency thus increases. In particular, we expect a higher rate of large flares. The increased flare rate produces higher X-ray luminosity because chromospheric evaporation produces more EM; at the same time, the plasma is heated to higher temperatures in larger flares [46].

8.5.6 Electron Densities in Stellar Coronae

Coronal electron densities control radiative losses from the coronal plasma; observationally, they can in principle also be used in conjunction with EMs to derive approximate coronal source volumes. The spectroscopic derivation of coronal densities is subtle, however. Two principal methods are available.

Densities from Fe line ratios: The emissivities of many transitions of Fe ions in the EUV range are sensitive to densities in the range of interest to coronal research [71]. The different density dependencies of different lines of the same Fe ion then also make their *line-flux ratios*, which (apart from blends) are easy to measure, useful diagnostics for the electron density.

Densities from He-like line triplets: The He-like triplets of C v, N vi, O vii, Ne ix, Mg xi, and Si xiii provide another interesting density diagnostic for stellar coronae. Two examples are shown in Fig. 8.8 (right). The spectra show, in order of increasing wavelength, the resonance, the intercombination, and the forbidden line of the O vii triplet. The corresponding transitions are depicted in the left panel of the figure. The ratio between the fluxes in the forbidden line and the intercombination line is sensitive to density [33] for the following reason: if the electron collision rate is sufficiently high, ions in the upper level of the forbidden transition, $1s2s\ ^3S_1$, do

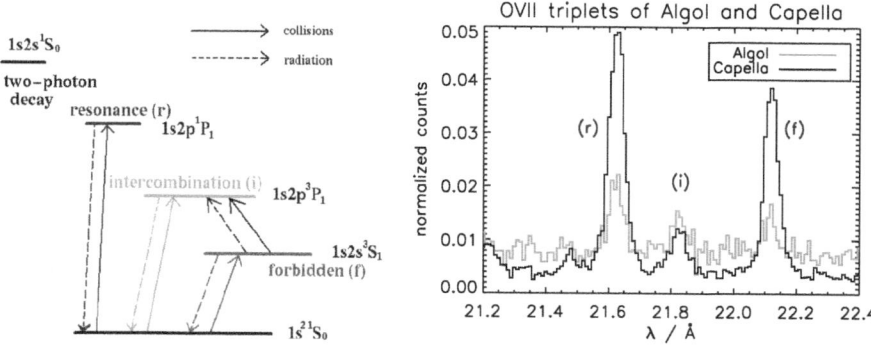

Fig. 8.8 *Left*: Term diagram for transitions in He-like triplets. The resonance, intercombination, and forbidden transitions are marked. The transition from $^3S_1$ to $^3P_1$ re-distributes electrons from the *upper level* of the forbidden transition to the *upper level* of the intercombination transition, thus making the f/i line-flux ratio density sensitive. In the presence of a strong UV field, however, the same transition can be induced by radiation as well. *Right*: He-like triplet of O VII for Capella (*black*) and Algol (*gray*). The resonance (r), intercombination (i), and forbidden (f) lines are marked. The f/i flux ratio of Algol is suppressed probably due to the strong UV radiation field of the primary B star (data from *Chandra*; both figures courtesy of J.-U. Ness)

not return to the ground level, $1s^2\ {}^1S_0$; instead the ions are collisionally excited to the upper levels of the intercombination transitions, $1s2p\ {}^3P_{1,2}$, from where they decay radiatively to the ground state. They thus enhance the flux in the intercombination line and weaken the flux in the forbidden line. The measured ratio $R = f/i$ of the forbidden to the intercombination line flux can be written as

$$R = \frac{R_0}{1 + n_e/N_c} = \frac{f}{i},$$ (8.32)

where R_0 is the limiting flux ratio at low densities and N_c the critical density at which R drops to $R_0/2$. The parameters R_0 and N_c are slightly dependent on the electron temperature in the emitting source. A few useful parameters are given in Table 8.2. A systematic problem with He-like triplets is that the critical density N_c increases with the formation temperature of the ion, i.e., higher Z ions measure only high densities at high T, while the lower density analysis based on C v, N vi, O vii, and Ne ix is applicable only to cool plasma.

A review of the literature shows a rather unexpected segregation of coronal densities into two realms at different temperatures. The cool coronal plasma measured by C, N, and O lines in inactive stars is typically found at low, solar-like densities of order 10^9 cm^{-3} to 10^{10} cm^{-3}. In *active* stars, the cooler components may show elevated densities up to several times 10^{10} cm^{-3}, but it is the hotter plasma component that apparently reveals extreme values up to $>10^{13}$ cm^{-3} [31, 97]. A basic concern with these latter measurements is that most of the reported densities are only slightly above the low-density limits for the respective ratios, and upper limits have equally been reported, sometimes resulting in conflicting statements for different line ratios

Table 8.2 Density-sensitive He-like triplets[a]

Ion	$\lambda(r, i, f)$ (Å)	R_0	N_c	$\log n_e$ range[b]	T range[c] (MK)
C v	40.28/40.71/41.46	11.4	6×10^8	7.7–10	0.5–2
N vi	28.79/29.07/29.53	5.3	5.3×10^9	8.7–10.7	0.7–3
O vii	21.60/21.80/22.10	3.74	3.5×10^{10}	9.5–11.5	1.0–4.0
Ne ix	13.45/13.55/13.70	3.08	8.3×10^{11}	11.0–13.0	2.0–8.0
Mg xi	9.17/9.23/9.31	2.66^d	1.0×10^{13}	12.0–14.0	3.3–13
Si xiii	6.65/6.68/6.74	2.33^d	8.6×10^{13}	13.0–15.0	5.0–20

[a]Data derived from Porquet et al. [88] at maximum formation temperature of ion

[b]Range where R is within approximately [0.1,0.9] times R_0

[c]Range of 0.5–2 times maximum formation temperature of ion

[d]For measurement with *Chandra* HETGS-MEG spectral resolution

in the same spectrum [77, 86, 79]. Several authors have concluded that the extremely high densities found in some active stars are spurious and perhaps not representative of coronal features. The observational situation is clearly unsatisfactory at the time of writing. The resolution of these contradictions requires a careful reconsideration of atomic physics issues.

8.6 The Structure of Stellar Coronae

The magnetic structure of stellar coronae is one of the central topics in the stellar coronal research discipline. The extent and predominant locations of magnetic structures currently hold the key to our understanding of the internal magnetic dynamo. All X-ray inferences of coronal magnetic structure in stars other than the Sun are so far indirect, while direct imaging, although at modest resolution, is available at radio wavelengths.

8.6.1 Magnetic Loop Models

Closed magnetic loops are the fundamental "building blocks" of the solar corona. When interpreting stellar coronae of any kind, we assume that this concept applies as well, although caution is in order. Even in the solar case, loops come in a wide variety of shapes and sizes (Fig. 8.9) and appear to imply heating mechanisms and heating locations that are poorly understood; see, for example, Aschwanden et al. [7]. Nevertheless, simplified loop models offer an important starting point for coronal structure studies and possibly for coronal heating diagnostics. A short summary of some elementary properties follows.

Rosner et al. [92] (RTV) have modeled hydrostatic loops with constant pressure (i.e., the loop height is smaller than the pressure scale height). They also assumed constant cross section, uniform heating, and absence of gravity, and found two scaling laws relating the loop semi-length L (in cm), the volumetric heating rate σ (in erg cm^{-3} s^{-1}), the pressure p (in dynes cm^{-2}), and the loop apex temperature T_a (in K),

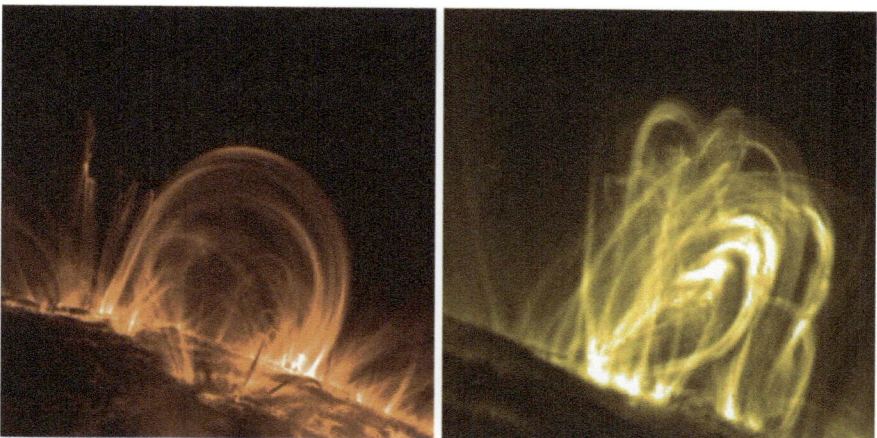

Fig. 8.9 *Left*: Example of a solar coronal loop system observed by *TRACE*. *Right*: Flaring loop system (observation by *TRACE* at 171Å). Although these images show the emission from relatively cool coronal plasma, they illustrate the possible complexity of magnetic fields

$$T_a = 1400(pL)^{1/3}, \qquad \sigma = 9.8 \times 10^4 p^{7/6} L^{-5/6}. \qquad (8.33)$$

Serio et al. [98] extended these scaling laws to loops exceeding the pressure scale height s_p, whereby, however, the limiting height at which the loops grow unstable is $(2–3)s_p$:

$$T_a = 1400(pL)^{1/3} e^{-0.04L(2/s_H + 1/s_p)}, \qquad \sigma = 10^5 p^{7/6} L^{-5/6} e^{0.5L(1/s_H - 1/s_p)}, \qquad (8.34)$$

where s_H is the heat deposition scale height. For loops with an area expansion factor $\Gamma > 1$, Vesecky et al. [114] found numerical solutions that approximately follow the scaling laws given by Schrijver et al. [96]:

$$T_a \approx 1400\Gamma^{-0.1}(pL)^{1/3}, \qquad T_a = 60\Gamma^{-0.1} L^{4/7} \sigma^{2/7}. \qquad (8.35)$$

There are serious disagreements between some *solar*-loop observations and the RTV formalism so long as simplified quasi-static heating laws are assumed, the loops being more isothermal than predicted by the models. There is, however, only limited understanding of possible remedies, such as heating that is strongly concentrated at the loop footpoints or dynamical processes in the loops (see, for example, a summary of this debate in [95]).

8.6.2 Coronal Structure from Loop Models

When we interpret stellar coronal spectra, we assume, to first order, that some physical loop parameters map on our measured quantities, such as temperature and EM (and possibly density), in a straightforward way. In the simplest approach, we assume that the observed luminosity L_X is produced by an ensemble of identical coronal loops with characteristic half-length L, surface filling factor f, and an apex temperature T used for the entire loop; then, on using Eq. (8.33) and identifying $L_X = \sigma V$, we obtain

$$L \approx 6 \times 10^{16} \left(\frac{R_*}{R_\odot} \right)^2 \frac{f}{L_X} T^{3.5} \quad \text{[cm]}. \tag{8.36}$$

This relation can only hold if L is smaller than the pressure scale height. As an example, for an active solar analog ($R = R_\odot$, $L_X = 10^{30}$ erg s^{-1}, $T = 10^7$ K) we obtain $L \approx 2 \times 10^{11} f$ cm. The coronal volume is approximately $V \approx 8R^2 fL$ and the electron density $n_e = (EM/V)^{1/2}$. Further, $L_X \approx 2 \times 10^{-23} EM$ erg s^{-1}. For the solar analog, thus, we find $n_e f \approx 2.5 \times 10^9$ cm^{-3}. The luminous, hot plasma component in magnetically active stars therefore seems to invariably require either very large, moderate-pressure loops with a large filling factor or solar-sized high-pressure compact loops with a very small ($\lesssim 1\%$) filling factor.

While the above interpretational work identifies spectral-fit parameters such as T or EM with parameters of theoretical loop models, a physically more appealing approach involves full hydrostatic models whose calculated emission spectra are directly fitted to the observations.

Such studies [104, 35, 67, 113] have found that the cooler component at ≈ 1–2 MK requires loops of small length ($L \ll R_*$) but high pressure ($p > p_\odot$), whereas the high-T component at ≈ 5–10 MK must be confined by very compact loops with extremely high base pressures (up to hundreds of dynes cm^{-2}) and small ($<1\%$) filling factors. These parameters are suspiciously "flare-like" – the observed hot plasma is perhaps indeed related to multiple, very compact flaring regions.

The most essential conclusion from these exercises is perhaps that, within the framework of such simplistic models, the loop heating rate required for magnetically active stars may exceed values for typical solar loops by orders of magnitude, pointing toward some enhanced heating process reminiscent of the energy deposition in flares. The compactness of the hot loops and the consequent high pressures also set these coronal structures clearly apart from any non-flaring solar coronal features.

8.6.3 Coronal Structure from Densities

Spectroscopically measured densities provide, in conjunction with the EM, important estimates of emitting volumes. If the trend suggested from density-sensitive line-flux ratios holds, namely that for increasing temperature, the pressures become progressively higher, then progressively smaller volumes are a consequence. The volume required for a luminosity of 10^{30} erg s^{-1} at 10 MK and with $n_e = 10^{13}$ cm^{-3} is $V = L_X/(2 \times 10^{-23} n_e^2)$ cm$^3 = 5 \times 10^{26}$ cm^3 (where the coefficient 2×10^{-23} is from Eq. (8.5), appropriate for $T = 10$ MK), corresponding to a layer of only

80 m height around a solar-like star, or 8 km for a filling factor of only 1%! Such scales are much smaller than chromospheric scale heights and therefore problematic. Still smaller filling factors must be assumed for a star of this kind. Similarly, from the RTV loop scaling law in Eq. (8.33), if applicable, a loop height $h = 2L/\pi = 8.5 \times 10^5 T^2/n_e \approx 80$ km is found, again an unreasonably small size.

The confinement of plasma at such high densities in compact sources would also require coronal magnetic field strengths of order $B > (16\pi n_e kT)^{1/2} \approx 1$ kG, i.e., field strengths like those very close to and just above (sun-)spots. In that case, the typical magnetic dissipation time is only a few seconds for $n_e \approx 10^{13}$ cm^{-3} if the energy is derived from the same magnetic fields, suggesting that the small, bright loops light up only briefly. In other words, the stellar corona would be made up of numerous ephemeral loop sources that cannot be treated as being in a quasi-static equilibrium [112].

8.6.4 X-ray Coronal Imaging: Overview

X-ray images of stellar coronae have been derived from eclipses in binaries or from rotational modulation in rapidly rotating stars. We keep in mind that any indirect imaging of this kind is highly biased by observational constraints (e.g., the volume that is subject to eclipses or self-eclipses or the accessible temperature range) and by the amount and density of plasma trapped in the magnetic fields. X-ray imaging captures strongly emitting plasma, *not* the entire magnetic field structure.

The "image" to be reconstructed consists of volume elements at coordinates (x, y, z) with *optically thin* fluxes $f(x, y, z)$ assumed to be constant in time. In the special case of negligible stellar rotation during the observation, the problem can be reduced to a 2-D projection onto the plane of the sky, at the cost of positional information along the line of sight (Fig. 8.10). In general, one thus seeks the geometric brightness distribution $f(x, y, z) = f_{ijk}$ (i, j, k) being the discrete number indices of the volume elements from a binned, observed light curve $F_s = F(t_s)$ that undergoes a modulation due to an eclipse or due to rotation.

8.6.5 Active-Region Modeling

In the most basic approach, the emitting X-ray or radio corona can be modeled by making use of a small number of simple, elementary building blocks that are essentially described by their size, their brightness, and their location. This approach is the 3-D equivalent to standard surface spot modeling. Preferred building block shapes are radially directed, uniformly bright, optically thin, radially truncated spherical cones with their apexes at the stellar center. Free parameters are their opening angles, their heights above the stellar surface, their radiances, and their central latitudes and longitudes. These parameters are then varied until the model fits the observed light curve.

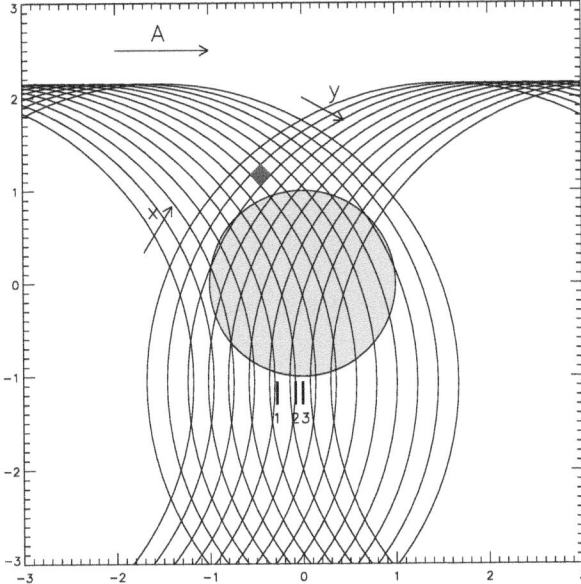

Fig. 8.10 Sketch showing the geometry of an eclipsing binary (in this case, α CrB, figure from [38]). The *large circles* illustrate the limbs of the eclipsing star that moves from *left* to *right* in front of the eclipsed star (shown in *light gray*). The limbs projected at different times during ingress and egress define a distorted 2-D array (x, y) of pixels (an example of a pixel is shown in *dark gray*)

A minimum solution can be found for a rotationally modulated star [47]. If a rotationally modulated feature is *invisible* during a phase interval φ of the stellar rotation, then all sources contributing to this feature must be confined to within a maximum volume, V_{max}, given by

$$\frac{V_{\mathrm{max}}}{R_*^3} = \frac{\psi}{3} - \frac{(2\pi - \varphi)(1 + \sin^2 i)}{6\sin i} + \frac{2\cot(\chi/2)}{3\tan i}, \qquad (8.37)$$

where $\tan\chi = \tan(\varphi/2)\cos i$, $\sin(\psi/2) = \sin(\varphi/2)\sin i$ with $0 \leq \psi/2 \leq \pi/2$, and χ and $\varphi/2$ lie in the same quadrant (i is the stellar inclination, $0 \leq i \leq \pi/2$). Together with the modulated fraction of the luminosity, lower limits to average electron densities in the modulated region follow directly.

8.6.6 Maximum Entropy Image Reconstruction

Maximum entropy methods (MEMs) are applicable both to rotationally modulated light curves and to eclipse observations. The standard MEM selects among all images f_{ijk} (defined in units of counts per volume element) that are compatible with the observation, the one that minimizes the Kullback contrast (relative entropy)

$$K = \sum_{i,j,k} f_{ijk} \ln \frac{f_{ijk}}{f_{ijk}^a}, \qquad (8.38)$$

with respect to an a priori image f_{ijk}^a, which is usually unity inside the allowed area or volume and vanishes where no brightness is admitted. Minimizing K thus introduces the least possible information while being compatible with the observation. The contrast K is minimum if f_{ijk} is proportional to f_{ijk}^a and thus flat inside the field of view, and it is maximum if the whole flux is concentrated in a single pixel (i, j, k). The compatibility with the observed count light curve is measured by χ^2,

$$\chi^2 = \sum_s \frac{(F_s^* - F_s)^2}{F_s^*}, \tag{8.39}$$

where F_s and F_s^* are, respectively, the observed number of counts and the number of counts predicted from f_{ijk} and the eclipse geometry. Poisson statistics usually requires more than 15 counts per bin. Finally, normalization is enforced by means of the constraint

$$N = \frac{\left(f^{\text{tot}} - \sum_{ijk} f_{ijk} \right)^2}{f^{\text{tot}}}, \tag{8.40}$$

where f^{tot} is the sum of all fluxes in the model.

The final algorithm minimizes the cost function

$$C = \chi^2 + \xi K + \eta N . \tag{8.41}$$

The trade-off between the compatibility with the observation, normalization, and unbiasedness is determined by the Lagrange multipliers ξ and η such that the reduced χ^2 is $\lesssim 1$, and normalization holds within a few percent.

8.6.7 Lucy/Withbroe Image Reconstruction

This method (after Refs. [66, 121]) iteratively adjusts fluxes in a given set of volume elements based on the mismatch between the model and the observed light curves in all time bins to which the volume element contributes. At any given time t_s during the eclipse, the observed flux F_s is the sum of the fluxes f_{ijk} from all volume elements that are unocculted:

$$F(t_s) = \sum_{i,j,k} f_{ijk} \mathbf{m}_s(i, j, k), \tag{8.42}$$

where $\mathbf{m}_s$ is the "occultation matrix" for the time t_s: it puts, for any given time t_s, a weight of unity to all visible volume elements and zero to all invisible elements (and intermediate values for partially occulted elements). Since F_s is given, one needs to solve Eq. (8.42) for the flux distribution, which is done iteratively as follows:

$$f_{ijk}^{n+1} = f_{ijk}^{n} \frac{\sum_{s} \dfrac{F_o(t_s)}{F_m^n(t_s)} \mathbf{m}_s(i,j,k)}{\sum_{s} \mathbf{m}_s(i,j,k)}, \tag{8.43}$$

where $F_o(t_s)$ and $F_m^n(t_s)$ are, respectively, the observed flux and the model flux (or counts) in the bin at time t_s, both for the iteration step n. Initially, a plausible, smooth distribution of flux is assumed, e.g., constant brightness or some r^{-p} radial dependence.

8.6.8 Backprojection and Clean Image Reconstruction

If rotation can be neglected during an eclipse, for example, in long-period detached binaries, then the limb of the eclipsing star is projected at regular time intervals onto the plane of the sky and therefore onto a specific part of the eclipsed corona, first during ingress, later during egress [38]. The two limb sets define a 2-D grid of distorted, curved pixels (Fig. 8.10). The brightness decrement during ingress or, respectively, the brightness increment during egress within a time step $[t_s, t_{s+1}]$ originates from within a region confined by the two respective limb projections at t_s and t_{s+1}. Ingress and egress thus each define a 1-D image by backprojection from the light curve gradients onto the plane of the sky. The relevant reconstruction problem from multiple geometric projections is known in tomography. The limiting case of only two independent projections can be augmented by a CLEAN step, as follows. The pixel with the largest *sum* of projected fluxes from ingress and egress is assumed to represent the location of a real source. A fraction, $g < 1$, of this source flux is then subtracted from the two projections and saved on a clean map, and the process is iterated until all flux is transferred onto the latter.

8.6.9 X-ray Coronal Structure Inferred from Eclipses

8.6.9.1 Extent of Eclipsed Coronal Features

Some shallow X-ray eclipses in tidally interacting binary systems of the RS CVn, Algol, or BY Dra type have provided important information on extended coronal structure. For example, Walter et al. [115] concluded that the coronae in the AR Lac binary components are bi-modal in size, consisting of compact, high-pressure (i.e., 50–100 dynes cm^{-2}) active regions with a scale height $<R_*$, while the subgiant K star is additionally surrounded by an extended $(2.7R_*)$ low-pressure corona. Further, a hot component pervading the entire binary system was implied from the absence of an eclipse in the hard ME detector on *EXOSAT* [117], and similar conclusions have been drawn from detailed light curve inversion analysis [89]; see Fig. 8.11.

X-ray dips or any periodic modulation have often been absent in X-ray observations of binaries. This again has been taken as evidence of a very extended (about

Fig. 8.11 Two examples of eclipses and the corresponding coronal image reconstructions. From *top* to *bottom*: Light curve of the YY Gem system (from [40], observation with *XMM-Newton* EPIC); light curve of the AR Lac system (after [102], observation with *ASCA* SIS); reconstructed image of the coronal structure of, respectively, YY Gem (at phase 0.375) and AR Lac (at quadrature). The latter figure shows a solution with intrabinary emission. (The light curve of AR Lac is phase-folded; the actual observation started around phase 0; data and image for AR Lac courtesy of M. Siarkowski.)

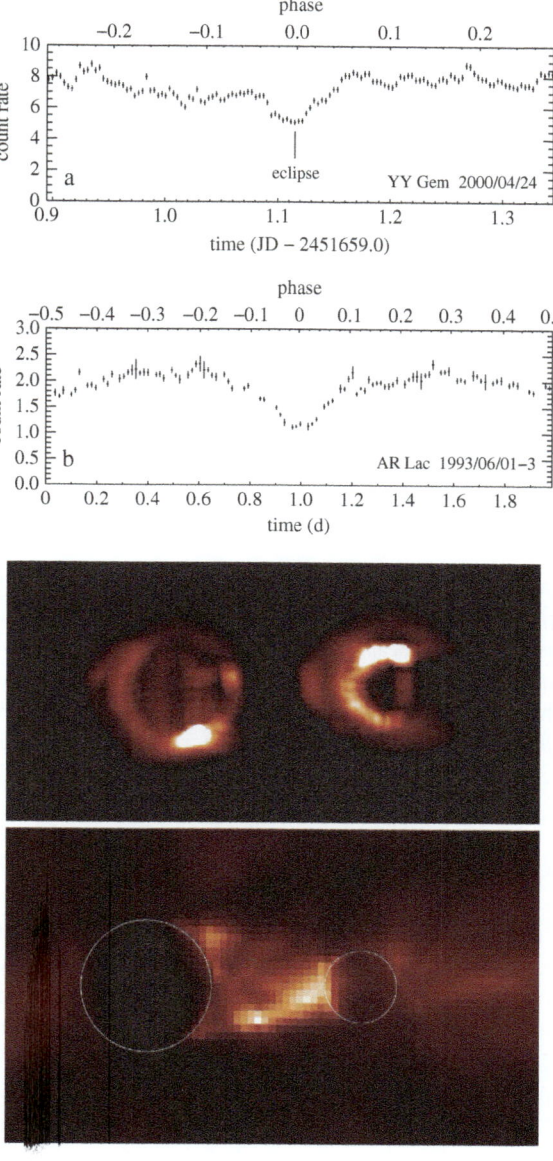

$1R_*$ in the case of Algol [116]) X-ray corona unless more compact structures sit at high latitudes where they remain uneclipsed.

Among wide, non-interacting eclipsing stars, α CrB provides a particularly well-suited example because its X-ray active, young solar analog (G5V) is totally eclipsed every 17 days by the optical primary, an A0 V star that is perfectly X-ray dark. Other parameters are ideal as well, such as the non-central eclipse, the eclipse timescale of a few hours, and the relatively slow rotation period of the secondary.

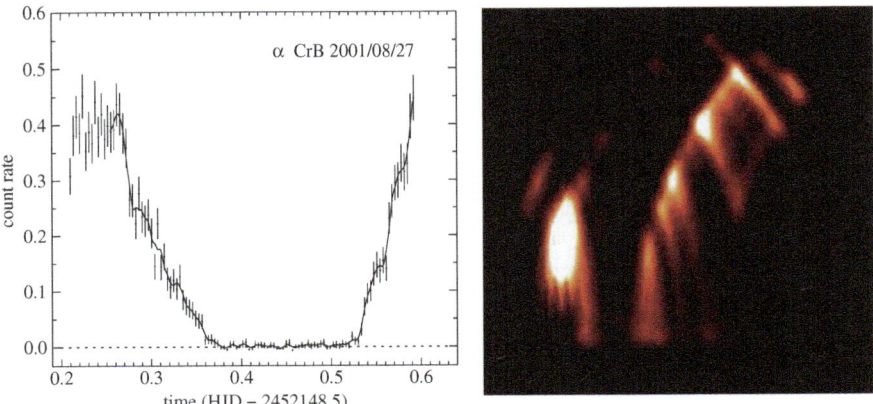

Fig. 8.12 Light curve and image reconstruction of the A+G binary α CrB. The *left panel* shows the light curve from observations with *XMM-Newton* and the *right panel* illustrates the reconstructed X-ray brightness distribution on the G star (after [38])

Image reconstructions from eclipse observations [38] reveal patches of active regions across the face of the G star; not much material is found significantly beyond its limb (Fig. 8.12). The structures tend to be of modest size ($\approx 5 \times 10^9$ cm), with large, X-ray faint areas in between, although the star's luminosity exceeds that of the active Sun by a factor of ≈ 30. These observations imply moderately high densities in the emitting active regions, reaching a few 10^{10} cm^{-3} in the brightest active regions.

8.6.10 Eclipsed X-ray Flares

Eclipses of flares have contributed very valuable information on densities and the geometric size of flaring structures. Only few reports are available, among them the following: Choi and Dotani [22] described a full eclipse of an X-ray flare in progress in the contact binary system VW Cep. During a narrow dip in the flare decay, the X-ray flux returned essentially to the pre-flare level. Geometric considerations then placed the flare near one of the poles of the primary star, with a size scale of order 5.5×10^{10} cm or somewhat smaller than the secondary star. The authors consequently inferred an electron density of 5×10^{10} cm^{-3}. A polar location was also advocated for a flare on Algol observed across an eclipse by Schmitt and Favata [93]. The flare emission was again eclipsed completely, and judged from the known system geometry, the flare was located above one of the poles, with a maximum source height of no more than approximately $0.5R_*$, implying a minimum electron density of 9.4×10^{10} cm^{-3} if the volume filling factor was unity. A more moderate flare was observed during an eclipse in the Algol system by Schmitt et al. [94]. In this case, the image reconstruction required an equatorial location, with a compact flare source of height $h \approx 0.1R_*$. Most of the source volume exceeded densities of

10^{11} cm^{-3}, with the highest values at $\approx 2 \times 10^{11}$ cm^{-3}. Because the *quiescent* flux level was attained throughout the flare eclipse, the authors argued that its source, in turn, must be concentrated near the polar region with a modest filling factor of $f < 0.1$ and electron densities of $\approx 3 \times 10^{10}$ cm^{-3}.

8.6.11 Radio Very Long Baseline Interferometry

By interferometrically combining radio telescopes over large distances, angular resolutions of as little as 1 milliarcsecond (mas) can be achieved in the microwave range. VLBI techniques have been very demanding for single late-type dwarf stars, owing both to low flux levels and small coronal sizes. Some observations with mas angular resolutions show unresolved quiescent or flaring sources, thus constraining the brightness temperature to $T_b > 10^{10}$ K (e.g., [17]), whereas others show evidence for extended coronae with coronal sizes up to several times the stellar size.

The dMe star UV Cet was found to be surrounded by a pair of giant synchrotron lobes, with sizes up to 2.4×10^{10} cm and a separation of $4-5$ stellar radii along the putative rotation axis of the star, suggesting very extended magnetic structures above the magnetic poles (Fig. 8.13a), perhaps arranged in a global dipole as sketched in Fig. 8.15b. I discuss this observation in some detail, following Benz et al. [18], to demonstrate the procedures with which we can characterize the magnetic field structure. Throughout, we assume, as detailed further above, that the emission is optically thin gyrosynchrotron emission from a power law population of accelerated electrons with a number density distribution in energy ε

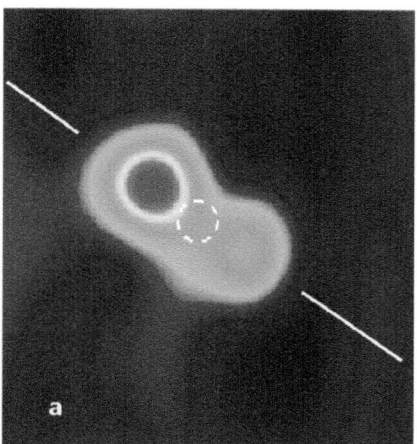
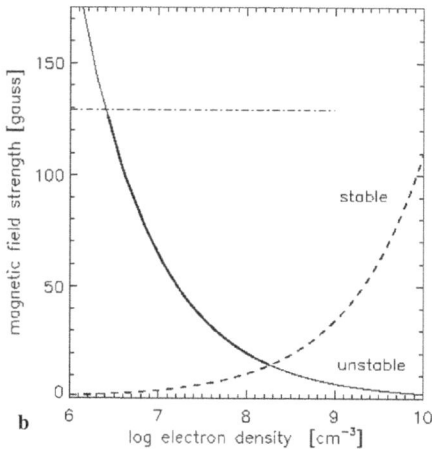

Fig. 8.13 (a) VLBA image of the dMe star UV Cet; the two radio lobes are separated by about 1.4 mas, while the best angular resolution reaches 0.7 mas. The *straight line* shows the orientation of the putative rotation axis, assumed to be parallel to the axis of the orbit of UV Cet around the nearby Gl 65 A. The *small circle* gives the photospheric diameter to size, although the precise position is unknown (after [18]). (b) Estimates of the magnetic field and the non-thermal electron density for the same observation (see text for details)

$$n(\varepsilon) = N(\delta - 1)\varepsilon_0^{\delta-1}\varepsilon^{-\delta} \quad [\text{cm}^{-3}\ \text{erg}^{-1}], \tag{8.44}$$

where $\varepsilon = (\gamma - 1)m_e c^2$ is the kinetic particle energy, γ the Lorentz factor, and $\delta > 1$ has been assumed so that N is the total non-thermal electron number density above ε_0.

We first need an expression for the emissivity: for isotropic pitch angle electron distributions according to Eq. (8.44) with $2 \lesssim \delta \lesssim 7$, for harmonics $10 \lesssim s \lesssim 100$, and for the x-mode [30]

$$\eta_\nu \approx 3.3 \times 10^{-24} BN\ 10^{-0.52\delta}(\sin\theta)^{-0.43+0.65\delta}\left(\frac{\nu}{\nu_c}\right)^{1.22-0.90\delta} \tag{8.45}$$

(in erg s^{-1} cm^{-3} Hz^{-1} sterad^{-1}). Here θ denotes the emission angle to the magnetic field and $\nu_c = eB/m_e c$ is the electron gyrofrequency. The coefficient of this expression is applicable if a power law cutoff has been set at 10 keV. This cutoff itself is of little relevance for gyrosynchrotron emission since electrons radiate weakly at such low energies.

We assume $\theta = \pi/2$ and $\delta \approx 2.5$ (the latter from modeling of M dwarf spectra, Sect. 8.4.3). Adding o- and x-mode (which have similar emissivities), the intensity is

$$I \approx 1.8\eta_\nu D, \tag{8.46}$$

which is to be compared with the observed intensity of 3.9×10^{-8} erg^{-1} cm^{-2} Hz^{-1} sterad^{-1}. Here, the fitted diameter of one of the radio blobs (0.3 mas or $D = 1.2 \times 10^{10}$ cm) was used. Combining Eqs. (8.45) and (8.46), we find

$$B \approx 2 \times 10^5 N^{-1/2}. \tag{8.47}$$

This relation is shown in Fig. 8.13b together with the condition that the particle pressure in the magnetic loops is less than the magnetic pressure,

$$N\bar{\varepsilon} \leq \frac{B^2}{8\pi}. \tag{8.48}$$

We find a lower limit to B of approximately 15 G and an upper limit to N of about 2×10^8 cm^{-3}.

At least part of the observed emission was slowly decaying, with a decay timescale of $\tau = 6650$ s (perhaps from a flare that filled the magnetospheric volume with electrons). Given the large size of the structure, the dominant energy decay process is likely to be due to synchrotron radiation loss given in Eq. (8.12). The average frequency of synchrotron emission can be expressed as

$$\bar{\nu} = 1.3 \times 10^6 B\gamma^2 \quad [\text{Hz}] \tag{8.49}$$

[70]. Using the observing frequency (8.4 GHz in this case) for $\bar{\nu}$, we find the decay time

$$\tau = \frac{\gamma}{\dot{\gamma}} \approx 8 \times 10^6 B^{-3/2} \quad [\text{s}], \tag{8.50}$$

and therefore $B = 113$ G. This defines an upper limit to the magnetic field strength because we have ignored other energy losses that might be present. The upper limit is also drawn in Fig. 8.13b. We have thus confined the magnetic field strength in the source to 15–130 G.

VLBA imaging and polarimetry of Algol reveal a similar picture with two oppositely polarized radio lobes separated along a line perpendicular to the orbital plane by more than the diameter of the K star ([75], Figs. 8.14 and 8.15b). Large-scale polarization structure further supports models that assume globally organized magnetic fields around active binary stars [16].

An important, early VLBI result for RS CVn and Algol-like binaries is evidence for a compact core plus extended halo radio structure of a total size that is comparable to the binary system size [74]. The basic idea here is the following: during quiescence, an optically thin, very large magnetosphere is filled with a power law distribution of electrons. The emission is essentially optically thin, with a flat radio spectrum. During an outburst, an active region injects a larger electron population into magnetic loops. The compact source (the "core"), unresolved by VLBI, becomes optically thick due to synchrotron self-absorption, and the radio spectral index consequently becomes positive, and the brightness temperature is thus equal to the effective electron temperature. Mutel et al. [74] measured a few times 10^{10} K

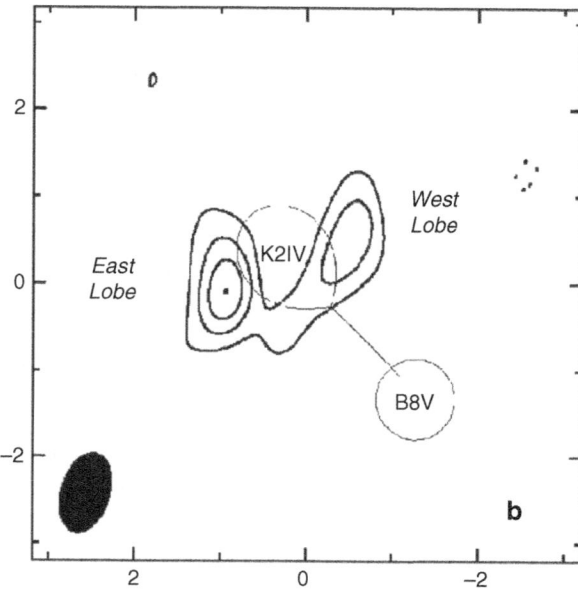

Fig. 8.14 VLBA observation of Algol at 8.4 GHz, resolving two lobes around the binary. The most likely configuration of the binary components is also drawn (after [75])

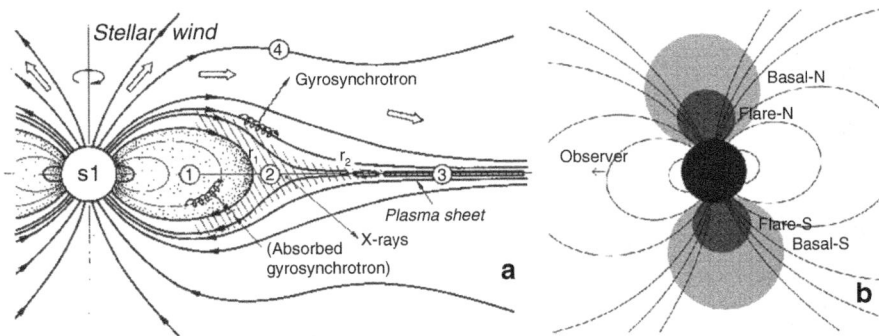

Fig. 8.15 (a) Equatorial model for the magnetosphere of the young B star S1 in ρ Oph [4]. (b) Sketch for radio emission from a global dipole consistent with the VLBI observation shown in Fig. 8.13b ([75]; reproduced with permission of the AAS)

for outbursts on RS CVn-type binaries. Once the electron injection has ceased, the lifetime of the energetic electron population is determined essentially by the synchrotron loss time. The final phase consists of magnetic loop expansion due to buoyancy, expanding the source until it merges with the pre-flare "halo" component. At the same time, the source becomes progressively more optically thin, developing a negative radio spectral index and mild circular polarization. Given the large size, the strength of the extended magnetic field must be rather moderate, of order 10 G. To emit radiation at frequency ν in the microwave range, high Lorentz factors are required. Because the spectral power for synchrotron emission is predominantly emitted at a harmonic of order γ^3 of the relativistic gyrofrequency

$$\nu_c = \frac{eB}{m_e \gamma c} = 2.8 \times 10^6 \frac{B}{\gamma},\qquad(8.51)$$

we require $\gamma \approx 6 \times 10^{-4}(\nu/B)^{1/2}$. For an observing frequency of 5 GHz, $\gamma \approx 10$. A model calculation of the spectral development of the microwave radiation was outlined in Sect. 8.4.3.

8.6.12 Radio Magnetospheric Models

VLBI observations of RS CVn and Algol binaries, T Tau stars, and magnetic Bp/Ap stars have shown some perplexing structures with sizes at least as large as the binary system, with polarization properties that suggest that the magnetic fields are globally ordered. This has led to a series of large-scale magnetic models for such systems. They have in common a global, dipole-like structure somewhat resembling the Earth's Van Allen belts (Fig. 8.15a). Stellar winds escaping along magnetic fields draw the field lines into a current sheet configuration in the equatorial plane. Particles can be accelerated in that region. They subsequently travel back to become trapped in the dipolar-like, equatorial magnetospheric cavity. Variants of this radiation-belt model, partly based on theoretical work of Havnes and Goertz [48], have been applied to RS CVn binaries (e.g., [73]), in an optically thick version to Bp/Ap stars [29, 63] and in an optically thin version to a young B star [4].

Polarization observations of RS CVn binaries support these models. The polarization degree seems to be anticorrelated with the luminosity for any given system, but the sense of polarization changes between lower and higher frequencies. For the entire binary sample, the polarization degree is inversely correlated with the stellar inclination angle such that low-inclination (pole-on) systems show the strongest polarization degrees, as expected for such global systems [76, 75, 73].

Flares still appear to originate in compact sources within such magnetospheres, probably close to the star. The "core plus halo" model then correctly describes the radio spectral properties – the halo corresponds to the extended magnetosphere. This is also supported by magnetic fields inferred to be stronger ($80-200$ G) in the flaring core and weaker ($10-30$ G) in the halo [109, 107, 75, 106]. Because the optical depth is frequency-dependent, the source is small at high frequencies (with a size $\approx R_*$, above 10 GHz) and large at small frequencies (with a size comparable to the binary system size at 1.4 GHz; [54, 52]). This effect explains the relatively flat, optically thick radio spectra seen during flares.

8.6.13 Extended or Compact Coronae?

As the previous discussions imply, we are confronted with mixed evidence for predominantly extended (source height $>R_*$) and predominantly compact ($\ll R_*$) coronal structures or a mixture thereof, results that variably come from radio or from X-ray astronomical observations. There does not seem to be unequivocal agreement on the type of structure that generally prevails. Several trends can be recognized, however, as summarized below.

Compact coronal structure. Steep (portions of the) ingress and egress light curves or prominent rotational modulation unambiguously argue in favor of short scale lengths perpendicular to the line of sight. Common to all are relatively high inferred densities ($\approx 10^{10}$ cm^{-3}). The pressures of such active regions may exceed pressures of non-flaring solar active regions by up to 2 orders of magnitude. Spectroscopic observations of high densities and loop modeling add further evidence for the presence of some rather compact sources. Flare modeling also provides modest sizes, often of order $0.1-1 R_*$, for the involved magnetic loops (Sect. 8.8).

Extended structure. Here, the arguments are less direct and are usually based on the absence of deep eclipses or very shallow ingress and egress curves. Caution is in order in cases where the sources may be located near one of the polar regions; in those cases, eclipses and rotational modulation may also be absent regardless of the source size. Complementary information is available from flare analysis (see Sect. 8.8) that in some cases does suggest quite large loops. The caveat here is that simple single-loop models may not apply to such flares. Clear evidence is available from radio interferometry that proves the presence of large-scale, globally ordered magnetic fields. The existence of prominent extended, closed magnetic fields on scales $> R_*$ is therefore *also* beyond doubt for several active stars.

The most likely answer to the question on coronal structure size is therefore an equivocal one: coronal magnetic structures follow a size distribution from very compact to extended ($\gtrsim R_*$) with various characteristic densities, temperatures, non-thermal electron densities, and surface locations. Quite different loop systems may be responsible when measuring cool X-ray lines or hot lines, or when observing in the microwave range. This is no different from what we see on the Sun even though various features observed on magnetically active stars stretch the comparison perhaps rather too far for comfort.

8.7 Stellar Radio Flares

Flares arise as a consequence of a sudden energy release and relaxation process of the magnetic field in solar and stellar coronae. Present-day models assume that the energy is accumulated and stored in non-potential magnetic fields prior to an instability that most likely implies reconnection of neighboring antiparallel magnetic fields. The energy is brought into the corona by turbulent footpoint motions that tangle the field lines at larger heights. The explosive energy release becomes measurable across the electromagnetic spectrum and, in the solar case, as high-energy particles in interplanetary space as well.

Flares are ubiquitous among coronal stars of all types, with very few exceptions. They have, of course, prominently figured in solar studies, and it is once again solar physics that has paved the way to the interpretation of stellar flares, even if not all features are fully understood yet. The complexity that flares reveal to the solar astronomer is inaccessible in stellar flares, especially in the absence of spatially resolved observations. Simplified concepts, perhaps tested for solar examples, must suffice. The following sections summarize the "stellar astronomer's way" of looking at flares.

8.7.1 Incoherent Radio Bursts

Active stars reveal two principal flare types at radio wavelengths, similar to the solar case: incoherent flares evolve on timescales of minutes to hours; they show broadband spectra and moderate degrees of polarization. These bursts are fully equivalent to solar microwave bursts. Like the latter, they show evidence for the presence of mildly relativistic electrons. The emission is therefore interpreted as gyrosynchrotron radiation from coronal magnetic fields. Many flares on single F/G/K stars are of this type, as are almost all radio flares on M dwarfs above 5 GHz [12] or on RS CVn binaries [74].

8.7.2 Coherent Radio Bursts

The second type of stellar equivalents to coherent solar radio bursts (showing high brightness temperature, short durations, perhaps small bandwidth, and perhaps high polarization degree) has been observed already in the early days of radio astronomy, given their sometimes extremely high fluxes. As for the Sun, they come in a bewildering variety which has made a clear identification of the ultimate cause difficult [13]. Coherent bursts are frequent on late-type main-sequence stars, but have also been reported from RS CVn binaries [118].

Such bursts carry important information in high-time resolution light curves. Radio "spike" rise times as short as 5–20 ms have been reported, requiring, from the light-time argument, source sizes of $R < c \Delta t \approx 1500$–6000 km. With

$$S = \frac{2k T_b \nu^2}{c^2} \frac{\pi R^2}{d^2} \tag{8.52}$$

and measured fluxes up to 1 mJy for sources at a distance of a few parsec, we derive brightness temperatures of order $T_b \approx 10^{16}$ K, a clear proof of the presence of a coherent mechanism [59, 60, 45, 15].

8.7.3 Radio Dynamic Spectra

The standard means to study solar coherent bursts are "dynamic spectra" (flux vs. frequency and time). If the elementary frequencies relevant for the coherent emission process, ν_p and ν_c, evolve in the source, or if the radiating source itself travels across density or magnetic field gradients, the emission leaves characteristic traces on the dynamic spectrum [13]. The study of drifts, decay times, harmonic structures, etc., may then help identify the emission process and thus infer magnetic field strengths, electron densities, electron energies, beam velocities, etc.

Applying the same technology to stellar observations has turned out to be extremely challenging, and only few successful dynamic spectra have been recorded to date (Fig. 8.16). A rich phenomenology has been uncovered, including: (a) short, highly polarized bursts with structures as narrow as $\Delta \nu / \nu = 0.2\%$, suggesting either plasma emission from a source of size $\sim 3 \times 10^8$ cm or a cyclotron maser in magnetic fields of ~ 250 G [14, 45, 15]; (b) evidence for spectral structure with positive drift rates of 2 MHz s^{-1} around 20 cm wavelength, taken as evidence for a disturbance propagating "downward" in the plasma emission interpretation [49]; and (c) in solar terminology, rapid broadband pulsations, "sudden (flux) reductions", and positive and negative drift rates of 250–1000 MHz s^{-1} [15, 2, 1].

Spectral bandwidths that are 1% of the emitting frequency were found for some bursts. If we conservatively assume a magnetic scale height of $L_B = 1 R_*$ (assuming that the emission is gyromagnetic), the source size can be estimated to be

$$r \approx \frac{\Delta \nu}{\nu} L_B, \tag{8.53}$$

which is $\approx$ a few 1000 km for $L_B = R_*$ of an M dwarf, again implying very high T_b, compatible with the above light-time argument [61, 45, 15].

Recent developments have permitted the recording of quite broad regions of the stellar burst spectrum, allowing a much better characterization of the burst types.

Fig. 8.16 Radio dynamic
spectra of M dwarf flares.
The *upper three panels* show
a flare on AD Leo, recorded
with the Arecibo (*top*),
Effelsberg (*middle*), and
Jodrell Bank (*bottom*)
telescopes in different
wavelength ranges (see also
[45]). The *bottom three
panels* show flares on AD
Leo (*top* and *middle*) and YZ
CMi (*bottom*) observed at
Arecibo (after [15],
reproduced with permission
of the AAS)

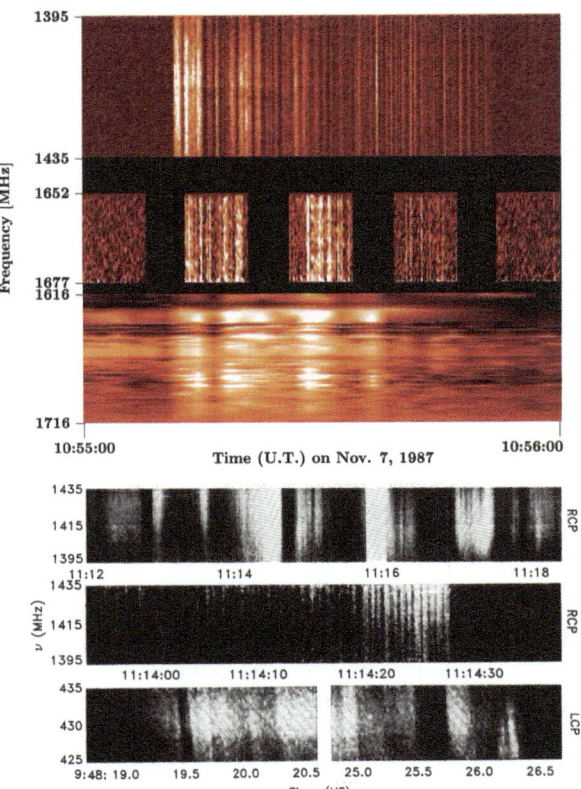

Osten and Bastian [78] obtained burst spectra from AD Leo recorded in the 1120–
1620 MHz range with a time resolution of 10 ms and a spectral resolution of
0.78 MHz. They find significant frequency–time drifts in short subbursts (Fig. 8.17).
The main characteristics of these structures are (i) durations of about 30 ms, (ii)
high polarization (>90%), frequency bandwidths of $\Delta\nu/\nu\approx5\%$, and inverse drift
rates (time interval per drift in frequency) that are symmetric around zero, with a
characteristic width corresponding to a frequency drift of 2.2 GHz s^{-1}. These characteristics are very close to those of narrowband *solar* decimetric "spike bursts" but
set them apart from type III bursts associated with electron beams in the solar corona
(type III bursts have considerably longer durations, occupy broader frequency intervals, and are at best moderately polarized).

There is one significant difference between the stellar and solar bursts: the duration of the solar examples is about three times shorter. We can relate this to Coulomb
collisional damping [43, 78] which is determined by the electron–ion collision time
τ_{ei}, from which we find

$$T = 8100\nu_{\text{MHz}}^{4/3}\tau^{2/3}. \tag{8.54}$$

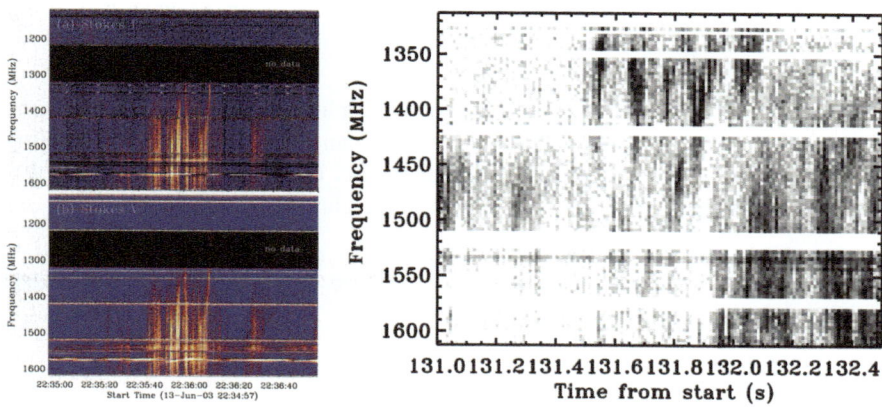

Fig. 8.17 Dynamic spectra of radio bursts observed in AD Leo [78]. The *left figure* shows the entire event and the *right figure* shows an extract revealing frequency drifts and spectral structure

We now identify τ with the observed characteristic event duration, and ν_{MHz} with the radio frequency in MHz. For solar events at 1.5 GHz, $T \approx 5$ MK [43], but for the longer events on AD Leo, $T \approx 13$ MK [78], which can be explained by the higher coronal temperature of AD Leo [39]. Conversely, the higher temperature of AD Leo's corona increases the gyroresonance absorption coefficient for underlying electron cyclotron maser emission significantly:

$$\kappa_{gr} \propto T^{s-1},\tag{8.55}$$

where $s = 2, 3$ is the harmonic of the absorbing layer, making the escape of the preferred harmonics of the maser difficult. This is not the case for plasma radiation for which free–free absorption is relevant:

$$\kappa_{ff} \propto T^{-3/2}\tag{8.56}$$

implying that the higher temperature of AD Leo's corona clearly favors a plasma emission process as long as $\omega_p > \Omega_c$. The latter condition is easily met as it requires $B < 500$ G. This is a nice example where a stellar observation helps identify the relevant emission mechanism that remains equivocal under solar conditions.

8.8 Stellar X-ray Flares

8.8.1 Cooling Physics

Flares cool through radiative, conductive, and possibly also volume expansion processes. We define the flare decay phase as the episode when the net energy loss by cooling exceeds the energy gain by heating, and the total thermal energy of the flare plasma decreases. The thermal energy decay timescale τ_{th} is defined as

$$\tau_{\text{th}} = \frac{E}{\dot{E}}, \tag{8.57}$$

where $E \approx 3n_e kT$ (for $n_e \approx n_{\text{H}}$) is the total thermal energy density in the flaring plasma of electron density n_e and temperature T, and $\dot{E}$ is the volumetric cooling loss rate (in erg cm^{-3} s^{-1}). For conduction parallel to the magnetic fields along a temperature gradient, the mean loss rate per unit volume is

$$\dot{E}_c = \frac{1}{L}\kappa_0 T^{5/2}\frac{dT}{ds} \approx \frac{4}{7L^2}\kappa_0 T^{7/2}, \tag{8.58}$$

where s is the coordinate along the field lines and the term $\kappa_0 T^{5/2}dT/ds$ is the conductive flux in the approximation of Spitzer [103], to be evaluated near the loop footpoint where T drops below 10^6 K, with $\kappa_0 \approx 9 \times 10^{-7}$ erg cm^{-1} s^{-1} K$^{-7/2}$. Equations (8.57) and (8.58) define the *conductive timescale* $\tau_{\text{th}} \equiv \tau_c$. The second equation in (8.58) should be used only as an approximation for non-radiating loops with a constant cross section down to the loss region and with uniform heating (or for time-dependent cooling of a constant pressure loop without heating; for the factor of 4/7, see [27, 56]). We have used L for the characteristic dimension of the source along the magnetic field lines, for example, the half-length of a magnetic loop. Strictly speaking, energy is not lost by conduction but is redistributed within the source; however, we consider energy lost when it is conducted to a region that is below X-ray-emitting temperatures, e.g., the transition region/chromosphere at the magnetic loop footpoints.

Radiative losses are by bremsstrahlung (dominant for $T \gtrsim 20$ MK), 2-photon continuum, bound-free, and line radiation. We note that the plasma composition in terms of element abundances can modify the cooling function $\Lambda(T)$, but the correction is of minor importance because stellar flares are usually rather hot. At relevant temperatures, the dominant radiative losses are by bremsstrahlung, which is weakly sensitive to modifications of the heavy-element abundances. The energy loss rate is

$$\dot{E}_r = n_e n_H \Lambda(T). \tag{8.59}$$

For $T \geq 20$ MK, $\Lambda(T) = \Lambda_0 T^\chi \approx 10^{-24.66} T^{1/4}$ erg cm^3 s^{-1} (after [110, 71]). Equations (8.57) and (8.59) define the *radiative timescale* $\tau_{\text{th}} \equiv \tau_r$.

8.8.2 Interpretation of the Decay Time

Equations (8.57), (8.58), and (8.59) describe the decay of the thermal energy, which in flare plasma is primarily due to the decay of temperature (with a timescale τ_T) and density. In contrast, the observed light curve decays (with a timescale τ_d for the *luminosity*) primarily due to the decreasing EM and, to a lesser extent, due to the decrease of $\Lambda(T)$ with decreasing temperature above ≈ 15 MK. From the energy equation, the thermal energy decay timescale τ_{th} is found to be

$$\frac{1}{\tau_{\mathrm{th}}} = \left(1 - \frac{\chi}{2}\right)\frac{1}{\tau_T} + \frac{1}{2\tau_d},$$ (8.60)

where the right-hand side is usually known from the observations (see [111] for a derivation). The decay timescale of the EM then follows as $1/\tau_{\mathrm{EM}} = 1/\tau_d - \chi/\tau_T$. Pan et al. [82] derived somewhat different coefficients in (8.60) for the assumption of constant volume or constant mass, including the enthalpy flux. In the absence of measurements of τ_T, it is often assumed that $\tau_{\mathrm{th}} = \tau_d$ although this is an inaccurate approximation.

In Eq. (8.60), τ_{th} is usually set to be τ_r or τ_c or, if both loss terms are significant $(\tau_r^{-1} + \tau_c^{-1})^{-1}$, taken at the beginning of the flare decay (note again that a simple identification of τ_r with τ_d is not accurate). If radiative losses dominate, the density immediately follows from Eqs. (8.57) and (8.59)

$$\tau_{\mathrm{th}} \approx \frac{3kT}{n_e \Lambda(T)},$$ (8.61)

and the characteristic size scale ℓ of the flaring plasma or the flare-loop semi-length L for a sample of $\mathcal{N}$ identical loops follow from

$$\mathrm{EM} = n_e n_H (\Gamma + 1)\pi \alpha^2 \mathcal{N} L^3 \approx n^2 \ell^3,$$ (8.62)

where α is the loop aspect ratio (ratio between loop cross-sectional diameter at the base and total length $2L$) and Γ is the loop expansion factor. The loop height for the important case of dominant radiative losses is [116, 111]

$$H = \left(\frac{8}{9\pi^4}\frac{\Lambda_0^2}{k^2}\right)^{1/3}\left(\frac{\mathrm{EM}}{T^{3/2}}\tau_r^2\right)^{1/3}\left(\mathcal{N}\alpha^2\right)^{-1/3}(\Gamma + 1)^{-1/3}.$$ (8.63)

A lower limit to H is found for $\tau_r \approx \tau_c$ in the same treatment:

$$H_{\min} = \frac{\Lambda_0}{\kappa_0 \pi^2}\frac{\mathrm{EM}}{T^{3.25}}\left(\mathcal{N}\alpha^2\right)^{-1}.$$ (8.64)

$\mathcal{N}$, α, and Γ are usually unknown and treated as free parameters within reasonable bounds. Generally, a small $\mathcal{N}$ is compatible with dominant radiative cooling.

8.8.3 Quasi-static Cooling Loops

Van den Ooord and Mewe [110] derived the energy equation of a cooling magnetic loop in such a way that it is formally identical to a static loop [92], by introducing a slowly varying flare heating rate that balances the total energy loss, and a possible constant heating rate during the flare decay. This specific solution thus proceeds through a sequence of different (quasi-)static loops with decreasing temperature.

The general treatment involves continued heating that keeps the cooling loop at coronal temperatures. If this constant heating term is zero, one finds for free quasi-static cooling

$$T(t) = T_0(1 + t/3\tau_{r,0})^{-8/7}, \tag{8.65}$$

$$L_r(t) = L_{r,0}(1 + t/3\tau_{r,0})^{-4}, \tag{8.66}$$

$$n_e(t) = n_{e,0}(1 + t/3\tau_{r,0})^{-13/7}, \tag{8.67}$$

where L_r is the total radiative loss rate and $\tau_{r,0}$ the radiative loss timescale given by Eq. (8.61) at the beginning of the flare decay.

This prescription is equivalent to requiring a constant ratio between radiative and conductive loss times, i.e., in the approximation of $T \gtrsim 20$ MK ($\Lambda \propto T^{1/4}$):

$$\frac{\tau_r}{\tau_c} = \text{const}\frac{T^{13/4}}{\text{EM}} \approx 0.18. \tag{8.68}$$

Accordingly, the applicability of the quasi-static cooling approach can be supported or rejected based on the run of T and EM during the decay phase. Note, however, that a constant ratio in Eq. (8.68) is not a sufficient condition to fully justify this approach.

8.8.4 Cooling Loops with Continued Heating

Whether or not flaring loops indeed follow a quasi-static cooling path is best studied on a density–temperature diagram (Fig. 8.18). Usually, characteristic values $T = T_a$ and $n_e = n_{e,a}$ as measured at the loop apex are used as diagnostics. For a magnetic loop in hydrostatic equilibrium, with constant cross section assumed, the RTV scaling law in Eq. (8.33) requires stable solutions (T, n_e) to be located where $T^2 \approx 7.6 \times 10^{-7} n_e L$ (for $n_e = n_i$). On a diagram of $\log T$ vs. $\log n_e$, all solutions are therefore located on a straight line with slope $\zeta = 0.5$. Figure 8.18 shows the path of a hydrodynamically simulated flare. The initial rapid heating leads to a rapid increase of T, inducing increased losses by conduction. As chromospheric evaporation grows, radiation helps to balance the heating. The flare decay sets in once the heating rate drops. At this moment, depending on the amount of ongoing heating, the magnetic loop is too dense to be in equilibrium, and the radiative losses exceed the heating rate, resulting in a thermal instability. In the limit of no heating during the decay, that is, an abrupt turnoff of the heating at the flare peak, the slope of the path becomes

$$\zeta \equiv \frac{d\ln T}{d\ln n_e} \equiv \frac{\tau_n}{\tau_T} = 2, \tag{8.69}$$

implying $T(t) \propto n_e^\zeta(t) = n_e^2(t)$ (see [99] for further discussion). Here, τ_T and τ_n are the e-folding decay times of the temperature and the electron density, respectively,

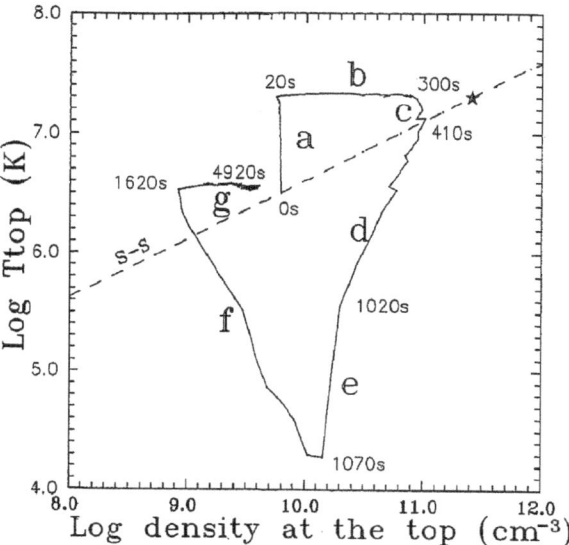

Fig. 8.18 Density–temperature diagram of a hydrodynamically simulated flare. The flare loop starts from an equilibrium (S-S, steady-state loop according to [92]); (*a*) and (*b*) refer to the heating phase; at (*c*), the heating is abruptly turned off, after which the loop cools rapidly (*d*, *e*), and only slowly recovers toward a new equilibrium solution (*f*, *g*) due to constant background heating (from [51])

under the assumption of exponential decay laws. Only for a non-vanishing heating rate does the loop slowly recover and eventually settle on a new equilibrium locus (Fig. 8.18). In contrast, if heating continues and is very gradually reduced, the loop decays along the static solutions ($\zeta = 0.5$). Observationally, this path is often followed by large solar flares [51].

Applying this concept to stellar astronomy, we replace n_e by the observable $\sqrt{\mathrm{EM}}$ and thus assume a constant flare volume, and further introduce the following generalization. In the freely cooling case after an abrupt heating turnoff, the entropy per particle at the loop apex decays on the thermodynamic decay time

$$\tau_{td} = 3.7 \times 10^{-4} \frac{L}{T_0^{1/2}} \quad [s], \tag{8.70}$$

where T_0 is the flare temperature at the beginning of the decay [99]. When heating is present, we introduce a correction term $F(\zeta) \equiv \tau_{LC}/\tau_{td}$ ([90]; τ_{LC} is the observed light curve decay time)

$$\tau_{LC} = 3.7 \times 10^{-4} \frac{L}{T_0^{1/2}} F(\zeta) \quad [s]. \tag{8.71}$$

$F(\zeta)$ is therefore to be numerically calibrated for each X-ray telescope. With known F, Eq. (8.71) can be solved for L. This scheme thus offers (i) an indirect method to study flaring loop geometries (L), (ii) a way of determining the rate and decay timescale of continued heating via $F(\zeta)$ and τ_{td}, and (iii) implications for the density decay time via $\tau_n = \zeta \tau_T$. Conditions of applicability include $\zeta \geq 0.3$ (most values are found in the range of $\sim 0.3 - 1$, [90]) and a resulting loop length L of less than one pressure scale height.

8.8.5 Two-Ribbon Flare Models

An approach that is entirely based on continuous heating (as opposed to cooling) was developed for the two-ribbon (2-R) class of solar flares. An example of this flare type is shown in Fig. 8.19. The 2-R flare model devised initially by Kopp and Poletto [55] is a parameterized magnetic energy release model. The time development of the flare light curve is completely determined by the amount of energy available in non-potential magnetic fields, and by the rate of energy release as a function of time and geometry as the fields reconnect and relax to a potential field configuration. It is assumed that a portion of the total energy is radiated into the observed X-ray band, while the remaining energy will be lost by other mechanisms. 2-R flares are well established for the Sun (Fig. 8.19); they often lead to large, long-duration flares that may be accompanied by mass ejections.

The magnetic fields are, for convenience, described along meridional planes on the star by Legendre polynomials P_n of order n, up to the height of the neutral point; above this level, the field is directed radially, that is, the field lines are "open". As time proceeds, field lines nearest to the neutral line move inward at coronal levels and reconnect at progressively larger heights above the neutral line. The reconnection point thus moves upward as the flare proceeds, leaving closed magnetic-loop systems underneath. One loop arcade thus corresponds to one N–S-aligned lobe between two zeros of P_n in latitude, axisymmetrically continued over some longitude in E–W direction. The propagation of the neutral point in height, $y(t)$, with a time constant t_0, is prescribed by (y in units of R_*, measured from the star's center)

$$y(t) = 1 + \frac{H_m}{R_*} \left(1 - e^{-t/t_0} \right) \tag{8.72}$$

$$H(t) \equiv [y(t) - 1] R_* \tag{8.73}$$

and the total energy release of the reconnecting arcade per radian in longitude is equal to the magnetic energy lost by reconnection,

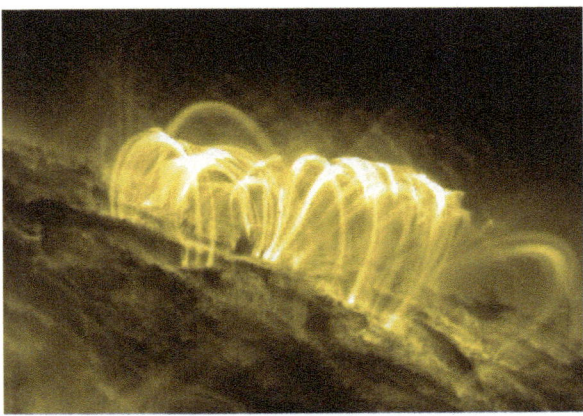

Fig. 8.19 *TRACE* image of a flaring magnetic loop arcade

$$\frac{dE}{dy} = \frac{1}{8\pi} 2n(n+1)(2n+1)^2 R_*^3 B^2 I_{12}(n) \frac{y^{2n}(y^{2n+1}-1)}{[n+(n+1)y^{2n+1}]^3} \qquad (8.74)$$

$$\frac{dE}{dt} = \frac{dE}{dy}\frac{dy}{dt} \qquad (8.75)$$

[87]. In Eq. (8.72), H_{m} is the maximum height of the neutral point for $t \to \infty$; typically, H_{m} is assumed to be equal to the latitudinal extent of the loops, i.e.,

$$H_{\mathrm{m}} \approx \frac{\pi}{n+1/2} R_* \qquad (8.76)$$

for $n > 2$ and $H_{\mathrm{m}} = (\pi/2)R_*$ for $n = 2$. Here, B is the surface magnetic field strength at the axis of symmetry and R_* is the stellar radius. Finally, $I_{12}(n)$ corresponds to $\int[P_n(\cos\theta)]^2 d(\cos\theta)$ evaluated between the latitudinal borders of the lobe (zeros of $dP_n/d\theta$) and θ is the co-latitude.

The free parameters are B and the efficiency of the energy-to-radiation conversion, q, both of which determine the normalization of the light curve; the timescale of the reconnection process, t_0, and the polynomial degree n determine the duration of the flare; and the geometry of the flare is fixed by n and therefore the asymptotic height H_{m} of the reconnection point. The largest realistic 2-R flare model is based on the Legendre polynomial of degree $n = 2$; the loop arcade then stretches out between the equator and the stellar poles. Usually, solutions can be found for many larger n as well. However, because a larger n requires larger surface magnetic field strengths, a natural limit is set to n within the framework of this model. Once the model solution has been established, further parameters, in particular the electron density n_e, can be inferred.

8.8.6 A Magnetohydrodynamic Model

Some scaling laws have been obtained from simulations based on the full set of magnetohydrodynamic equations [100, 101] (see Chap. 2). For the flare peak temperature T, the loop magnetic field strength B, the pre-flare loop electron density n_0, and the loop semi-length L, one finds, under the condition of dominant conductive cooling (appropriate for the early phase of a flare),

$$T \approx 1.8 \times 10^4 B^{6/7} n_0^{-1/7} L^{2/7} \ [\mathrm{K}]. \qquad (8.77)$$

The law follows from the balance between conduction cooling ($\propto T^{7/2}/L^2$, Eq. (8.58)) and magnetic reconnection heating ($\propto B^3/L$). Assuming loop filling through chromospheric evaporation and balance between thermal and magnetic pressure in the loop, two further "pressure-balance scaling laws" follow:

$$\mathrm{EM} \approx 3 \times 10^{-17} B^{-5} n_0^{3/2} T^{17/2} \ [\mathrm{cm}^{-3}], \qquad (8.78)$$

$$\mathrm{EM} \approx 2 \times 10^8 L^{5/3} n_0^{2/3} T^{8/3} \ [\mathrm{cm}^{-3}]. \qquad (8.79)$$

An alternative scaling law applies if the density development in the initial flare phase is assumed to follow balance between evaporation enthalpy flux and conduction flux, although the observational support is weaker,

$$EM \approx 1 \times 10^{-5} B^{-3} n_0^{1/2} T^{15/2} \text{ [cm}^{-3}].$$ (8.80)

Additionally, a steady solution is found for which the radiative losses balance conductive losses. This scaling law applies to a steady loop,

$$EM \approx \begin{cases} 10^{13} T^4 L & \text{[cm}^{-3}] & \text{for} & T < 10^7 \text{ K} \\ 10^{20} T^3 L & \text{[cm}^{-3}] & \text{for} & T > 10^7 \text{ K} \end{cases}$$ (8.81)

and is equivalent to the RTV scaling law in Eq. (8.33).

The advantage of these scaling laws is that they make use exclusively of the flare-peak parameters T, EM, B (and the pre-flare density n_0) and do not require knowledge of the time evolution of these parameters.

8.8.7 Observations of Stellar X-ray Flares

One of the main results that have come from extensive modeling of stellar X-ray flares is that extremely large stellar flares require large volumes under all realistic assumptions for the flare density, i.e., flaring complexes of magnetic loops must either be high or must spread across a large surface area. This is because, first, the energy released comes from the non-potential portion of the magnetic fields that are probably no stronger than a few 100 G in the corona; and second, small-loop models require higher pressure to produce the observed luminosity, hence requiring excessively strong magnetic fields.

There is an extensive literature discussing individual flares observed on a variety of stars. I will not discuss these results individually here but the interested reader may consult the compilation of results by Güdel [37] and the references given therein. This section summarizes some systematic trends found in stellar flares.

When the flare energy release evaporates plasma into the corona, heating and cooling effects compete simultaneously, depending on the density and temperature profiles in a given flare. It is therefore quite surprising to find a broad correlation between peak temperature T_p and peak emission measure EM_p, as illustrated in Fig. 8.20 for the sample reported by Güdel [37]. A regression fit gives (for 66 entries)

$$EM_p \propto T_p^{4.30 \pm 0.35}.$$ (8.82)

The correlation overall indicates that *larger flares are hotter*. A similar relation was reported previously for solar flares [32].

It is interesting to note that this correlation is similar to the $T - L_X$ correlation for the "non-flaring" coronal stars in Fig. 8.7 at cooler temperatures. This same sample

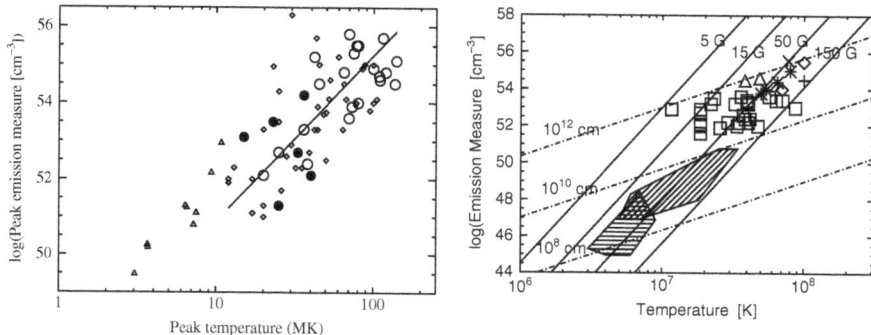

Fig. 8.20 *Left*: Peak temperatures and EMs of the flares (from [37] and references therein). Key to the symbols: *Filled circles*: *XMM-Newton* observations. *Open circles*: *ASCA* or *BeppoSAX* observations. *Small diamonds*: observations from other satellites. The *solid line* shows a regression fit given by Eq. (8.82). *Triangles* represent non-flaring parameters of G stars, referring to the hotter plasma component in 2-T spectral fits to *ROSAT* data. *Right*: Theoretical EM–T relations based on the reconnection model by Shibata and Yokoyama, showing *lines* of constant loop length L and lines of constant magnetic field strength B. Hatched areas are loci reported for solar flares and other symbols refer to individual stellar flares in star-forming regions (figure courtesy of K. Shibata and T. Yokoyama, after [101])

is plotted as triangles in Fig. 8.20, again only for the hotter plasma component. The stars follow approximately the same slope as the flares, albeit at cooler temperatures, and for a given temperature, the EM is higher. This trend may suggest that flares systematically contribute to the hot plasma component, although we have not temporally averaged the flare temperature and EM for this simple comparison.

In the context of the magnetohydrodynamic scaling laws presented in Eqs. (8.77) and (8.78), the observed loci of the flares require loop magnetic field strengths similar to solar flare values ($B \approx 10 - 150$ G) but the loop lengths must increase toward larger flares. This is seen in Fig. 8.20 where lines of constant L and B are plotted for this flare model. Typical loop lengths are thus of order $L \approx 10^{11}$ cm in this interpretation.

8.8.8 The "Neupert Effect"

In the framework of chromospheric evaporation, we assume that the emitted radio emission or its flux at the Earth, $F_R(t)$ is at any time proportional to the deposition rate of kinetic energy by non-thermal electrons into the plasma of the chromosphere. This plasma is thereby heated and "evaporates" into the corona. We define the conversion factor α as the ratio between the *thermal* energy flux being deposited in the chromosphere by non-thermal electrons and the observed radio flux density. Conversion into other forms of energy (mechanical, turbulence, etc.) occurs in parallel. Our simplification consists, first, in the assumption that these other energy transformations do not interact so that α remains constant over the course of the relatively short radio flare; second, we will assume that the dominant losses of the *thermal* energy occur via radiation; we will neglect energy loss into cool, non-X-ray-emitting plasma, e.g., by conduction, and cooling by adiabatic expansion. Also, we assume that there is no direct heating in parallel to the chromospheric evaporation. Third, we will use one temperature parameter for the flare plasma at any given time: the temperature T can be considered as describing an isothermal plasma dominating the total losses from the flaring loops.

The rate of change of the total thermal energy E in a plasma of volume V with electron density n_e is determined by the influx of kinetic energy and by radiation, hence the energy conservation equation for the *thermal* plasma is

$$\frac{d}{dt}(3n_e k T V) = \alpha F_R(t) - n_e^2 V \Lambda(T), \tag{8.83}$$

where $\Lambda(T)$ is the total luminosity of a plasma with unit emission measure (EM) at a temperature of T. Note again that Eq. (8.83) does not, by definition of α, describe the total energy budget of the flare, but merely the energy conversion of interest here (see results from RHESSI in Chap. 5). For large flare temperatures ($\gg 20$ MK), Λ is dominated by bremsstrahlung losses, roughly scaling as $T^{1/2}$. For somewhat more moderate temperatures ($\gtrsim 10$–20 MK), losses via line emission become important, with the (isothermal) approximation for Λ

$$\Lambda(T) = 1.86 \cdot 10^{-25} T^{1/4} \ [\text{erg s}^{-1} \ \text{cm}^3] \tag{8.84}$$

(see after Eq. (8.59) – for simplicity, we define here EM $= n_e^2 V$ with $n_H = 0.85 n_e$ and a corresponding correction in Λ). In the absence of kinetic energy influx $\alpha F_R(t)$, the thermal energy decays radiatively (neglecting conduction), and hence we define an e-folding decay time τ for the thermal energy as a function of the two independent variables, temperature T and density n_e, at a given instant,

$$\left.\frac{dE(t)}{dt}\right|_{F_R=0} = \frac{E(t)}{\tau(n_e(t), T(t))}, \tag{8.85}$$

so that

$$\tau(n_e(t), T(t)) = \frac{E(t)}{L_{\text{rad}}(t)} = \frac{3kT}{n_e \Lambda(T)}, \tag{8.86}$$

where L_{rad} is the total luminosity (approximately equal to the X-ray luminosity, L_X, but also with contributions from lower temperatures, e.g., in the ultraviolet, which we neglect here). Then,

$$\frac{dE(t)}{dt} = \alpha F_R(t) - L_{\text{rad}}(t) = \alpha F_R(t) - \frac{E(t)}{\tau(t)}. \tag{8.87}$$

The general solution of the inhomogeneous, linear differential equation (8.87) reads

$$E(t) = e^{-\int_{t_0}^t \tau(y)^{-1}dy} \left(E_0 + \alpha \int_{t_0}^t F_R(u)e^{+\int_{t_0}^u \tau(y)^{-1}dy}du \right),$$
(8.88)

where $E_0 = E(t_0)$ for a fixed t_0 before the flare start. The integration over τ^{-1} is along the time axis. If the initial thermal energy content can be neglected ($E_0 = 0$), then

$$E(t) = \alpha \int_{t_0}^t F_R(u)e^{-\int_u^t \tau(y)^{-1}dy}du.$$
(8.89)

We define an average decay constant for any time interval $[u, t]$ by

$$\bar{\tau}^{-1} = \bar{\tau}^{-1}(u, t) = \frac{\int_u^t \tau(y)^{-1}dy}{t - u}.$$
(8.90)

Then

$$E(t) = \alpha \int_{t_0}^t F_R(u)e^{-(t-u)/\bar{\tau}(t,u)}du,$$
(8.91)

i.e., for constant τ the energy profile is the *convolution* of the kinetic energy influx with an exponential function. Equation (8.91) is a generalized form of the *Neupert effect*. In the limit $\tau \to \infty$, Eqs. (8.87) and (8.91) become

$$\frac{dE(t)}{dt} = \alpha F_R(t),$$
(8.92)

$$E(t) = \alpha \int_{t_0}^t F_R(u)du,$$
(8.93)

i.e., the total energy content of the plasma is the integral of the kinetic energy influx (and radiation is inhibited). Equations (8.92) and (8.93) remind us of the classical formulation of the Neupert effect, with the observed X-ray losses replaced by the thermal energy content of the plasma. These two equations are applicable only for the increasing portion of the soft X-ray light curve or, correspondingly, for the time interval where $F_R \neq 0$. On using Eq. (8.86), we obtain the generalized Neupert effect for the light curve,

$$L_{rad}(t) = \frac{\alpha}{\tau(t)} \int_{t_0}^t F_R(u)e^{-(t-u)/\bar{\tau}(t,u)}du.$$
(8.94)

Note the importance of the thermal *energy* decay time $\tau = 3kT/(n_e \Lambda(T))$. Serio et al. [99] and Jakimiec et al. [51] find $Tn^{-2} = const$ for the radiative cooling phase. If the temperatures are very high and the bremsstrahlung approximation applies, then $\tau = T^{1/2}n^{-1}$ and thus $\tau = \text{const}$ in Eq. (8.94). However, in the case of more moderate temperatures in Eq. (8.84), only for $T^{3/4}n^{-1} = \text{const}$ does the classical Neupert effect approximately apply to the light curve. The actual functional dependence of $\psi(T)$ is more complicated. Thus, in general, the relevant parameters to be investigated for the Neupert effect are the *energies*. Further, $L_{\text{rad}}(t)$ describes all radiative losses across the electromagnetic spectrum; the observed X-ray luminosity L_X generally depends on the selected bandpass and therefore constitutes only a lower limit to L_{rad}.

Despite these caveats, it is surprising that the Neupert effect is often well observed, at least qualitatively, in solar radio and X-ray light curves. Instead of radio emission, bursts can be monitored in the U band: because U-band emission is likely to be a prompt reaction to the bombardment of the chromosphere by the same electron population that induces gyrosynchrotron emission, the Neupert effect should hold as well. A stellar example is shown in Fig. 8.21. These observations give clear evidence that at least some giant flares on active stars are subject to similar evaporation physics that occurs in the solar corona.

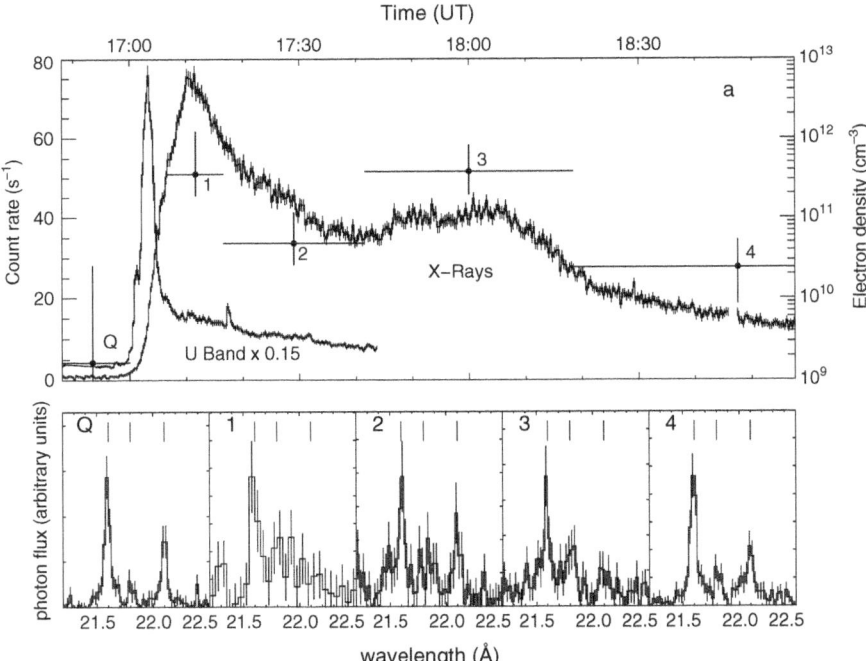

Fig. 8.21 Flare on Proxima Centauri, observed with *XMM-Newton*. The *top panel* shows the X-ray light curve and the much shorter U band flare (around 17 UT). The *bottom panel* shows the O VII He-like triplets observed during various time intervals of the flare. The locations of the *r*, *i*, and *f* lines are marked by *vertical lines*. The resulting electron densities are given in the *top panel* by the crosses, where the *horizontal arm* lengths indicate the time intervals over which the data were integrated and the right axis gives the logarithmic scale (after [41])

8.9 The Statistics of Flares

The study of stellar coronal structure confronts us with several problems that are difficult to explain by a scaling of solar coronal structure: (i) characteristic coronal temperatures increase with increasing magnetic activity. (ii) Characteristic coronal densities are typically higher in active than in inactive stars, and pressures in hot loops can be exceedingly high. (iii) The maximum stellar X-ray luminosities exceed the levels expected from complete coverage of the surface with solar-like active regions by up to an order of magnitude. (iv) Radio observations reveal a persistent population of non-thermal high-energy electrons in magnetically active stars even if the lifetime of such a population should only be tens of minutes to about an hour under ideal trapping conditions in coronal loops and perhaps much less due to efficient scattering of electrons into the chromosphere. Several of these features are reminiscent of flaring, as are some structural elements in stellar coronae. If flares are important for any of the above stellar coronal properties indeed, then we must consider the effects of frequent flares that may be unresolved in our observations but that may make up part, if not all, of the "quiescent" emission.

8.9.1 Stochastic Variability – What is "Quiescent Emission"?

The problem has been attacked in several dedicated statistical studies. Early statistical investigations (fluctuation analysis in light curves) remained ambiguous, reporting a significant amount of variability, although not necessarily being due to flares, or statistical absence of low-level flaring within the sensitivity limits [3, 24, 81]. There are indications in newer data that M dwarfs are continuously variable on short ($\lesssim 1$ day) timescales, and that the luminosity distribution is very similar to the equivalent distribution derived for solar flares, which suggests that the overall stellar light curves of dM stars are variable in the same way as a statistical sample of solar flares [68].

Very long light curves obtained from the *EUVE* satellite reveal an astonishing level of continuous variability in active M dwarfs (Fig. 8.22). Some of those data can be used to investigate statistical properties of the occurrence rate of flares as a function of total emitted energy. The increased sensitivity of *XMM-Newton* and *Chandra* is now revealing extreme levels of activity. Some X-ray light curves show no steady time interval exceeding a few tens of minutes within the sensitivity limit. In the day-long light curve in Fig. 8.23, no more than 30%, and probably much less, of the average X-ray emission of UV Cet can be attributed to any sort of steady emission, even outside the obvious, large flares. On the contrary, almost the entire light curve is resolved into frequent, stochastically occurring flares of various amplitudes.

In the *solar* corona, the flare rate increases steeply toward lower radiative energies, with no evidence (yet) for a lower threshold (e.g., [58]). Figure 8.24 shows an example of a *GOES* light curve in the 1.5–12 keV range, purposely selected

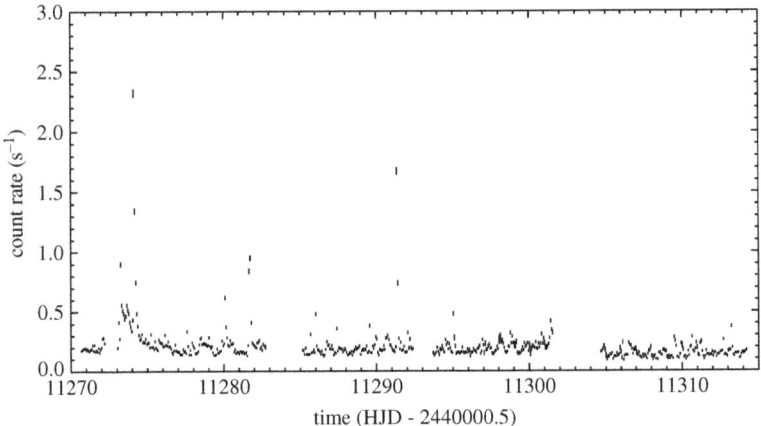

Fig. 8.22 A long light curve of the dMe star AD Leo, obtained by the DS instrument on *EUVE*. Most of the discernible variability is due to flares (after [39])

during an extremely active period in November 2003. While the *GOES* band is harder than typical bands used for stellar observations, it more clearly reveals the level of the underlying variability (a typical detector used for stellar observations would see much less contrast). If the solar analogy has any merit in interpreting stellar coronal X-rays, then low-level emission in stars that does show flares *cannot* be truly quiescent, that is, constant or slowly varying exclusively due to long-term evolution of active regions, or due to rotational modulation. A measure of *flare rates* is therefore not meaningful unless it refers to flares above a given luminosity or energy threshold. This is – emphatically – not to say that steady emission is absent

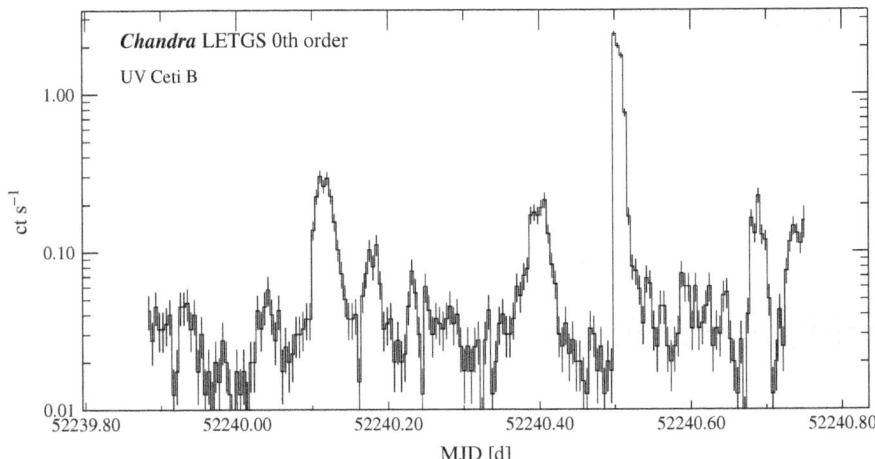

Fig. 8.23 Light curve of UV Ceti B, observed with the *Chandra* LETGS/HRC over about 1 day. Note the logarithmic flux axis (figure courtesy of M. Audard, after [11])

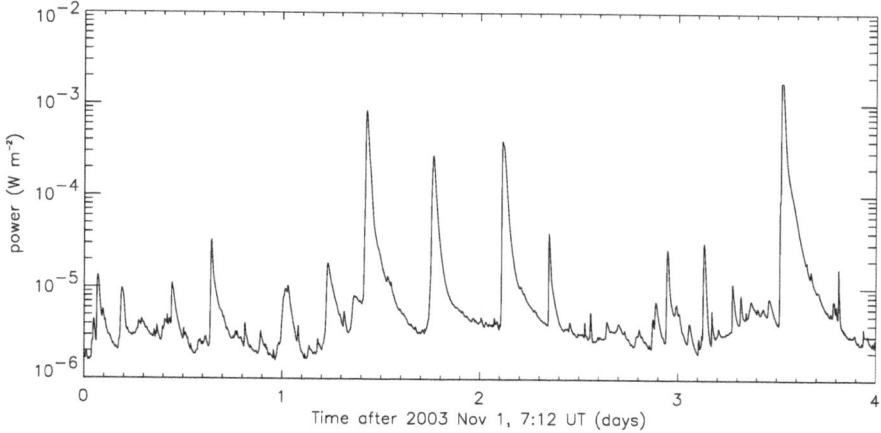

Fig. 8.24 *GOES* full-disk solar X-ray light curve, observed in the 1.5–12 keV band in November 2003. The abscissa gives time after November 1, 2003, 7:12 UT in days

in magnetically active stars. However, once we accept the solar analogy as a working principle, the question is not so much about the presence of large numbers of flares, but to what extent they contribute to the overall X-ray emission from coronae.

8.9.2 The Flare-Energy Distribution

The suggestion that stochastically occurring flares may be largely responsible for coronal heating is known as the "microflare" or "nanoflare" hypothesis in solar physics [83]. Observationally, it is supported by evidence for the presence of numerous small-scale flare events occurring in the solar corona at any time (e.g., [62]). Their distribution in energy is a power law,

$$\frac{dN}{dE} = kE^{-\alpha}, \tag{8.95}$$

where dN is the number of flares per unit time with a total energy in the interval $[E, E + dE]$ and k is a constant. If $\alpha \geq 2$, then the energy integration (for a given time interval) diverges for $E_{\min} \rightarrow 0$, that is, by extrapolating the power law to sufficiently small flare energies, *any* energy release power can be attained. This is not the case for $\alpha < 2$. Solar studies have repeatedly resulted in α values of 1.6–1.8 for ordinary solar flares [25] (see Chap. 5), but some recent studies of low-level flaring suggest $\alpha = 2.0–2.6$ [58, 84].

Relevant stellar studies have been rare (see Table 8.3). Early investigations lumped several stars together to produce meaningful statistics. Full forward modeling of a superposition of stochastic flares was applied to EUV and X-ray light curves by Kashyap et al. [53] and Güdel et al. [39] based on Monte Carlo simulations and by Arzner and Güdel [6] based on an analytical formulation. The results of

Table 8.3 Stellar radiative flare-energy distributions

Star sample	Photon energies [keV]	log (Flare energies)[a]	α	References
M dwarfs	0.05–2	30.6 – 33.2	1.52±0.08	Collura et al. [24]
M dwarfs	0.05–2	30.5 – 34.0	1.7±0.1	Pallavicini et al. [81]
RS CVn binaries	EUV	32.9 – 34.6	1.6	Osten and Brown [80]
Two G dwarfs	EUV	33.5 – 34.8	2.0–2.2	Audard et al. [10]
F-M dwarfs	EUV	30.6 – 35.0	1.8–2.3	Audard et al. [9]
Three M dwarfs	EUV	29.0 – 33.7	2.2–2.7	Kashyap et al. [53]
AD Leo	EUV&0.1–10	31.1 – 33.7	2.0–2.5	Güdel et al. [39]
AD Leo	EUV	31.1 – 33.7	2.3 ± 0.1	Arzner and Güdel [6]

[a] Total flare-radiated X-ray energies used for the analysis (in ergs).

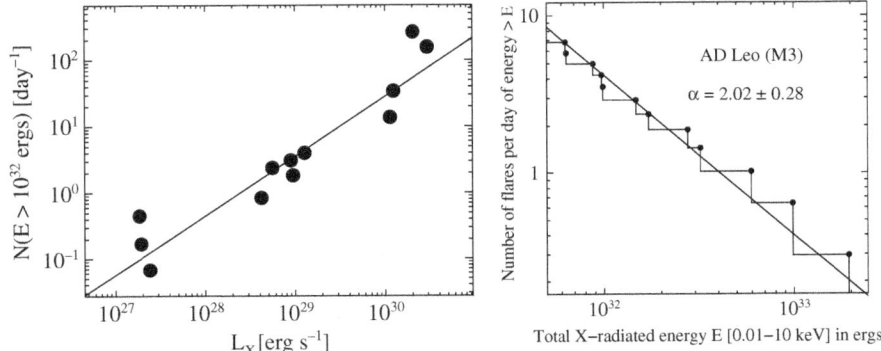

Fig. 8.25 *Left:* The rate of flares above a threshold of 10^{32} erg in total radiated X-ray energy is plotted against the low-level luminosity for several stars, together with a regression fit. *Right:* Flare energy distribution for AD Leo, using a flare identification algorithm for an observation with *EUVE* (both figures courtesy of M. Audard, after [9])

these investigations are in full agreement, converging to $\alpha \approx 2.0-2.5$ for M dwarfs (Table 8.3, Fig. 8.25). If the power law flare energy distribution extends by about 1–2 orders of magnitude below the actual detection limit in the light curves, then the *entire* emission could be explained by stochastic flares. The coronal heating process in magnetically active stars would – in this extreme limit – be one solely due to *time-dependent* heating by flares, or, in other words, the X-ray corona would be an entirely hydrodynamic phenomenon rather than an ensemble of hydrostatic loops.

8.9.3 Microflaring at Radio Wavelengths

Quiescent radio emission can apparently persist for quite long periods. Losses by collisions (see Eq. (8.11)) require a very low ambient electron density to maintain the electron population. Alternatively, electrons could be frequently injected at many coronal sites. Based on spectral observations, White and Franciosini [118] suggest that the emission around 1.4 GHz shown in Fig. 8.26 is composed of

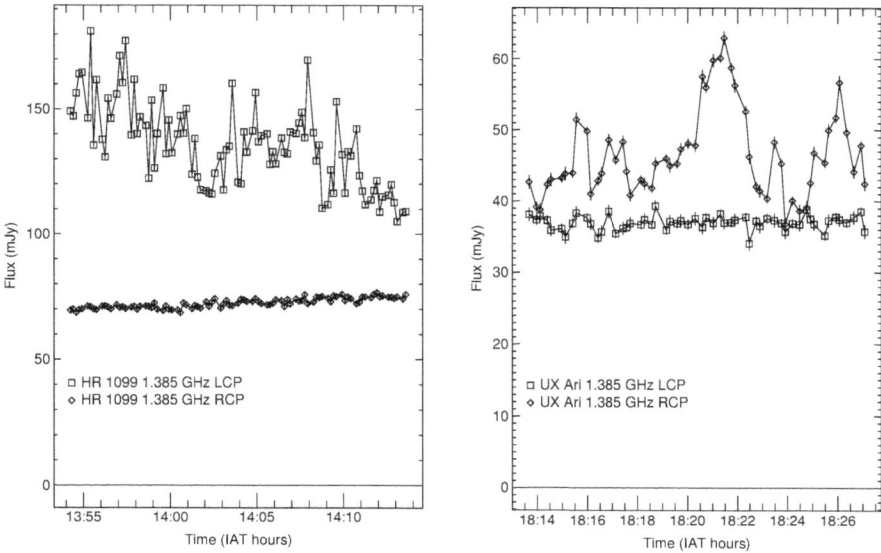

Fig. 8.26 Light curves of HR 1099 (*left*) and UX Ari (*right*) obtained in the two senses of circular polarization. The brighter of the two polarized fluxes varies rapidly and has been interpreted as 100% polarized coherent emission superimposed on a gradually changing gyrosynchrotron component (from [118]; figures from S. M. White)

a steady, weakly polarized broadband gyrosynchrotron component plus superimposed, strongly and oppositely polarized, fluctuating plasma emission that is perceived as quasi-steady but that may occasionally evolve into strong, polarized flare emission. Continual flaring may thus also reflect in radio light curves.

8.10 A Flare-Heating Approach

The quiescent emission of synchrotron radiation requires the continuous presence of non-thermal electrons, while X-rays are emitted by the thermal bulk at typically 10^6 to a few times 10^7 K. For optically thin radio emission ($\nu \gtrsim 5$ GHz), the radio luminosity is

$$L_R = 4\pi \eta_\nu V_R \ [\text{erg s}^{-1} \text{ Hz}^{-1}] \,, \qquad (8.96)$$

where V_R is the radio source volume. The density of non-thermal electrons will be assumed to be distributed in energy according to a power law,

$$n(\varepsilon) = \frac{(\delta - 1)}{\varepsilon_0} \, N \left(\frac{\varepsilon}{\varepsilon_0} \right)^{-\delta} \ [\text{cm}^{-3} \text{ erg}^{-1}], \qquad (8.97)$$

with the lower cutoff at $\varepsilon_0 \approx 10$ keV $= 1.6 \cdot 10^{-8}$ erg being compatible with most acceleration processes proposed for stellar coronae and with solar flare observations. We keep ε_0 fixed at 10 keV in the following considerations. It is not a sensitive value for synchrotron emission since the latter becomes appreciable only at higher energies. We assume a homogeneous source for simplicity. The gyrosynchrotron emissivity is approximately given by Dulk [30] (sum of x and o modes, see Eq. (8.46))

$$\eta \approx 1.8 \cdot 3.3 \times 10^{-24} \, BN \, 10^{-0.52\delta} (\sin\theta)^{-0.43+0.65\delta} \left(\frac{\nu}{\nu_B} \right)^{1.22-0.90\delta} \quad (8.98)$$

$$\equiv 1.8 \cdot \vartheta(B, \delta, \nu, \theta) \, N \qquad [\text{erg s}^{-1} \, \text{Hz}^{-1} \, \text{cm}^{-3} \, \text{sterad}^{-1}], \quad (8.99)$$

where ν_B is the electron gyrofrequency and θ is the angle between the line of sight and the magnetic field. In a steady-state situation, the density of non-thermal electrons is given by

$$n(\varepsilon) = \frac{\dot{n}_{\text{in}}(\varepsilon)}{V_R} \tau(\varepsilon), \qquad (8.100)$$

where $\dot{n}_{\text{in}}(\varepsilon)$ is the total number of electrons of energy ε accelerated per unit time and $\tau(\varepsilon)$ is the electron lifetime. Let

$$\dot{n}_{\text{in}}(\varepsilon) = \dot{n}_{0,\text{in}} \left(\frac{\varepsilon}{\varepsilon_0} \right)^{-\kappa}, \qquad (8.101)$$

$$\tau(\varepsilon) = \tau_0 \left(\frac{\varepsilon}{\varepsilon_0} \right)^{\alpha}. \qquad (8.102)$$

With Eq. (8.100), the power law index of the electrons in Eq. (8.97) is $\delta \approx \kappa - \alpha$.

Let a be the fraction of the energy that goes into accelerated particles, and b the fraction of the total coronal energy ultimately radiated into the observed X-ray band. Since some of the thermal energy is lost by conduction and other processes, $b < 1$. Then, X-rays are related to the total energy input

$$\dot{E} = \frac{1}{a} \int_{\varepsilon_0}^{\infty} \dot{n}_{\text{in}}(\varepsilon)\varepsilon \, d\varepsilon = \frac{1}{b} L_X. \qquad (8.103)$$

Using Eqs. (8.96, 8.97, 8.98, 8.99, 8.100, 8.101, 8.102, 8.103), the relation between L_R and L_X becomes

$$L_R = 1.8 \cdot 4\pi \vartheta(B, \nu, \theta, \delta) \frac{a}{b} \varepsilon_0^{-1} \tau_0 \left(\frac{\alpha - 1}{\delta - 1} + 1 \right) L_X, \qquad (8.104)$$

where we require $\delta > 1$ and $\alpha > 2 - \delta$ for convergence. Equation (8.104) is general and includes different possible scenarios. Let us select typical parameters for stellar observations, viz. $\delta = 3$ (implying $\alpha > -1$), $\nu = 5$ GHz, and $\theta = 30°$; then

$$L_R = 3.5 \cdot 10^{-22} B^{2.48} \frac{a}{b} \tau_0 (\alpha + 1) L_X. \qquad (8.105)$$

Late-type main-sequence and subgiant stars appear to follow a linear relation between L_R and L_X: $L_X \approx 10^{15.5} L_R$ (the coefficient is 0.5–1 dex smaller for RS

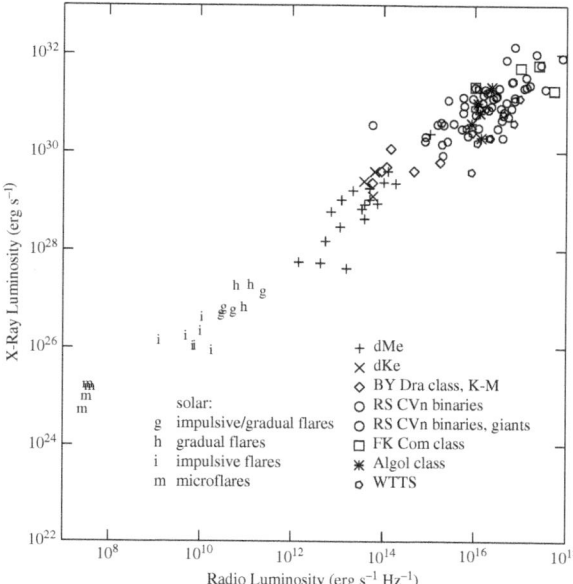

Fig. 8.27 Correlation between radio and X-ray luminosities for magnetically active stars (from [36])

CVn-type binaries [44]; see Fig. 8.27). We thus suggest numerical relations between a, b, τ, and B for active stars:

$$B^{2.48}\frac{a}{b}\tau_0(\alpha+1) \approx \begin{cases} 9.0 \cdot 10^5 & dMe, dKe, BYDra \\ 5.4 \cdot 10^6 & RSCVn, Algols, PTTS, FKCom. \end{cases} \tag{8.106}$$

Two scenarios are possible: If the acceleration efficiency a is close to unity, the non-thermal electrons first emit a small fraction of the total kinetic energy as synchrotron radiation before they lose most of their energy by collisions, thereby heating the X-ray-emitting corona ("causal relation" between non-thermal and thermal energy). It is unknown whether fully efficient accelerators are realized in nature.

The other scenario implies a common energy release, most of it heating the corona by thermal processes (e.g., Ohmic heating), and a much smaller fraction a accelerating particles. Some of the latter energy may also end up as heat, but this is negligible. Therefore, the thermal plasma and non-thermal electrons radiate independently ("common origin" scenario). This scenario can explain the relation found by assuming that coronal heating in active stars *necessarily* implies electron acceleration to relativistic energies. The fraction between thermal and non-thermal energy release is the crucial, but unknown number. It is unclear whether it can be constant over many different types of stars.

The lifetime τ (or τ_0 and α), which is relevant here for the gyrosynchrotron emitting high-energy particles, depends on the process that scatters energetic electrons into the loss-cone of the trapped particle velocity distribution, such as collisions, whistler wave instability, or cyclotron maser action. Their effects on the relation between trapping time and particle energy are opposite. As an example, let us assume $\tau(\varepsilon) = $ const (i.e., $\alpha = 0$) and use solar active region and flare values for B ($\approx$100 G) and a/b (order of unity; [26, 122]). Equation (8.106) then would require $\tau \approx 10$ s for dM(e), dK(e), and BY Dra stars. The lifetime is an average over the time of flight of the particles that are lost immediately (moving parallel to the field lines) and the trapped particles temporarily residing in a lower density plasma. The estimated values for τ are compatible with the minimum observed timescale of variations in the "quiescent" emission (e.g., [50]). For the halo of RS CVn binaries, B can be as low as $\approx$10 G [74]. Then, from Eq. (8.106), τ is of the order of many hours, compatible with observed long timescales (e.g., [69]).

Acknowledgments It is a pleasure to thank the organizers for inviting me to the Montegufoni summer school. I thank several colleagues for providing me figure material, which is reproduced with permission of the publishers. In particular, some of the material presented here has been reprinted, with permission, from the *Annual Review of Astronomy and Astrophysics*, Volume 40, © 2002 by Annual Reviews, www.annualreviews.org (Fig. 1b, 2, 13–16, 22, 26, 27; Güdel 2002). Some material has been reprinted, with permission, from the *Astronomy and Astrophysics Review*, Vol. 12 (Figs. 1a, 4, 6–12, 18–25; Güdel 2004). Research at PSI has been supported by the Swiss National Science Foundation.

References

1. M. Abada-Simon, A. Lecacheux, M. Aubier, J.A. Bookbinder: Astron. Astrophys., **321**, 841 (1997)
2. M. Abada-Simon, A. Lecacheux, P. Louarn, G.A. Dulk, L. Belkora, et al.: Astron. Astrophys., **288**, 219 (1994)
3. C.W. Ambruster, S. Sciortino, and L. Golub: Astrophys. J. Supp., **65**, 273 (1987)
4. P. André, T. Montmerle, E.D. Feigelson, P.C. Stine, K.-L. Klein: Astrophys. J., **335**, 940 (1988)
5. S.K. Antiochos: Astrophys. J., **241**, 385 (1980)
6. K. Arzner, M. Güdel: Astrophys. J., **602**, 363 (2004)
7. M.J. Aschwanden, R.W. Nightingale, D. Alexander: Astrophys. J., **541**, 1059 (2000)
8. M. Aschwanden, C.J. Schrijver: Astrophys. J. Supp., **142**, 269 (2002)
9. M. Audard, M. Güdel, J.J. Drake, V. Kashyap: Astrophys. J., **541**, 396 (2000)
10. M. Audard, M. Güdel, E.F. Guinan: Astrophys. J. Lett., **513**, L53 (1999)
11. M. Audard, M. Güdel, S.L. Skinner: Astrophys. J., **589**, 983 (2003)
12. T.S. Bastian: Solar Phys., **130**, 265 (1990)
13. T.S. Bastian, A.O. Benz, D.E. Gary: Ann. Rev. Astron. Astrophys., **36**, 131 (1998)
14. T.S. Bastian, J.A. Bookbinder: Nature, **326**, 678 (1987)
15. T.S. Bastian, J. Bookbinder, G.A. Dulk, M. Davis: Astrophys. J., **353**, 265 (1990)
16. A.J. Beasley, M. Güdel: Astrophys. J., **529**, 961 (2000)
17. A.O. Benz, W. Alef, M. Güdel: Astron. Astrophys., **298**, 187 (1995)
18. A.O. Benz, J. Conway, M. Güdel: Astron. Astrophys., **331**, 596 (1998)
19. R.J. Bray, L.E. Cram, C.J. Durrant, E.E. Loughhead: *Plasma Loops in the Solar Corona*. Cambridge University Press, Cambridge (1991)
20. P.J. Cargill: Astrophys. J., **422**, 381 (1994)
21. F. Chiuderi Drago, E. Franciosini: Astrophys. J., **410**, 301 (1993)
22. C.S. Choi, T. Dotani: Astrophys. J., **492**, 761 (1998)

23. E.W. Cliver, B.R. Dennis, A.L. Kiplinger, S.R. Kane, D.F. Neidig, N.R. Sheeley Jr., M.J. Koomen: Astrophys. J., **305**, 920 (1986)
24. A. Collura, L. Pasquini, J.H.M.M. Schmitt: Astron. Astrophys., **205**, 197 (1988)
25. N.B. Crosby, M.J. Aschwanden, B.R. Dennis: Solar Phys., **143**, 275 (1993)
26. B.R. Dennis: Solar Phys., **118**, 49 (1988)
27. J.F. Dowdy Jr., R.L. Moore, S.T. Wu: Solar Phys., **99**, 79 (1985)
28. J.J. Drake, G. Peres, S. Orlando, J.M. Laming, A. Maggio: Astrophys. J., **545**, 1074 (2000)
29. S.A. Drake, D.C. Abbott, T.S. Bastian, J.H. Bieging, E. Churchwell, et al.: Astrophys. J., **322**, 902 (1987)
30. G.A. Dulk: Ann. Rev. Astron. Astrophys., **23**, 169 (1985)
31. A.K. Dupree, N.S. Brickhouse, G.A. Doschek, J.C. Green, J.C. Raymond: Astrophys. J. Lett., **418**, L41 (1993)
32. U. Feldman, J.M. Laming, G.A. Doschek: Astrophys. J. Lett., **451**, L79 (1995)
33. A.H. Gabriel, C. Jordan: MNRAS, **145**, 241 (1969)
34. D.E. Gary, J.L. Linsky: Astrophys. J., **250**, 284 (1981)
35. M.S. Giampapa, R. Rosner, V. Kashyap, T.A. Fleming, J.H.M.M. Schmitt, J.A. Bookbinder: Astrophys. J., **463**, 707 (1996)
36. M. Güdel: Ann. Rev. Astron. Astrophys., **40**, 217 (2002)
37. M. Güdel: Astron. Astrophys. Rev., **12**, 71 (2004)
38. M. Güdel, K. Arzner, M. Audard, R. Mewe: Astron. Astrophys., **403**, 155 (2003b)
39. M. Güdel, M. Audard, V.L. Kashyap, J.J. Drake, E.F. Guinan: Astrophys. J., **582**, 423 (2003a)
40. M. Güdel, M. Audard, H. Magee, E. Franciosini, N. Grosso, F.A. Cordova, R. Pallavicini, R. Mewe: Astron. Astrophys., **365**, L344 (2001a)
41. M. Güdel, M. Audard, S.L. Skinner, M.I. Horvath: Astrophys. J., **580**, L73 (2002a)
42. M. Güdel, A.O. Benz: Astron. Astrophys., **211**, L5 (1989)
43. M. Güdel, A.O. Benz: Astron. Astrophys., **231**, 202 (1990)
44. M. Güdel, A.O. Benz: Astrophys. J. Lett., **405**, L63 (1993)
45. M. Güdel, A.O. Benz, T.S. Bastian, E. Fürst, G.M. Simnett, R.J. Davis: Astron. Astrophys., **220**, L5 (1989)
46. M. Güdel, E.F. Guinan, S.L. Skinner: Astrophys. J., **483**, 947 (1997)
47. M. Güdel, J.H.M.M. Schmitt: Röntgenstrahlung from the Universe. In: H.U. Zimmermann, J.E. Trümper, H. Yorke (eds.) MPE, München, p. 35, (1995)
48. O. Havnes, C.K. Goertz: Astron. Astrophys., **138**, 421 (1984)
49. P.D. Jackson, M.R. Kundu, S.M. White: Astrophys. J. Lett., **316**, L85 (1987)
50. P.D. Jackson, M.R. Kundu, S.M. White: Astron. Astrophys., **210**, 284 (1989)
51. J. Jakimiec, B. Sylwester, J. Sylwester, S. Serio, G. Peres, F. Reale: Astron. Astrophys., **253**, 269 (1992)
52. K.L. Jones, R.T. Stewart, G.J. Nelson: MNRAS, **274**, 711 (1995)
53. V. Kashyap, J.J. Drake, M. Güdel, M. Audard: Astrophys. J., **580**, 1118 (2002)
54. K.-L. Klein, F. Chiuderi-Drago: Astron. Astrophys., **175**, 179 (1987)
55. R.A. Kopp, G. Poletto: Solar Phys., **93**, 351 (1984)
56. R.A. Kopp, G. Poletto: Astrophys. J., **418**, 496 (1993)
57. M.R. Kundu, R.K. Shevgaonkar: Astrophys. J., **297**, 644 (1985)
58. S. Krucker, A.O. Benz: Astrophys. J. Lett., **501**, L213 (1998)
59. K.R. Lang, J. Bookbinder, L. Golub, M.M. Davis: Astrophys. J. Lett., **272**, L15 (1983)
60. K.R. Lang, R.F. Willson: Astrophys. J., **305**, 363 (1986)
61. K.R. Lang, R.F. Willson Astrophys. J., **326**, 300(1988)
62. R.P. Lin, R.A. Schwartz, S.R. Kane, R.M. Pelling, K. Hurley: Astrophys. J., **283**, 421 (1984)
63. J.L., Linsky, S.A. Drake, T.S. Bastian: Astrophys. J., **393**, 341 (1992)
64. J.L. Linsky, D.E. Gary: Astrophys. J., **274**, 776 (1983)
65. J.L. Linsky, B.M. Haisch: Astrophys. J. Lett., **229**, L27 (1979)
66. L.B. Lucy: Astron. J., **79**, 745 (1974)
67. A. Maggio, G. Peres: Astron. Astrophys., **325**, 237 (1997)

68. A. Marino, G. Micela, G. Peres: Astron. Astrophys., **353**, 177 (2000)
69. M. Massi, F. Chiuderi-Drago: Astron. Astrophys., **253**, 403 (1992)
70. D.B. Melrose: *Plasma Astrophysics*. Gordon and Breach, New York (1980)
71. R. Mewe, E.H.B.M. Gronenschild, G.H.J. van den Oord: Astron. Astrophys. Supp., **62**, 197 (1985)
72. R. Mewe, J.S. Kaastra, G.H.J. van den Oord, J. Vink, Y. Tawara: Astron. Astrophys., **320**, 147 (1997)
73. D.H. Morris, R.L. Mutel, B. Su: Astrophys. J., **362**, 299 (1990)
74. R.L. Mutel, J.-F. Lestrade, R.A. Preston, R.B. Phillips: Astrophys. J., **289**, 262 (1985)
75. R.L. Mutel, L.A. Molnar, E.B. Waltman, F.D. Ghigo: Astrophys. J., **507**, 371 (1998)
76. R.L. Mutel, D.H. Morris, D.J. Doiron, J.-F. Lestrade: Astron. J., **93**, 1220 (1987)
77. J.-U. Ness, M. Güdel, J.H.M.M. Schmitt, M. Audard, A. Telleschi: Astron. Astrophys., **427**, 667 (2004)
78. R.A. Osten, T.S. Bastian: Astrophys. J., **637**, 1016 (2006)
79. R.A. Osten, T.R. Ayres, A. Brown, J.L. Linsky, A. Krishnamurthi: Astrophys. J., **582**, 1073 (2003)
80. R.A. Osten, A. Brown: Astrophys. J., **515**, 746 (1999)
81. R. Pallavicini, G. Tagliaferri, L. Stella: Astron. Astrophys., **228**, 403 (1990)
82. H.C. Pan, C. Jordan, K. Makishima, R.A. Stern, K. Hayashida, M. Inda-Koide: MNRAS, **285**, 735 (1997)
83. E.N. Parker: Astrophys. J., **330**, 474 (1988)
84. C.E. Parnell, P.E. Jupp: Astrophys. J., **529**, 554 (2000)
85. V. Petrosian: Astrophys. J., **299**, 987 (1985)
86. K.J.H. Phillips, M. Mathioudakis, D.P. Huenemoerder, D.R. Williams, M.E. Phillips, F.P. Keenan: MNRAS, **325**, 1500 (2001)
87. G. Poletto, R. Pallavicini, R.A. Kopp: Astron. Astrophys., **20**, 93 (1988)
88. D. Porquet, R. Mewe, J. Dubau, A.J.J. Raassen, J.S. Kaastra: Astron. Astrophys., **376**, 1113 (2001)
89. P. Preś, M. Siarkowski, J. Sylwester: MNRAS, **275**, 43 (1995)
90. F. Reale, R. Betta, G. Peres, S. Serio, J. McTiernan: Astron. Astrophys., **325**, 782 (1997)
91. F. Reale, S. Serio, G. Peres: Astron. Astrophys., **272**, 486 (1993)
92. R. Rosner, W.H. Tucker, G.S. Vaiana: Astrophys. J., **220**, 643 (1978)
93. J.H.M.M. Schmitt, F. Favata: Nature, **401**, 44 (1999)
94. J.H.M.M. Schmitt, J.-U. Ness, G. Franco: Astron. Astrophys., **412**, 849 (2003)
95. C.J. Schrijver, M.J. Aschwanden: Astrophys. J., **566**, 1147 (2002)
96. C.J. Schrijver, J.R. Lemen, R. Mewe: Astrophys. J., **341**, 484 (1989)
97. C.J. Schrijver, R. Mewe, G.H.J. van den Oord, J.S. Kaastra: Astron. Astrophys., **302**, 438 (1995)
98. S. Serio, G. Peres, G.S. Vaiana, L. Golub, R. Rosner: Astrophys. J., **243**, 288 (1981)
99. S. Serio, F. Reale, J. Jakimiec, B. Sylwester, J. Sylwester: Astron. Astrophys., **241**, 197 (1991)
100. K. Shibata, T. Yokoyama: Astrophys. J. Lett, **526**, L49 (1999)
101. K. Shibata, T. Yokoyama: Astrophys. J., **577**, 422 (2002)
102. M. Siarkowski, P. Preś, S.A. Drake, N.E. White, K.P. Singh: Astrophys. J., **473**, 470 (1996)
103. L. Spitzer: *Physics of Fully Ionized Gases*. Interscience, New York (1962)
104. R.A. Stern, S.K. Antiochos, F.R. Harnden Jr.: Astrophys. J., **305**, 417 (1986)
105. A. Telleschi, M. Güdel, K.R. Briggs, M. Audard, J.-U. Ness, S.L. Skinner: Astrophys. J., **622**, 653 (2005)
106. C. Trigilio C.S. Buemi, G. Umana, M. Rodonò, P. Leto, et al.: Astron. Astrophys., **373**, 181 (2001)
107. G. Umana, P. Leto, C. Trigilio, R.M. Hjellming, S. Catalano: Astron. Astrophys., **342**, 709 (1999)
108. G. Umana, C. Trigilio, S. Catalano: Astron. Astrophys., **329**, 1010 (1998)

109. G. Umana, C. Trigilio, R.M. Hjellming, S. Catalano, M. Rodonò: Astron. Astrophys., **267**, 126 (1993)
110. G.H.J. van den Oord, R. Mewe: Astron. Astrophys., **213**, 245 (1989)
111. G.H.J. van den Oord, R. Mewe, A.C. Brinkman: Astron. Astrophys., **205**, 181 (1988)
112. G.H.J. van den Oord, C.J. Schrijver, M. Camphens, R. Mewe, J.S. Kaastra: Astron. Astrophys., **326**, 1090 (1997)
113. R. Ventura, A. Maggio, G. Peres: Astron. Astrophys., **334**, 188 (1998)
114. J.F. Vesecky, S.K. Antiochos, J.H. Underwood: Astrophys. J., **233**, 987 (1979)
115. F.M. Walter, D.M. Gibson, G. Basri: Astrophys. J., **267**, 665 (1983)
116. N.E. White, J.L. Culhane, A.N. Parmar, B.J. Kellett, S. Kahn, G.H.J. van den Oord, J. Kuijpers: Astrophys. J., **301**, 262 (1986)
117. N.E. White, R.A. Shafer, K. Horne, A.N. Parmar, J.L. Culhane: Astrophys. J., **350**, 776 (1990)
118. S.M. White, E. Franciosini: Astrophys. J., **444**, 342 (1995)
119. S.M. White, M.R. Kundu, P.D. Jackson: Astron. Astrophys., **225**, 112 (1989)
120. S.M. White, J. Lim, M.R. Kundu: Astrophys. J., **422**, 293 (1994)
121. G.L. Withbroe: Solar Phys., **45**, 301 (1975)
122. G.L. Withbroe, R.W. Noyes: Ann. Rev. Astron. Astrophys., **15**, 363 (1977)

GPSR Compliance

*The European Union's (EU) General Product Safety Regulation (GPSR)
is a set of rules that requires consumer products to be safe and our
obligations to ensure this.*

*If you have any concerns about our products, you can contact us on
ProductSafety@springernature.com*

In case Publisher is established outside the EU, the EU authorized
representative is:

Springer Nature Customer Service Center GmbH
Europaplatz 3
69115 Heidelberg, Germany

Batch number: 09491090

Printed by Printforce, the Netherlands